EXTREME STATES OF MATTER

EXTREME STATES OF MATTER

Editor

Helfried Rothbauer

scitus academics

Extreme States of Matter

Edited by **Helfried Rothbauer**

This Book is Printed on Acid-Free Paper. ♾

ISBN: 9781681178530

Library of Congress Control Number: 2017948880

SCITUS Academics LLC,

616, Corporate Way, Suite 2, 4766,

Valley Cottage, NY 10989

www.scitusacademics.com

Printed in United States of America.

Preface

In physics, four states of matter are apparent in daily life: solid, liquid, gas, and plasma. Many other states are known to exist only in extreme situations, such as Bose–Einstein condensates, neutron-degenerate matter, and quark-gluon plasma, which only occur in situations of extreme cold, extreme density, and extremely high-energy color-charged matter respectively. Some other states are believed to be possible but remain theoretical for now. The aim of high energy physics is to determine the most fundamental building blocks of matter and to understand the interactions between these particles. The research effort of the high energy theory group covers a wide range of fields, including quantum field theory, string theory, quantum gravity models in various dimensions, and the theory of turbulence, particle cosmology, phenomenology of the Standard Model and beyond, and also computer simulations of problems that arise in these areas. Black hole theory provides an important testing ground for the quantum theory of gravity and in recent work significant progress has been achieved in explaining black hole entropy and Hawking radiation from a more fundamental point of view. Work on quantum black holes has led to new relations between strings and non-Abelian gauge theory. This application of string theory has already provided new insights into strongly coupled gauge theories, and it continues to be an exciting area.

This book gives intriguing approaches into the unusual forms and behavior of matter under extremely high pressures and temperatures, devoted to more specific topics which arise when nuclear collisions are considered as a tool for the experimental study of QCD thermodynamics. It also presents the studies into the behavior of substances at ultimately high pressures and temperatures obtainable by way of kinetic or electromagnetic energy cumulation in laboratory conditions. Also considered are the diversified states of matter and the processes occurring under gravitational forces and thermonuclear energy release. The book will be of interest to students, novice researchers and all the many young scientists starting their scientific research in this field, providing them with a general, self-sufficient introduction that highlight in particular the basic concepts and ideas.

Table of Contents

CHAPTER 1

Matter under Extreme Conditions Experiments at the Linac Coherent Light Source

S H Glenzer[1], L B Fletcher[1], E Galtier[1], B Nagler[1]

[1] *SLAC National Accelerator Laboratory, 2575 Sand Hill Road, MS 19, Menlo Park, CA 94025, USA*

[2] *Université de Bordeaux-CNRS-CEA, Centre Lasers Intenses et Applications (CELIA), Talence, F-33405, France*

ABSTRACT

The matter in extreme conditions end station at the Linac Coherent Light Source (LCLS) is a new tool enabling accurate pump–probe measurements for studying the physical properties of matter in the high-energy density (HED) physics regime. This instrument combines the world's brightest x-ray source, the LCLS x-ray beam, with high-power lasers consisting of two nanosecond Nd:glass laser beams and one short-pulse Ti:sapphire laser. These lasers produce short-lived states of matter with high pressures, high temperatures or high densities with properties that are important for applications in nuclear fusion research, laboratory astrophysics and the development of intense radiation sources. In the first experiments, we have performed highly accurate x-ray diffraction and x-ray Thomson scattering measurements on shock-compressed matter resolving the transition from compressed solid matter to a co-existence regime and into the warm dense matter state. These complex charged-particle systems are dominated by strong correlations and quantum effects. They exist in planetary interiors and laboratory experiments, e.g., during high-power laser interactions with solids or the compression phase of inertial confinement fusion implosions. Applying record peak brightness x-rays resolves the ionic interactions at atomic (Ångstrom) scale lengths and measure the static structure factor, which is a key quantity for determining equation of state data and important transport coefficients. Simultaneously, spectrally resolved measurements of plasmon features provide dynamic structure factor information that yield temperature and density with unprecedented precision at micron-scale resolution in dynamic compression experiments. These studies have demonstrated our ability to measure fundamental thermodynamic properties that determine the state of matter in the HED physics regime.

1. INTRODUCTION

Matter in extreme conditions (MEC) is central to scientific research aimed at demonstrating nuclear fusion in the laboratory [1–5], for developing intense radiation sources with unique properties [6–11], and to many fundamental physics processes in astrophysics [12–18]. The capabilities of the linac coherent light source (LCLS) allow unprecedented explorations in this area of research enabling ultrafast probing with high-repetition rates and high x-ray photon energies [19]. It is particularly suited for a class of experiments that produces short-lived states of matter with extreme conditions using intense high-power laser beams where penetrating high-peak brightness x-rays resolve the dynamic evolution of the target [20]. The x-rays visualize structural changes and determine the physical properties with spectrally resolved x-ray scattering [21, 22], x-ray imaging [23] and x-ray diffraction [22–25]. High repetition rates allow averaging and summation over many experiments to produce unprecedented data quality with large signal-to-noise ratios and high confidence in the results [21]. For this purpose, the LCLS x-ray beam has recently been combined with high-power lasers and a suite of optical and x-ray diagnostics at the MEC instrument located in the far experimental hall (FEH) of LCLS.

An areal view of LCLS and the location of the near and FEH are shown in figure 1. LCLS delivers approximately 10^{11}–10^{12} x-ray photons in a micron-scale focal spot and when operated in a seeded mode [26] allows measurements with high spectral resolution of $\Delta E/E = 10^{-4}$, high wave-number resolution of $\Delta k/k = 10^{-2}$, and high temporal resolution of 20–50 fs. The MEC instrument is equipped with one ultra-short pulse laser and two nanosecond laser beams that drive the material into extreme matter conditions, accessing a large phase space of high energy density (HED) physics coupled with the unique LCLS x-ray beam. At present, temperatures of 100 eV at solid density or a few eV at 3 times solid density have been reported [21, 27]. Recently, high-power ultra-short pulse laser–gas interaction experiments [28] have begun exploring much higher temperatures at sub-atmospheric target densities.

Figure 1. Areal view of the linac coherent light source. Electrons are injected at the 2 km point of the Linac that accelerates electron bunches over 1 km to energies ranging from 4.3 to 14 GeV. A 130 m long undulator is then producing the free electron laser x-ray beam through self amplified spontaneous emission of radiation (SASE) or through self-seeding. Energies are ranging from 1.1 Å to 15 Å at 120 Hz for experiments in the near and far experimental halls. The MEC is the 3rd instrument in the far experimental hall that is reached after a 200 m long x-ray transport line. At MEC, the x-ray beam can deliver x-ray energies from 480 to 11 000 eV with pulse durations of approximately 20 to 500 fs and up to 2 mJ energy per pulse.

Figure 2 shows the temperature–density parameter space that is currently accessible for ultrafast high-resolution pump–probe studies together with the expected increase after upgrading the short pulse laser to 100–200 TW peak power. Importantly, the parameter space overlaps with areas in phase space where the conditions access important physics areas and processes. For example at low temperatures, conditions can be produced that are found in giant planets, planetary interiors, brown dwarfs or meteor impacts. Both isochoric heating using free electron laser radiation [21, 29–32] or a high-power laser-produced radiation sources [33–37] and dynamic compression of matter by laser ablation pressure are suitable techniques [38–47]. The resulting HED physics conditions are often characterized by material pressures of 1 mega bar (100 GPa) and above [48]. In the high-density part of the phase space there is also significant overlap with conditions produced at the largest facilities such as the National Ignition Facility (NIF), the Omega laser, or the *Z* pulsed power facility (e.g. [49]).

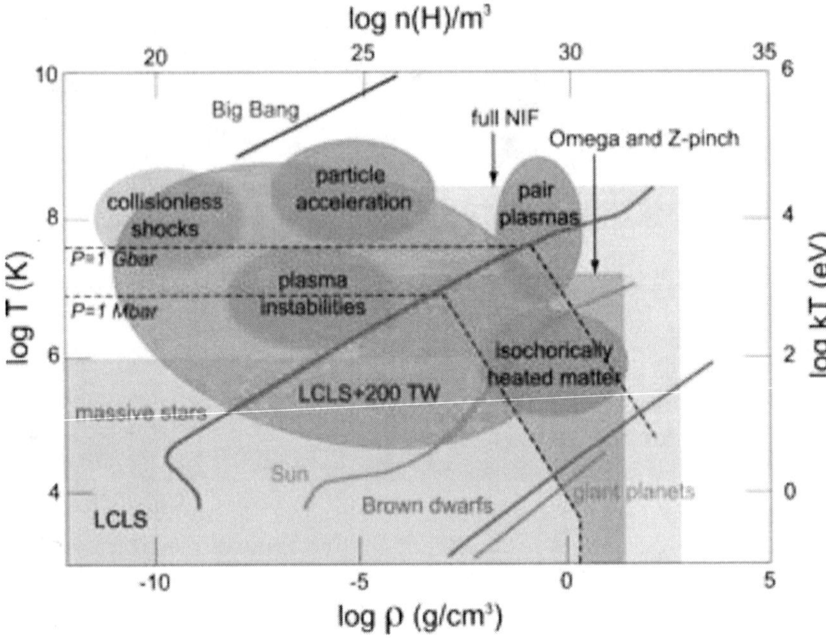

Figure 2. The range in temperature–density phase space is shown that can be produced with LCLS (beige), with LCLS and long pulse lasers, and with a 200 TW high intensity ultra-short pulse laser coupled to LCLS (green). The combination with the LCLS x-ray beam provides world unique experimental capability that will allow studies of laser–plasma interactions, coupling and heating of fusion plasmas, and the demonstration of the physics mechanism for the generation of intense radiation and particle sources by high power lasers. Also shown is the phase space covered by the Omega laser, the *Z* pulsed power facility, and the National Ignition Facility (NIF) laser facility (adopted from [49]).

Fusion research aims towards producing confined burning plasmas with temperatures and densities where heavy isotopes of hydrogen fuse to helium nuclei and neutrons producing 17.6 MeV energy per fusion reaction. This process powers the Sun and has led to all life on earth. The present worldwide effort generally pursues two complementary efforts where hot dense plasmas are confined by magnetic fields or by inertia and heated to fusion temperatures that are typically in the range of 5–50 keV (1 eV = 11 605 K) [50]. The understanding of the physical properties of fusion plasmas and the material and target conditions involved in producing and confining the plasma is of critical importance for the successful demonstration of a laboratory fusion concept and potential future developments as energy source. Key to these research efforts is the application of computer simulations that rely on equation of state (EoS) data and physics models describing thermal transport, thermodynamic equilibrium and hydrodynamic instabilities [51]. The physics becomes important at various stages of the formation of the plasma that follows trajectories through the phase space accessible at the MEC instrument [40].

In addition, to direct LCLS x-ray heating or laser dynamic compression techniques, a 30 TW Ti:sapphire short pulse laser beam has recently become available at MEC. This laser will be further upgraded to a 200 TW class system. This capability is of particular interest for studies of conditions including those relevant to energetic phenomena in astrophysics [12, 18], for producing sources of intense radiation and for studies of their interaction with matter. Specifically, high-power short pulse lasers produce high-temperature plasmas and relativistic electrons that propagate along filaments and lead to energetic particle beams [9]. Extremely high magnetic fields surround the filaments and affect energy transport and ionization. The target interior is also an intense source of positrons and high-energy photons, x- and γ-rays. LCLS will allow us to probe these conditions and to optimize their use as sources for interaction physics studies.

Central to studies of the high-power laser beam interactions with matter is the on going development of accurate and precise experimental pump–probe and diagnostic techniques. This is required to provide critical experimental tests of simulations and theory of matter in the HED state and further to uncover the underlying physics mechanisms that determine the coupling of ultra intense photon and particle beams with matter.

Figure 3 shows schematics of MEC studies with LCLS. High-energy x-ray pulses are required to penetrate through dense plasmas and to provide *in situ* measurements of the target conditions. The x-ray beam must provide sufficiently large numbers of photons in femtosecond pulses to allow scattering experiments in single shots during the short-lived states of the dynamically evolving plasma. Further, focusing of the beam to small volumes will allow unprecedented spatial resolution in studies of matter at HED and will resolve previously unexplored energy and particle transport properties. Novel combinations of high-power laser beams, laser-produced radiation, and multiple x-ray pulses with multiple x-ray energies at sub-nanosecond delay will open up novel pump–probe capabilities for resolving the evolution of extreme matter states towards or away from equilibrium.

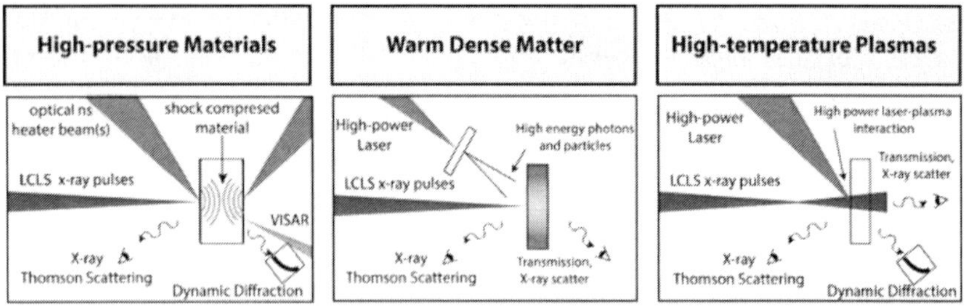

Figure 3. Examples of geometries are shown for investigations of matter in extreme conditions with ultrafast pump–probe experiments at LCLS.

The method of production of extreme matter states can vary from the direct irradiation of a sample by the LCLS or a high-power short pulse laser of thin (1 nm–10 000 nm) samples, the irradiation of thicker samples by long pulse lasers to compress samples, or the interaction of energetic particle beams produced by short pulse laser focused to high intensity (>10^{16} W cm^{-2}).

The first experiments on the MEC instrument have investigated laser-compressed solids. They address a wide variety of topics in high-pressure research including shock waves, equation-of-state, shock-induced chemical reactions, strength kinetics, dislocation dynamics, high strain rate phenomena, or plasmon and phonon dispersion [52–54]. Dynamic high-pressure experiments utilize nanosecond long pulse laser beams focused onto a target, to spot sizes of 50–400 μm using phase plate beam smoothing conditioning. The beams compress solid targets by ablation pressures in the mega bar pressure range, see figure 3 (left). X-ray scattering and dynamic diffraction measurements determine the properties of the HED state and can be complemented by shock and particle velocity characterization in the laser-driven target using a velocity interferometer system for any reflector (VISAR) [55], and a Fourier domain interferometer (FDI) [56]. These optical measurements characterize reflective surface motions from shock or compression waves and target surface expansion.

The recent addition of the high-power short pulse laser to MEC greatly enhances the capabilities for ultrafast pump–probe studies allowing direct x-ray probing of high-power laser interaction with matter or optical probing of x-ray heated systems, see figure 3 (right). In addition, secondary photon and particle beams are now becoming available for interaction studies with warm dense matter (WDM) or to study the dynamic heating into the WDM state using the LCLS x-ray beam, see figure 3 (middle). WDM is a state of matter that is close to solid density, but with temperatures ranging from a couple of eV to tens of eV. In this state, standard plasma theories are not valid as the matter is too dense and cold, and neither are solid-state theories because the thermal energy is of the same order as the interaction energies between atoms and perturbation theory breaks down. Despite its transient small-scale character in laboratory experiments in, e.g., laser- or beam-produced plasma experiments, WDM is found abundantly in nature, for example in steady state conditions in interiors of large planets and accretion disks, or in transient conditions during large (meteor) impact events.

This paper reviews the first LCLS experiments that have investigated high-pressure states in single shot mode. Of particular importance for successful studies of HED matter are independent measurements of temperatures, densities and ionization state. For this purpose, MEC provides the capability for spectrally resolved x-ray Thomson scattering (XRTS) measurements. This technique was first developed using backscatter geometries accessing the non-collective Compton scattering regime that provides information on the microscopic physics by measuring the free electron distribution function [20, 35]. At low-temperatures and small degrees of ionization, the free–free Compton feature blends with bound-free scattering and the data interpretation becomes complex requiring additional information of finite-temperature bound wave functions and continuum lowering. On the other hand, the unique high-peak brightness properties of the LCLS x-ray laser are well suited for collective forward scattering on plasmons whose spectra yield temperatures and densities with unprecedented precision at micron-scale resolution in dynamic compression experiments. The plasmon scattering spectrum is of fundamental interest because it holds promise to determine plasma parameters and the physical properties from first principles using the plasmon dispersion and the principle of detailed balance. These techniques have been developed at LCLS utilizing high repetition rates of 120 Hz measuring quasi noise-free plasmon spectra within 7 s [21, 57]. Subsequent single shot measurements in dynamic compression experiments [22] have measured temperatures and densities providing data needed to infer pressures and compression.

Our experiments are precise enough to critically test theory of dense plasmas achieving an accuracy of the elastic scattering feature data that can clearly distinguish between the results of simulations and various theoretical models [58–63]. The LCLS beam visualizes the evolution of the compressed material lattice across the melting line and the coexistence regime into a WDM state [22]. The x-ray diffraction measurements provide structural measurements in these different regimes. In particular for the WDM states, the data reveal a narrow scattering peak that can be interpreted as the combined effects of screening and short-range repulsion (SRR) in the ion–ion interaction potential [58]. The measurements test predictions from density functional theory coupled to many particle molecular-dynamics (DFT-MD) simulations [60–64]. Up to now, these simulations were used to predict physical properties derived from optical observations of particle and shock velocities. Here, structural and physical properties that are sensitive to many-particle electron–ion (e–i) and ion–ion interaction physics have been obtained. An independent theoretical analysis using pseudo-atom molecular dynamics [65, 66] has shown excellent agreement with the LCLS data when using Kohn–Sham DFT. However, the Thomas–Fermi (orbital free) form of DFT was found to show significant discrepancies, while the hyper-netted chain (HNC) approximation for the Ornstein–Zernike equations was found to be marginal [67].

The experimental data further allow a direct calculation of macroscopic material properties. Especially equation-of-state data can be derived on the shock Hugoniot together with new off-Hugoniot data that occur by shock coalescence. We find that they approach an isentrope and are in agreement with the DFT-MD simulations performed for the measured temperature and density values of the experiment

2. OVERVIEW OF THE MEC INSTRUMENT AT LCLS

2.1. LCLS x-ray beam

LCLS utilizes one third of the 3 km long SLAC Linac [68] to accelerate electrons with a single bunch charge of 150 pC up to approximately 13.6 GeV electron beam energy. During a single pass through a long (100 m) undulator, this pulse will initiate fee electron laser (FEL) action at the undulator fundamental wavelength, which will be tunable between 15 Å and 1.5 Å. The FEL will either operate in self-amplified spontaneous emission (SASE) mode or in a seeded mode. In SASE mode, the FEL radiation field will arise from amplification of the spontaneous undulator radiation created at the start of the long undulator resulting in a bandwith of approximately 40 eV at 8 keV. The seeded LCLS x-ray beam will result in a bandwidth of about 1 eV at 8 keV. Thus, the spectral resolution in seeded-beam LCLS scattering experiments will be limited by the x-ray spectrometer.

To resolve the electron plasma (Langmuir) oscillations [69], we operate LCLS in the seeded x-ray mode a just recently invented tool [26]. In this study, we developed 8 keV peak brightness radiation of 2.7×10^{34} photons $s^{-1}mm^{-2}$ $mrad^{-2}$ 0.1% bandwidth at LCLS, see figure (4). Within the first 15 undulators the Linac produces SASE with 1 GW power. The electron beam is then passed through a 4 m long chicane while the x-ray beam traverses a diamond crystal that Bragg reflects a narrow energy range of x-rays with a bandwidth of $\Delta E/E = 0.5 \times 10^{-4}$ in the center of the broad SASE spectrum. The diamond (400) reflection was chosen as the wake monochromator for suitable crystal perfection, angular width and low absorption. Thus, the transmitted x-ray spectrum creates a 5 MW trailing monochromatic seed pulse, which has been amplified to ultrahigh peak brightness and total power of 10–15 GW in the subsequent 17 undulators. The resulting record peak brightness x-ray beam is then transported and focused into the dense laser-driven target.

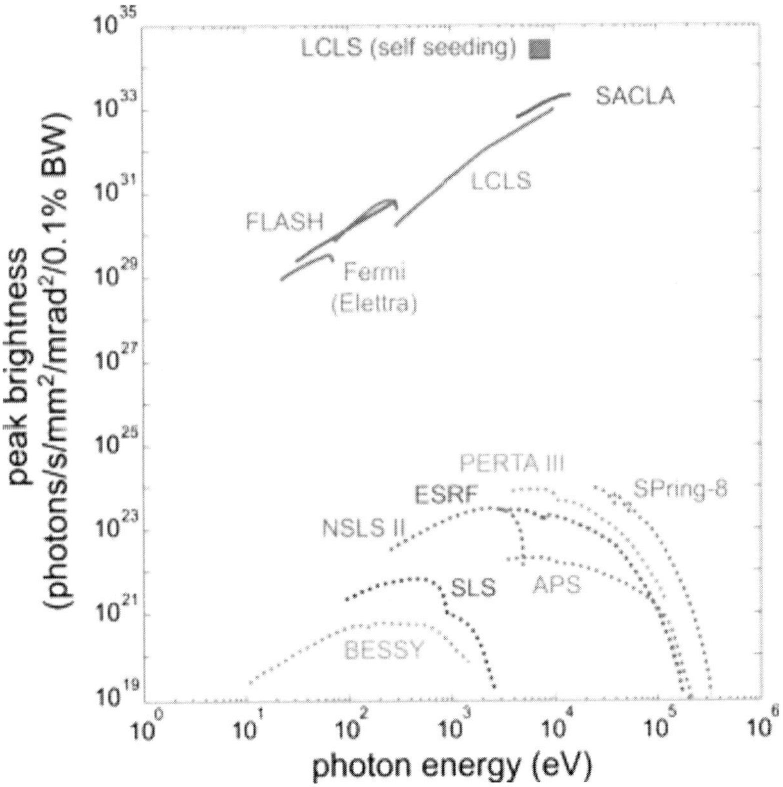

Figure 4. Comparison of peak brightness of various x-ray light sources is shown. An internally self-seeded beam at record peak brightness provides the required photon flux to measure scattering spectra and diffraction data in single dynamic compression experiments and provides the spectral bandwidth to resolve plasmons spectral features.

The operational phase has shown that the normal operation of hard x-ray self-seeding performs better with the nominal 150 pC charge and the corresponding 50 fs pulse duration. The stability is improved by roughly a factor of two compared to low charge 20 pC operation with a similar reduction in the per pulse energy fluctuations. In this mode, the pulse energy of the seeded beam is significantly improved with the longer pulse (higher charge) mode. Recently, multi-pulse, multi-color x-ray pulses have been delivered in this mode [70]. The seeded beam normally gives a 2–4 times improvement in time-averaged x-ray power compared with a post-monochromator of similar bandwidth with similar pulse durations and shot-to-shot intensity fluctuations. The narrow seeded line, 0.4–1.1 eV full-width at half maximum, for a 50 fs pulse duration typically contains an average pulse energy of 0.3 mJ, with occasional shots up to 1 mJ. The peak brightness of 2.7×10^{34} photons s^{-1} mm^{-2} $mrad^{-2}$ 0.1% BW is calculated from the measured pulse duration and a mean pulse energy of 0.3 mJ.

2.2. MEC x-ray instrumentation

The particular strength of the MEC instrument [23] is to combine the LCLS x-ray laser beam with high power optical laser beams [71], and a suite of dedicated diagnostics tailored for HED science. They include two x-ray scattering spectrometers for forward collective and backward non-collective scattering, emission spectrometers and interferometer systems. While the vacuum target chamber makes the end-station very versatile, it has been designed to service key scientific areas including WDM physics, high-pressure studies, and laser–plasma interactions.

MEC is the third instrument in the FEH delivering a focused x-ray beam in the energy range $4 < E_{\text{x-ray}}$ 1st < 11 keV for 1st harmonic experiments or $10 < E_{\text{x-ray}}$ 3rd < 25 for experiments with 3rd harmonic beams. A selection of compound refractive lenses can be inserted in the beam to provide focusing at target chamber center for the selected x-ray energy. By locating the sample at or out of focus, the beam size at the sample can be selected from about ~3 μm to several tens of microns. Figure 5 indicates the location of the instrument in the FEH and a schematic of the x-ray focusing and short pulse laser systems.

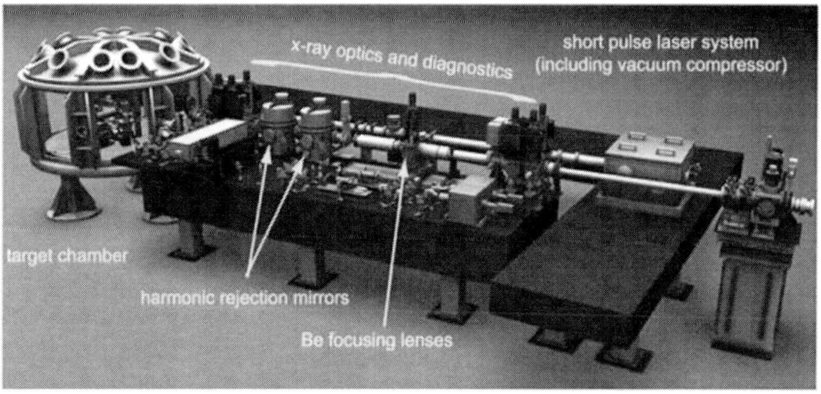

Figure 5. (Left) The x-ray beam is transported 200 m between the near and far experimental halls and combined with high-power lasers at the MEC instrument at the end of the facility. (Right) The x-ray beam is focussed by compound beryllium lenses into the center of the target chamber. The beamline further consists of third harmonic rejection mirrors, x-ray beam intensity, profile, and positioning monitors. Also shown is the path of the short pulse laser system. The long pulse lasers enter the target chamber at an angle of 90° to the x-ray beam axis and can be redirected inside the target chamber using 2" optics (not shown).

Figure 6 shows a schematic of the x-ray beam characterization tools in MEC and examples of x-ray near and far field images from CCD and imprint data. The x-ray focusing capability is a prerequisite for high spatial resolution x-ray imaging and for controlled x-ray scattering measurements in a well-defined target volume. Further, the high photon flux allows direct isochoric x-ray heating of matter that can be probed by, e.g., two-color x-ray probing or ultrafast optical laser measurements. The latter requires accurate cross timing of x-ray and optical laser beams, which can be achieved with an uncertainty of approximately 50 fs.

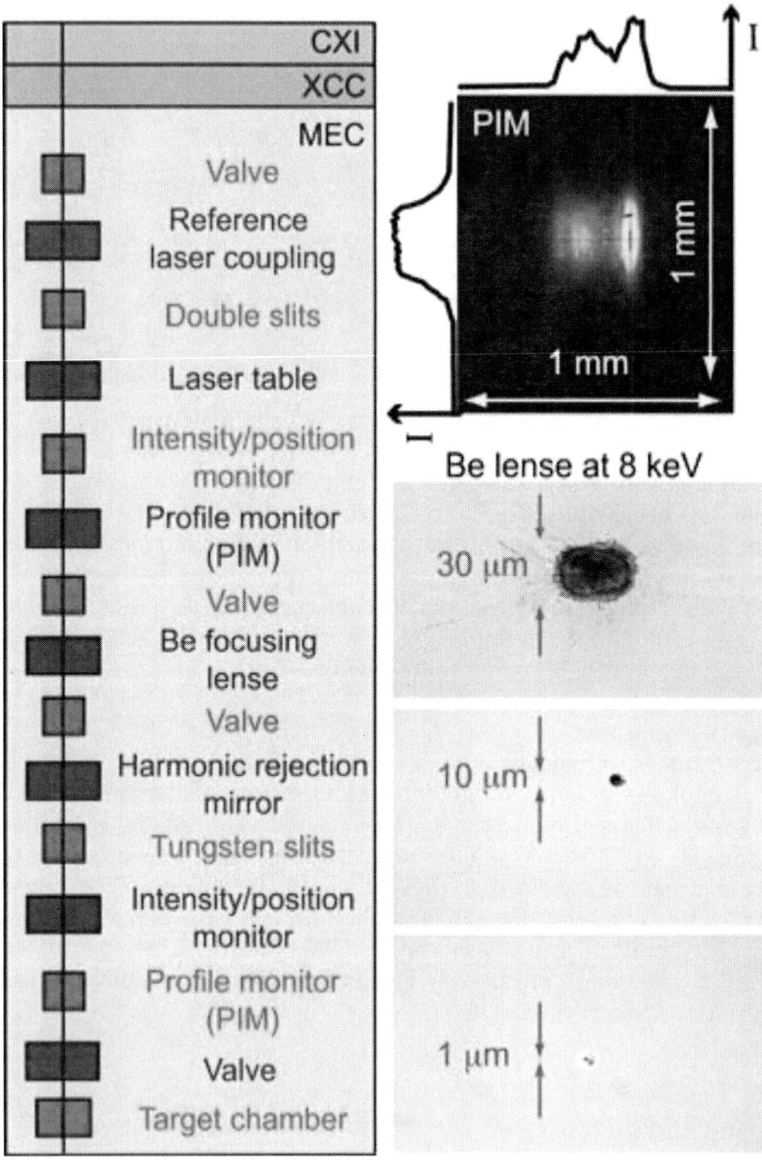

Figure 6. (Left) A schematic is shown of the x-ray beam transport optics and x-ray characterization tools at MEC. (Right) Example of x-ray near field intensity distribution is shown before entering the target chamber. Also shown are examples of x-ray focal spot measurements using imprints on targets coated with 100 nm thick Au layers.

2.3. Target configurations

The existing target chamber at MEC allows for great flexibility of the x-ray focus location, the optical laser mirror and lens locations, diagnostic instruments, and targets options. Ongoing experiments at the Titan [39] and Trident [72] laser facilities have previously demonstrated a subset of experimental configurations and target types in the same target chamber architecture. Here, the LCLS x-ray beam will be a new feature increasing the complexity of the experimental platforms and laser set up. Further, the high repetition rates of LCLS of 120 Hz require new developments in target delivery to fully utilize the x-ray capabilities. Figure 7 shows two examples of target capabilities that have been successfully fielded at MEC.

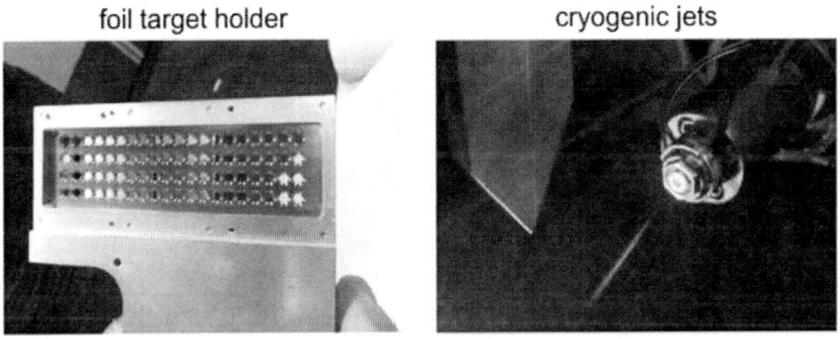

Figure 7. (Left) The target holder is shown for solid target experiments. The long pulse lasers heat the foil that is mounted in one opening in a single dynamic compression experiment. Subsequent experiments use a new foil target in the next opening. (Right) An example of a continuous cryogenic jet is shown. These targets allow experiments at high repetition rates. Pump–probe studies using LCLS combined with the short pulse laser have been demonstrated at 5 Hz.

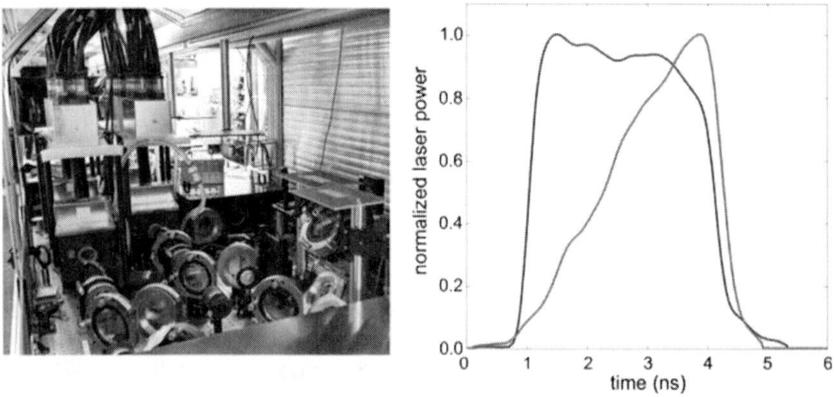

Figure 8. (Left) Picture of MEC Nd:glass laser system showing the 25 mm rod amplifiers. A laser pulse shaping system delivers about 4.5 J in 3 ns per beam on target at 527 nm. (Right) Examples of laser pulse shapes are shown for shock (blue) and ramp wave (red) compression experiments.

The solid-density target holder provides thin foils for experiments where strong shocks have been launched on either side by ablation pressure produced by the nanosecond laser beams. The shocked matter temperature and density conditions and the shock speeds will be probed with spectrally, temporally and wavenumber resolved x-ray scattering and optical measurements on targets irradiated on either side or by a single beam from one side.

For this class of experiments, the shot repetition rates are limited by (1) target alignment time and (2) by the cooling time of the glass rod amplifiers. Here, it should be noted that when using the x-ray beam alone, 120 Hz experiments have been demonstrated because the foil is heated by the x-ray beam locally over diameters of several 10 μm and can be rapidly translated orthogonal to the x-ray beam axis to deliver a new target surface for subsequent shots, see [21, 57].

More recently, continuous cryogenic liquid jets have been developed and fielded at MEC. First experiments used hydrogen for laser–plasma experiments with densities of 0.07 g cc^{-1} ($n_0 = 4.2 \times 10^{22}$ cm^{-3}), see figure 7 (right). This capability allows experiments at high repetition rates for structural investigations and ultrafast heating studies. Typical jet diameters are between 8 and 20 μm [31, 32], with recent developments towards smaller diameter jets being under way. Other demonstrated cryogenic jet targets include argon, CO_2, and deuterium.

2.4. Optical high-power laser systems

2.4.1. Long pulse Nd:glass laser

MEC is equipped with two frequency doubled Nd:Glass laser systems [23, 71] that each deliver green (527 nm) light on target with energies up to 18 J for pulse durations larger than 15 ns. For typical applications with pulse lengths of several nanoseconds, the peak laser power is 1–2 GW per beam with variable pulse shaping and rise times of 0.5–1 ns see figure 8. The arrival time of the optical laser beams with respect to the x-ray beam can be changed during the experiment, and is accurate within 25 ps rms. Typical repetition rates for the Nd:glass laser are 1 shot every 7 min when using both lasers simultaneously.

To access a broad range of interesting regimes by dynamic compression it is desirable to further improve the existing systems. At highest priority is increasing the total laser energy on target, to provide precision laser pulse shaping, and shot-by-shot pulse monitoring with the goal to drive larger samples with spatially flat topped intensity profiles for planar shocks. In addition, cross timing and equivalent plane focal spot monitoring will enable measurements of the laser-beam uniformity at the plane of the target. For this purpose, the monopotassium phosphate frequency doubling crystals may be replaced by lithium triborate crystals for a 30% increase of laser energy on target. In addition, for tailoring the focal spot profile on target a suite of phase plates are available to deliver spot diameters in the range of $50\,\mu\text{m} < d < 400\,\mu\text{m}$. To further reduce the effects of laser-beam non-uniformities experiments on solid targets have employed low-Z ablators of varying thickness from $2\,\mu\text{m} < d < 50\,\mu\text{m}$. Future improvements in laser technology and additional amplifier stages for delivering increased laser energy on target will be important to continue accessing important new regimes of the HED phase space, see figure 2.

2.4.2. Short pulse Ti:sapphire laser

The existing MEC short pulse laser system is shown schematically in figure 9. The system is capable of producing 1 J pulses with a temporal width of 40 fs, and thus delivers a peak power of 25 TW. When focused to $a \sim 5\,\mu\text{m}$ spot size on a target, this provides peak intensities in the relativistic regime where the electron quiver velocity approaches relativistic velocities, i.e. at intensities exceeding 10^{19} W cm^{-2}. The repetition rate of this system is determined by the final amplifier GAIA pump laser and is currently 5 Hz.

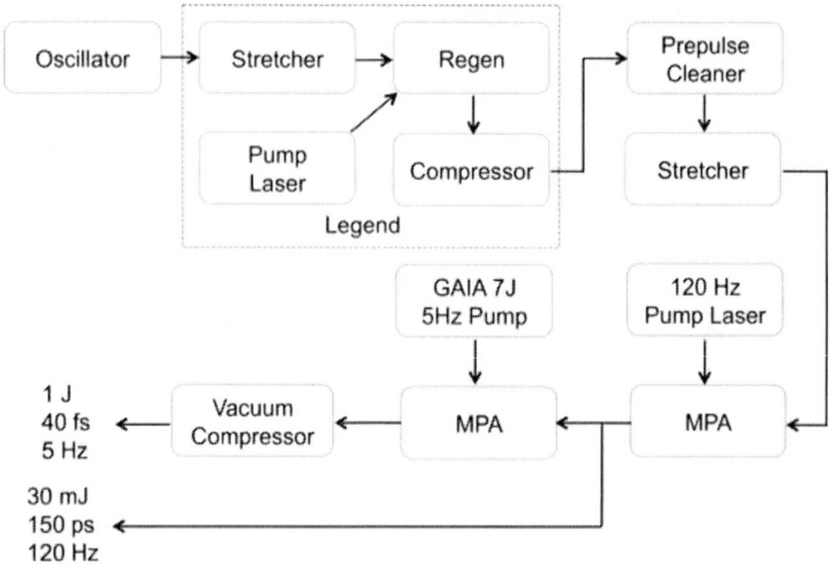

Figure 9. Schematic for two different modes of short pulse laser operations at MEC is shown. Pulses at 30 mJ can be delivered at 120 Hz to match the LCLS x-ray beam repetition rate. A higher 25 TW laser power option exists at 5 Hz.

At relativistic laser intensities, the contrast ratio between the main pulse and any possible pre-pulses or pedestal pulses is important. To reduce the pre-pulse intensity a cross-polarized wave generator (XPW) has been implemented. In this case, the pulse is compressed after the regenerative amplifier (Legend). The resulting short pulse can then be cleaned in a saturable absorber or XPW device. These pulse 'cleaning' techniques have demonstrated contrast ratios in the 10^{10} range, albeit at the loss of pulse energy. To compensate the loss an additional multi-pass amplifier (MPA) has been added. Because these MPA are pumped by CW diode-pumped Nd:YLF lasers, they provide an exceptionally stable, typically better than 0.2% rms energy stability, and a near perfect Gaussian spatial profile. Because these lasers run at 120 Hz, this will provide the option to have lower energy 30 mJ short pulses at 120 Hz to match the pulse rate of LCLS. The addition of a MPA that can be pumped by a 7 J, GAIA laser provides higher-energy high-power experiments at 25 TW albeit at reduced repetition rates of 5 Hz.

The system can be further upgraded to 200 TW peak laser power by adding a third MPA to the laser to be pumped by the MEC nanosecond laser. At this stage, larger diffraction gratings may be needed in the pulse vacuum compressor in order to accommodate the increased energy without damage. Also, a deformable mirror and wavefront sensor system will be fielded to optimize the wavefront for the tightest possible focal spot. This should provide 6–8 J of output energy in a ~45 fs pulse width with good spatial wave front and temporal contrast. This high-power laser will produce focused peak intensities in the 10^{20} W cm^{-2} range at a repetition rate that is defined by the nanosecond laser at 1 shot every 3–10 min. It may further be possible to develop a burst mode which will deliver a 0.01 Hz shot rate followed by an extended cool down period that may be used for other experimental activities.

2.5. Diagnostic systems

2.5.1. General description

The size of the MEC target chamber with a radius of 1 m provides sufficient space to place in-vacuum optical and x-ray diagnostics. An extensive suite of x-ray scattering diagnostics is available including highly efficient graphite crystal spectrometers for single shot measurements of scattering spectra, high-resolution spectrometers for x-ray emission and absorption measurements, and areal detectors for angularly resolved scattering measurements, and phase contrast imaging. The suite of x-ray diagnostics is complemented by optical measurement capabilities, FDI and VISAR. In addition, high magnification optical backlighting has been applied using the optical short pulse laser. For alignment purposes, questar long distance microscopes are available providing a field of view of 0.5–5 mm with a spatial resolution of 7 μm. Below, we discuss examples of x-ray and optical measurements applied during the first shock-compressed matter experiments at MEC.

2.5.2. Spectrally-resolved x-ray scattering

2.5.2.1. Highly annealed pyrolytic graphite (HAPG) spectrometer

To maximize the collection fraction, x-ray scattering experiments require spectrometers with high efficiency, but with high resolving power of $\Delta E/E < 10^{-3}$. In the multi-keV x-ray regime, artificial crystals from low-Z materials such as carbon with a mosaic spread of ~0.1° approach these requirements. To further improve collection efficiency the distances from the crystal to both source and detector are typically chosen to be the same, resulting in mosaic focusing and high spectral resolution [73]. In addition, a cylindrically bent crystal in von-Hamos geometry will further optimize collection by about a factor of 2 [74].

At MEC, HAPG crystals have been fielded [75, 76]. These crystals have been shown to provide advantages over previously used HOPG [74, 77]. The point-spread function is dominated by so-called depth broadening. Due to negligible photo-absorption (the attenuation length of 8 keV x-rays in C is about 1 mm), x-rays penetrate deep into the crystal until they encounter a crystallite that is properly oriented to reflect the ray. In von-Hamos geometry, a x-ray point source will therefore be broadened in the detector plane and in the dispersion plane by $2D/\tan(\theta_B)$ with the crystal thickness D and the Bragg angle θ_B. Ray tracing simulations for a point source and monochromatic 8 keV x-rays have been performed for HAPG crystals of 30 × 30 mm^2 size and 51.7 mm radius of curvature ($\theta_B = 13°$). The calculations show that the point spread function of the 40 μm thick crystal is about 6 eV wide with $E/\Delta E = 1300$ suitable for resolving plasmon scattering spectra in dense plasmas. This crystal has been used for the forward scattering measurements. In backscatter, a 100 μm thick crystal has been employed with lower spectral resolution of 14 eV and $E/\Delta E = 550$. This crystal results in a higher reflectivity, as determined by the integral over the reflection curve. Some decrease of reflectivity in the high-photon energy tail is observed due to photo-absorption in the crystal volume.

Figure 10 compares the instrument function for a single shot with the one for experiments that average over 700 shots with a seeded x-ray beam. Here, we observe that statistical fluctuation in the x-ray laser energy result in additional broadening of the effective profile. Even when taking these fluctuations into account the diagnostic is well suited for resolving the plasmon scattering spectrum and the plasmon frequency shift.

2.5.2.2. Plasmon measurements

Plasmon measurements were first validated at MEC utilizing high-resolution measurements from isochorically heated targets at 120 Hz repetition rates. Subsequently, the technique was applied for single shot measurements on shock-compressed aluminum. Figure 11 shows the experimental scattering spectra from solid-density aluminum from Fletcher *et al* [57]. Results are shown from the forward scattering and backward scattering spectrometers with the x-ray beam in seeded and SASE mode of operation. The seeded x-ray beam provides a bandwidth of 1 eV and combined with 8 eV spectrometer resolution resolves the plasmon that is down-shifted from the elastic 8 keV scattering feature by 19 eV.

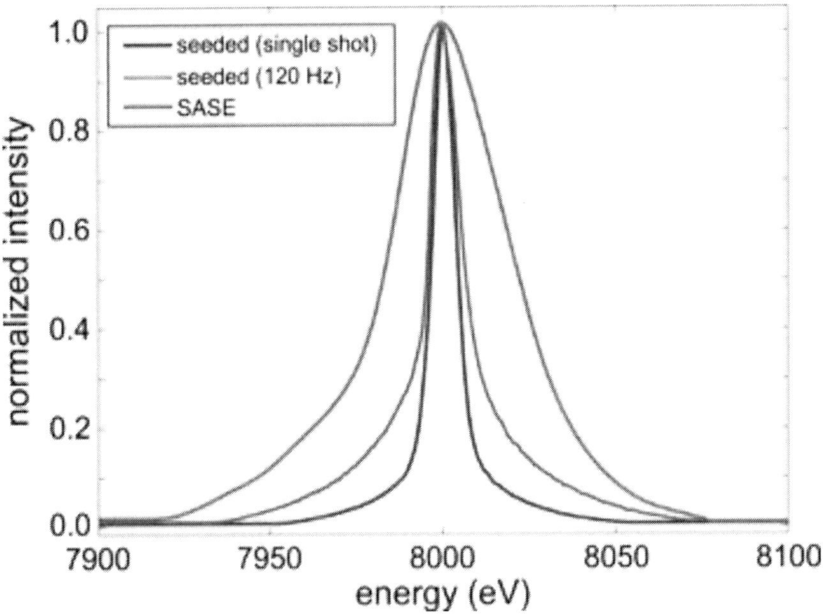

Figure 10. The HAPG instrument function is shown for a 40 μm thick crystal used for forward (plasmon) scattering experiments. The single shot profile is compared with the one obtained when averaging of 700 seeded beam shots with slightly fluctuating energy at a high-repetition rate. These spectra demonstrate a spectral resolution of 0.1%.

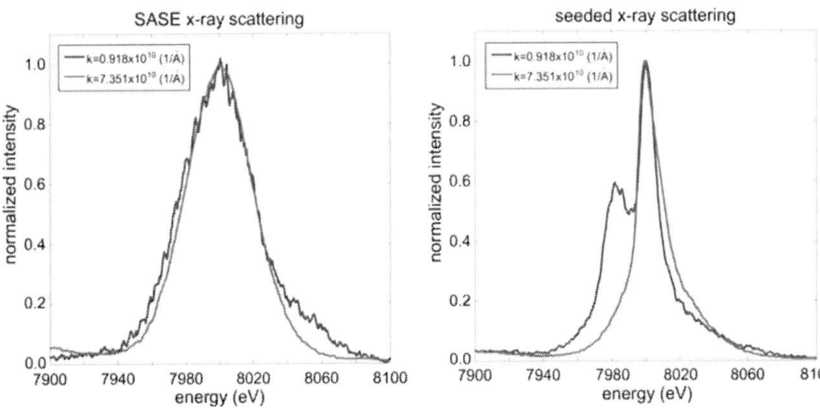

Figure 11. X-ray Thomson scattering data from solid density aluminum targets are shown from forward (plasmon) scattering and backscattering spectrometers for seeded (right) and SASE (left) x-ray beam operation [55]. These data are accumulated over 700 shots at 120 Hz. The data indicate 8 eV resolution for a seeded beam and 50 eV for SASE operations.

Further, figure 11 shows the results of the backward scattering spectrometer which observes elastic scattering at 8 keV reflecting the instrument function of the spectrometer convolved with a Gaussian profile that accounts for the statistical fluctuations of the x-ray energy of a seeded beam over 700 shots. These measurements do not indicate an inelastic scattering feature in the vicinity of the plasmon peak validating that bound-free scattering is not affecting the results of the forward scattering measurements; bound-free scattering features are predicted to be negligible in this energy range and no feature has been observed with the backscattering spectrometer. For these measurements, the crystal in the forward scattering spectrometer is 40 μm thick while in backscattering we employed a 100 μm thick crystal giving rise to slightly different instrument functions and consequently slight differences in the spectral shape of the elastic scattering feature.

Because the x-ray seed power is generated from the random SASE process that fluctuates in energy and amplitude, less than 1/3 of the shots provide sufficient photons with adequate spectrum for single shot x-ray scattering experiments. When summing over 700 shots the spectrum shows a broad foot, but the bandwidth is mostly preserved greatly improving the resolution when compared to the SASE spectrum. In forward direction, the x-ray scattering spectra with SASE operation show a slight broadening of the red wing of the scattering spectrum, see figure 11, but the SASE scattering spectrum is not suitable for inferring the dense plasma conditions.

Figure 12 shows the theoretical fit of the dynamic structure factor, see section 3, to the experimental forward scattering spectrum. The shift of the plasmon peak yields the electron density of $n_e = 1.8 \times 10^{23}$ cm$^{-3} \pm 5\%$. The electron temperature of the fit is 2 eV accounting for the isochoric heating of the aluminum by the x-ray laser pulse [22]. The ions are cold with temperatures of order 0.1 eV consistent with the observation of Debye–Scherrer rings from solid aluminum. Here, the plasmon frequency is extremely sensitive to the electron density, but temperature effects are too small to affect the plasmon energy shift. The accuracy of the electron density measurement is very high due to the sensitivity of the plasmon resonance to the plasma frequency. With $n_e = Z/A 6.03 \times 10^{23}\,\rho$ (the electron density n_e in cm^{-3} and the mass density ρ in g cm^{-3}), assuming the ionization state of $Z = 3$ and using the mass $A = 26.98\,u$, the measured electron density results in $\rho = 2.7$ g cm^{-3} as expected for solid aluminum.

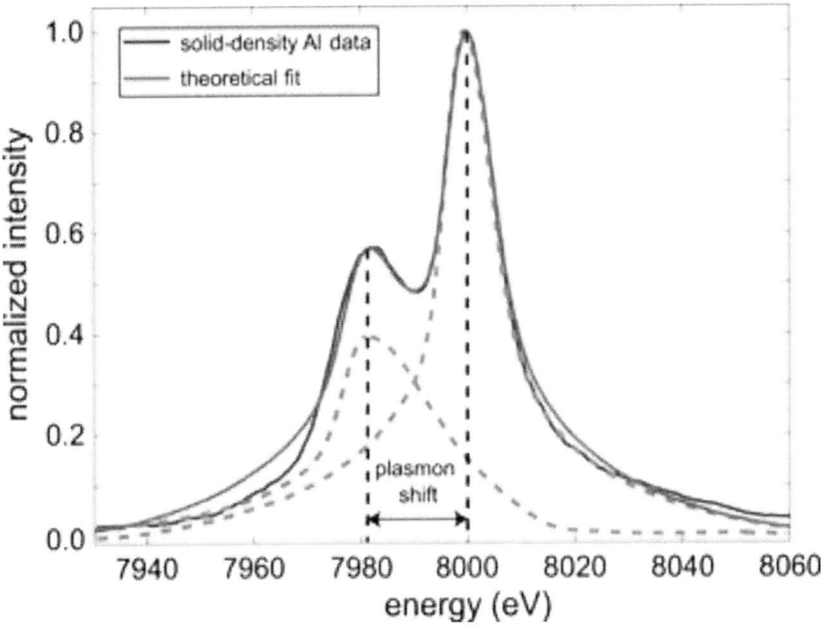

Figure 12. The theoretical fit to the experimental forward x-ray plasmon scattering data from solid-density aluminum is shown for the electron density of $n_e = 1.8 \times 10^{23}$ cm^{-3}. In these near Fermi-degenerate states, the theoretical fit of the plasmon feature is not sensitive to the electron temperature; however, the intensity of the elastic scattering feature at 8 keV depends on both electron and ion temperature. The dashed curve centered at 8000 eV shows the elastic scattering component and the dashed curve centered at 7981 eV represents the plasmon spectrum.

2.5.3. VISAR system

The MEC instrument is equipped with a line imaging VISAR system for velocity measurements of rapid surface motions that are typically in the range of 1–20 km s^{-1}. This diagnostic tool obtains accurate continuous velocity-time histories of surface motions. The VISAR measures velocities by using the Doppler shift of a green laser beam diffusely reflected from the moving surface. The reflected light is collected and collimated into a Michelson-type interferometer. The imaging field of view is 0.5–1 mm with a spatial resolution of 10 μm. The interferometer splits the light into equal parts so that the apparent optical path lengths of the two arms are equal to allow interference of spatially incoherent light, but at the same time with light from one arm delayed in time so that fringe shifts occur with wavelength changes. These requirements are satisfied by the use of an etalon in one arm. At MEC, several etalons of 25 mm diameter and with varying thickness between 5.072 mm–15.01 mm and of 50 mm diameter and with varying thickness in the range of 25.036 mm–75.04 mm are available. Two Hamamatsu streak cameras measure the fringes in two arms with different etalons providing a temporal window of 1 ns–1 ms and temporal resolution of approximately 20 ps. The system can operate at a repetition rate of about 0.1 Hz.

Figure 13 shows an example of a VISAR streak camera record when used in combination with the LCLS x-ray beam. This experiment uses a graphite target irradiated by a 1 TW frequency-doubled drive beam launching a shock wave that propagates through the target. The probe beam reflection from VISAR shows steady interference fringes from the backside of target that is coated with a reflective aluminum surface. The fringes disappear at the time of shock breakout when the optical reflectivity of the surface drops significantly. With exact cross timing of the VISAR probe laser, streak camera and drive laser these data yield the averaged shock velocity. In addition, step targets or wedge targets will provide more accurate measurements of the shock velocity. However, a unique feature can be observed at the time when the LCLS x-ray beam is fired (in this case about

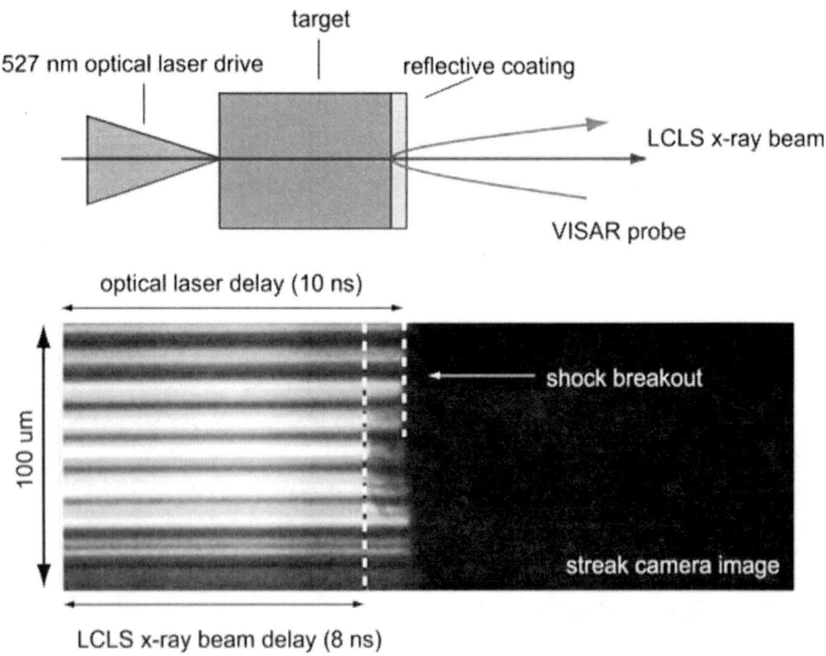

Figure 13. An example of a VISAR streak camera record shows the drop in the fringe intensity due to the heating of the rear surface by the LCLS x-ray beam. These measurements provide timing and location of the x-ray beam with respect to shock break out.

¾ through the optical laser drive). At that time, the reflectivity of the back coating is reduced providing an accurate measurement of both the x-ray probe time with respect to shock breakout and the exact location of the LCLS beam in the target.

3. THEORY AND SIMULATIONS

3.1. Dynamic compression using counter-propagating shock waves

3.1.1. Simulations of shock states

Figure 14 shows examples of radiation-hydrodynamic design calculations of shock-compressed solid aluminum experiments that utilize experimental laser powers, pulse shapes and phase-plate focal spots [78] available at MEC. A 2 μm thick CH coating was applied on the laser drive sides. The simulations use the radiation-hydrodynamic code HELIOS [79]. The nanosecond laser beams are frequency converted to green delivering 527 nm laser light on target with the pulse shape shown in blue in figure 8 in a focal spot of approximately 50 μm.

The simulations indicate that the shocks propagate through the solid with an averaged shock velocity of 14 km s^{-1}, traversing 25 μm thick aluminum plus 2 μm CH and break out or coalesce in the center at about 1.9 ns after the start of the laser irradiation. These shock waves initially compress the solid by a factor of 2–3 and during coalescence they reach 4 times solid density and multi mega bar pressures. These conditions can be compared with the Rankine–Hugoniot relations with $\rho_1/\rho_0 = D/(D - u)$, where D is the shock velocity, u is the particle velocity. For two-fold compression, i.e. $\rho_1/\rho_0 = 2$, we infer ½ $D = u$, and with a shock velocity of 14 km s^{-1} (20 km s^{-1}) we obtain pressures of $\Delta P = \rho_0 Du \sim$ 2.6 Mbar (5.4 Mbar). Here, the density and pressure define the Hugoniot curve of the shocked material. Experiments that will simultaneously measure the temperature will determine the thermodynamic state of the shock wave and will be important to estimate heat capacity and energy partition of the shocked material.

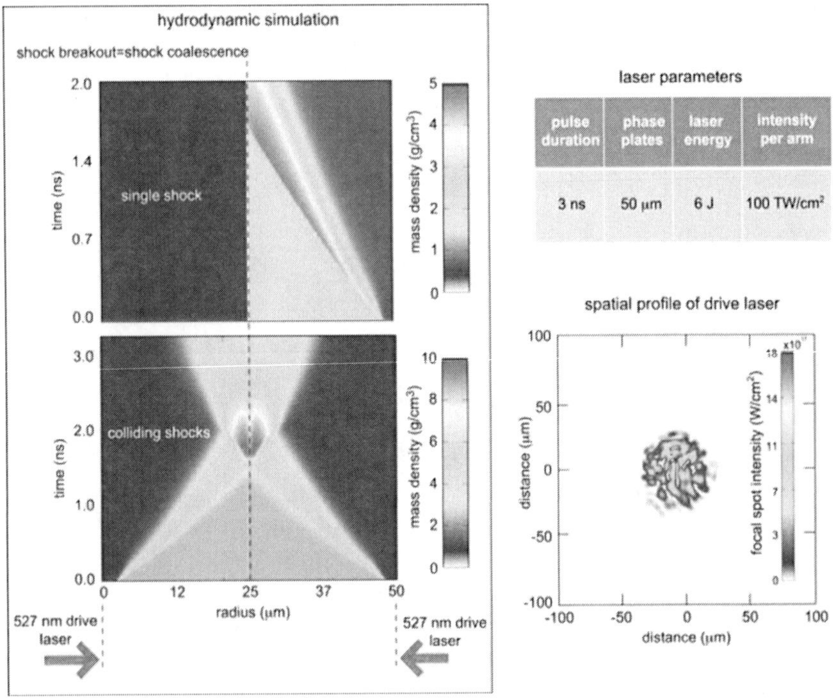

Figure 14. (Left) Radiation-hydrodynamic design of single shock wave experiments and counter-propagating shocks driven by two MEC nanosecond lasers with the same pulse shape. (Right) Laser intensities and focal spot data are obtained from phase plate characterization.

3.1.2. Counter-propagating shocks

To reach high compression states of solid targets by intense laser radiation in the laboratory, the technique of coalescing shocks is often applied [39, 44]. It is well known that a large number of shocks with a fixed pressure change results in adiabatic compression also described with negligible entropy changes. At MEC, the counterpropagating shock technique has been applied in experiments that reach several times solid density, temperatures of several 10 000 K, and pressures in excess of 1 Mbar. Here, the entropy largely determined by the conditions in the first shock. These extreme states of matter serve as model systems to investigate the thermodynamic state, the material structure, and the physical (microscopic) properties.

For these studies, the adiabatic exponent $\gamma = (\partial \ln p / \partial \ln \rho)_{S=\mathrm{const}}$ is a central quantity. For perfect gases, $\gamma = c_p/c_v$ is the ratio of heat capacities at constant pressure and volume, respectively. Furthermore, γ determines the sound speed, and the EoS [1, 47, 48, 80–82]. For the examples below we choose $\rho_0 = 2.7$ g cc^{-1}. Here and in the following, we denote conditions in the uncompressed material by the index 0, the single shock compressed matter by index 1, and the region of shock overlap by index 2.

We apply the Rankine–Hugoniot equations [47] to determine the polytropic index γ. We describe each shock by the Rankine–Hugoniot equation which relates the particle velocities u_0, u_1, pressures p_0, p_1 and mass densities ρ_0, ρ_1 in the uncompressed and in the compressed region, respectively and which follow from the conservation of total mass, energy, and momentum. One finds for the energy jump across the shock front

$$\begin{aligned} \mathcal{E}_0 - \mathcal{E}_1 &= u_0^2 - u_1^2 + \frac{p_0}{\rho_0} - \frac{p_1}{\rho_1} \\ &= \frac{1}{2}(p_0 + p_1)(\rho_1^{-1} - \rho_0^{-1}). \end{aligned} \tag{1}$$

These relations hold for each individual shock as well as for the two shock fronts that evolve after the shock collision between the single and the doubly shocked material and which travel in opposite directions away from each other. The adiabatic exponent γ is introduced to eliminate the internal energy $\varepsilon = p/(\gamma - 1)\,\rho$. Accounting for different values of γ before and after compression, one derives for the compression ratio

$$\frac{\rho_1}{\rho_0} = \frac{1 + (p_1/p_0)\dfrac{\gamma_1 + 1}{\gamma_1 - 1}}{p_1/p_0 + \dfrac{\gamma_0 + 1}{\gamma_0 - 1}}. \tag{2}$$

In the strong shock limit $p_1/p_0 \to \infty$, one recovers the familiar result

$$\lim_{p_1/p_0 \to \infty} \frac{\rho_1}{\rho_0} = \frac{\gamma_1 + 1}{\gamma_1 - 1}. \tag{3}$$

Note that the strong shock compression depends only on γ_1, i.e. the conditions in the compressed region. For a perfect monoatomic gas with $\gamma = 5/3$, one obtains $\lim_{p_1/p_0 \to \infty} \rho_1/\rho_0 = 4$. The pressure ratio is derived from equation (3) as

$$\frac{p_j}{p_i} = 1 + \frac{2\gamma_j - 2\gamma_i \dfrac{\rho_j}{\rho_i} \dfrac{\gamma_j - 1}{\gamma_i - 1}}{\dfrac{\rho_j}{\rho_i}(\gamma_j - 1) - \gamma_j - 1}. \tag{4}$$

The jump in particle velocities can be expressed as

$$(u_i - u_j)^2 = \left(1 - \frac{\rho_i}{\rho_j}\right)\frac{P_i}{\rho_i}\left(\frac{P_j}{P_i} - 1\right). \quad (5)$$

In the case of two colliding strong shocks, symmetry of the problem with respect to the contact surface between the two shock waves at the moment of collision competely determines the final compression [47]. In particular, the material at the contact surface is at rest, $u_2 = 0$. Furthermore, the material velocity in the unshocked material u_0 is negligible compared to u_1, the velocity in the first, strongly shocked material region. For a system with constant heat capacity ratio $\gamma_2 = \gamma_1 = \gamma_0$, one recovers the following result for the second compression:

$$\rho_2/\rho_1 = [2 + (\gamma + 1)(\rho_1/\rho_0 - 1)]/ \\ [2 + (\gamma - 1)(\rho_1/\rho_0 - 1)]. \quad (6)$$

For the ideal gas with $\gamma = 5/3$ one finds $\rho_2/\rho_1 = 2.5$. For a system where both density jumps are accurately measured one easily inverts the equations to obtain γ_1 and γ_2. For the example above with $\rho_1/\rho_0 = 2$ and $\gamma_2 = \gamma_1 = 2$ (see [44]), we find $\rho_2/\rho_1 = 5/3$ or $\rho_2/\rho_0 = 10/3$. Further, by varying the laser intensity on both sides of the foil and measuring the density jumps with, e.g, XRTS, a range of conditions can be accessed, and γ can be directly determined.

3.2. Spectrally-resolved x-ray scattering

3.2.1. General description

Spectrally resolved x-ray scattering has emerged as a powerful tool to measure the plasma conditions of dense compressed states of matter and as a core capability to determine the physical properties of MEC. XRTS provides the dynamic structure factor $S(\mathbf{k}, \omega)$, which is central for calculating e–i collisions for wave damping, electrical conductivity, thermal conductivity, free–free opacity, line-broadening, and plasma oscillations. Experiments on laser-produced dense plasmas utilize fast detection of the angular and frequency resolved x-ray scattering spectra. Previous studies at large kilojoule type laser facilities [20, 35–46] have demonstrated spectrally resolved scattering techniques measuring the x-ray Compton backscattering and the plasmon forward scattering spectra providing the temperature and density of the plasma. The temporal resolution achieved in previous studies is about 100 ps using fast framing camera detectors, or about 10 ps using a laser-produced K-α source.

The collimated high-energy LCLS x-ray beam enables novel investigations of HED states using the high photon flux in 50 fs pulses with photon energies of 4–8 keV that are unique to the MEC end station. Experiments with ultrafast 50 fs x-ray scattering techniques and high spectral resolution can now be performed at orders of magnitude higher repetition rates than previous studies so that high signal-to-noise ratios and precision spectral measurements will be achieved. Furthermore, using the LCLS x-ray beam as opposed to laser-produced plasma sources will provide a well-defined incident scattering vector; these facts combined with a seeded x-ray beam with 1 eV spectral bandwidth will provide precision spectra and highly accurate measurements of the scattering features. Combining this unique x-ray capability with powerful lasers allows investigating the microscopic and hydrodynamic physics properties of HED plasmas.

The x-ray scattering process is significantly different from optical scattering [20, 82] because the energy of the incident x-ray photon with frequency ω_0 is large enough to give a significant Compton shift to the frequency of the scattered radiation. During the scattering process, the incident photons transfer momentum $\hbar\mathbf{k}$ and energy $\hbar\omega = (\hbar\mathbf{k})^2/2m_e$ to the electrons. The magnitude of the **k**-vector is given by

$$k = |k| = \frac{4\pi E_0}{hc}\sin\left(\frac{\theta}{2}\right) \tag{7}$$

with $E_0 = \hbar\omega_0$ being the energy of the probe x-rays and θ the scattering angle. Momentum and energy are primarily transferred to the free and weakly bound electrons whose binding energy is less than the Compton energy $\hbar\omega$. In the non-collective (Compton scattering) regime, the contribution of the free electrons that carry the information on the Fermi energy (i.e., electron density) or the electron temperature of the dense plasma is measured in the inelastic scattering feature. Moreover, bound electrons with ionization energies larger than $\hbar\omega$(states deep in the Fermi sphere) cannot be excited, and no Compton energy transfer occurs during the scattering process.

Thus, the total XRTS spectrum results in elastic and inelastic spectral features that are described by the dynamic form factor, $S(\mathbf{k}, \omega)$ [83]. The ion structure is measured from the un-shifted elastic scattering component at E_0that is commonly referred to as Rayleigh peak. For the analysis of x-ray scattering experiments, we developed a comprehensive analysis tool [84] that is based on the Chihara formula [85]. The formula describes the contributions to elastic scattering and inelastic scattering from free (delocalized) and weakly bound electrons by the dynamic form factor

$$\begin{aligned} S(\mathbf{k},\omega) &= |f_1(\mathbf{k}) + q(\mathbf{k})|^2 S_{ii}(\mathbf{k}) + Z_f S_{ee}(k,\omega) \\ &\quad + Z_C \int S_{CE}(\mathbf{k},\omega-\omega')S_S(\mathbf{k},\omega)\mathrm{d}\omega' \end{aligned} \tag{8}$$

with Z_f and Z_C denote the number of free and bound electrons, respectively. It is primarily the intensity of the elastic (Rayleigh) peak described by the first term that provides the ionic structure while the intensity and spectral dispersion of the inelastic free-electron scattering described by the second term provides information on the electronic system. The last term of equation (8) includes inelastic scattering by weakly bound electrons, which arises from bound–bound and bound-free transitions to the continuum of core electrons within an ion, $S_{CE}(\mathbf{k}, \omega)$, modulated by the self-motion of the ions, represented by $S_S(\mathbf{k}, \omega)$. The corresponding spectrum of the scattered radiation is that of a Raman-type band. Recently, there has been much activity to improve the theoretical description of this term, e.g., [86, 87]. For plasmon scattering applied in this study, we find that this term is not important when compared to the free electron dynamic structure. Principally, the relative contributions of the bound-free feature (third term in equation (8)) and the free–free feature (second term in equation (8)) can be utilized to infer the number of bound electrons and hence the ionization degree and the density of the dense plasma [87].

We have developed a theoretical fitting code that calculates $S(\mathbf{k}, \omega)$. Where physics uncertainties exist, several theoretical models have been developed for comparisons and to establish sensitivities. They are available for comparisons with experimental data. There are three areas, (1) for term 1, four different models have been implemented for calculations of the ion–ion structure factor $S_{ii}(\mathbf{k})$ to describe the elastic (Rayleigh) scattering peak [88]. They include a Debye model, a one component plasma model and a screened one component plasma (SOCP) approximation with and without negative screening; these have been used previously with some success; (2) for term 2, calculations of the free–free contribution include e–i collisions in the Born–Mermin approximation (BMA) [89–91] together with local field corrections (LFC) [92, 93] which has been applied with good success to describe the dispersion [42] and width of plasmons [36]. These models show little effect for calculations of the Compton profile which can be tested by comparison with the random phase approximation (RPA) [94]; (3) for term 3, calculations of the bound-free contribution include the form factor approximation and the impulse approximation [84] that show differences for low-temperature HED plasmas with little ionization and which are presently under scrutiny for testing of continuum lowering models. Importantly, the present models include different continuum lowering physics and finite temperature effects.

LCLS provides new opportunities to develop novel physics areas using x-ray scattering. Among those are first-principal electron temperature measurements using detailed balance, observations of plasmon damping by collisions, LFC effects on the plasmon dispersion and effects of viscosity and damping on phonon modes on the ion feature. For high-pressure conditions, elastic and inelastic bound-free scattering can become important and will be utilized in a novel way to characterize ultrafast phase transitions, continuum lowering or ion–ion correlations.

3.2.2. Compton scattering feature

Initially, XRTS was developed in isochorically-heated matter. This path was chosen because the advantage of *a priori* known mass density allows the electron density to be directly inferred from the degree of ionization. The first experiments have tested calculations of the Compton scattering feature that shows significant broadening due to the thermal motion of the electrons resulting in factor of >2 larger width than scattering spectra from cold beryllium [35]. In many dynamic compression experiments, the conditions are close to the Fermi degenerate state. Here, the total width of the Compton peak depends on $\sqrt{E_{\mathrm{Fermi}}} \sim n_e^{1/3}$ in the partly Fermi degenerate regime with $E_{\mathrm{Fermi}} = \hbar^2(3\pi^2 n_e)^{2/3}/2m_e$. Thus, the width of the Compton scattering profile is fairly insentitive to the electron density increase during shock collision. Figure 15 shows an example for 8.5 keV x-ray probing. The challenge will be observing the whole scattering spectrum with sufficient energy range and sufficient resolution to measure this effect.

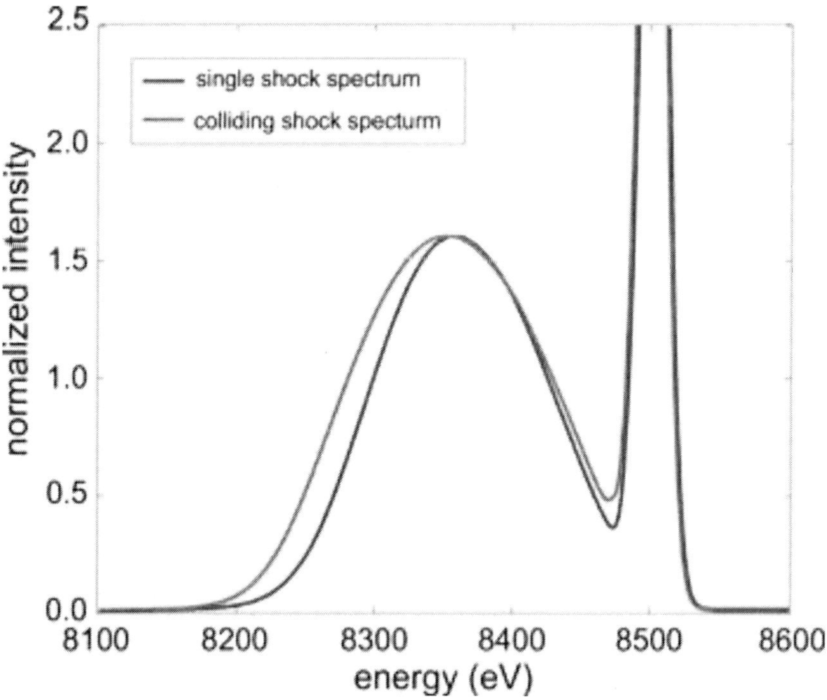

Figure 15. Calculated x-ray scattering spectra for 90° scattering of a 8.5 keV x-ray beam from a aluminum that is shocked by a single strong shock (red curve) or by two counter propagating shock waves (blue curve) with density determined by equation (6).

Finally, with experiments aimed towards producing high-temperature states, Compton scattering is the technique of choice to measure the transition from a Fermi parabolic distribution towards a Boltzmann Gaussian profile whose width provides the electron temperature with high accuracy.

3.2.3. Plasmon scattering

To access collective scattering requires that the scattering parameter must be larger than one:

$$\alpha = \frac{1}{k\lambda_S} = \frac{\text{Scattering length}}{\text{Screening length}} > 1. \tag{9}$$

The scattering length is defined by the scattering vector **k** (equation (7)) and is given by the probe energy and scattering angle: $\lambda = \lambda_0/[4\pi \sin(\theta/2)]$. In the limits of degenerate or classical plasmas the screening length assumes the Thomas–Fermi length or the Debye length, respectively. In case of $\alpha < 1$, the probe radiation with wavelength λ_0 is scattered from electron density fluctuations on a spatial scale λ smaller than the screening length λ_S. In this case, no collective effects like electron plasma oscillations can be observed and the scattered radiation shows the Compton scattering feature providing the electron velocity distribution function. In case of $\alpha > 1$, the scattering wavelength λ is larger than the screening length λ_S. In a weakly degenerate solid-density plasma with an electron temperature of the order of the Fermi temperature, $T_e \sim T_F = 15$ eV, collective scattering will occur in forward scattering geometry with $\theta < 40°$ and x-ray probe energies of $E_0 > 3$ keV. In this regime, the scattered spectrum shows collective effects corresponding to scattering resonances from electron plasma (Langmuir) oscillation, i.e., plasmons, and from ion-acoustic modes.

In this study, our measurements observe a plasmon shift of $\Delta E = 25$ eV that is approximated by the dispersion relation for plasmons in dense matter, often expressed in the familiar expression [95, 96]:

$$\omega^2 \sim \omega_{pe}^2 + 3(kv_{th})^2 + \left[\frac{\hbar k^2}{2m_e}\right]^2. \tag{10}$$

For HED states, the first term of equation (10) dominates the frequency shift. With ${\omega_{pe}}^2 = n_e e^2/\varepsilon_0 m_e$ the first term being sensitive to the electron density, n_e. The second term is known as the thermal correction from the Bohm–Gross dispersion relation and is of order <5% of the total shift. However, T_e will influence the damping of the plasmon and a rough estimate can be inferred from the fit to the width of the experimental data. The third term is the well-known Compton shift and exactly calculated with equation (1). Higher order terms have been estimated analytically; they are included in numerical results of e.g., [96].

Under the assumption that inter particle interactions are weak, so that the nonlinear interactions between different density fluctuations are negligible, the plasmon shift has been first calculated in the RPA. In the classical limit, it reduces to the usual Vlasov equation. In the limit of the RPA, strong coupling effects are not accounted for, thus limiting the model validity. To expand the theoretical modeling into the WDM regime, the BMA has been developed. The theory was subsequently improved to allow for different models for the dynamic collision frequency and for LFC. Its application to laser experiments on compressed boron has shown that the description of the plasmon dispersion is greatly improved with this theory [42]. However, at the time of this writing, the theoretical description of many-body effects in WDM is not complete.

Figure 16 demonstrates that for small forward scattering angles the calculated plasmon energy shift is not sensitive to the choice of the model for $S_{ee}(k, \omega)$. For a scattering angle of 13° the plasmon shift calculated with BMA and BMA–LFC agree over a range of densities. The shift reflects the relative increase in plasma frequency with density and is well understood compared to the absolute plasmon energy at arbitrary wavenumber. This is apparent in the calculations with increasing scattering angle that show discrepancies between these models. It becomes clear that MEC provides a great opportunity for future investigations of these and related phenomena when using larger scattering angles. For the purpose of the present analysis we conclude that it is possible to achieve an error in density of 5% when choosing conditions where differences in the theoretical approximations are minimized.

3.2.4. Detailed balance

For HED matter at significant temperature, the collective x-ray forward scattering spectrum will show two plasmon features. In addition to the downshifted feature, an up-shifted plasmon will occur on the high-energy wing with nearly the same frequency shift. Compared to the intensity of the downshifted plasmon, the up-shifted intensity is reduced by the Bose–Einstein function dictated by the principle of detailed balance. The intensity ratio of these plasmon features thus offers a first-principal method to measure the ultrafast electron temperature evolution of HED plasmas.

The relation that demonstrates this method is derived from the fluctuation–dissipation theorem. The density fluctuations in the plasma, described by the dynamic structure factor, are related to the dissipation of energy, described by the dielectric function $\varepsilon(\mathbf{k}, \omega)$

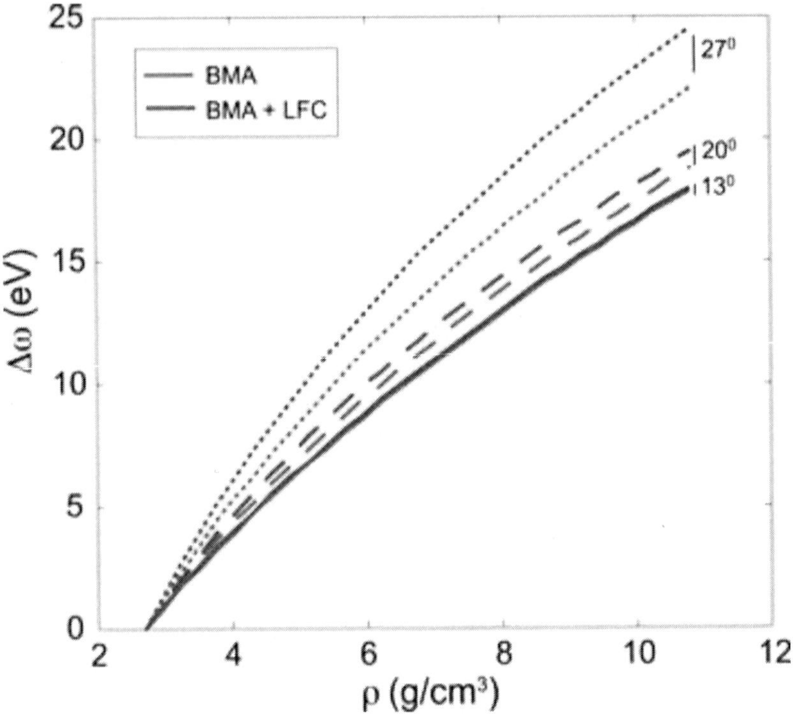

Figure 16. Calculations of the plasmon frequency shift are shown for compressed aluminium and for three scattering angles, temperature of 1.75 eV, and three delocalized electrons. Calculations in the Born–Mermin approximation with and without local field correction agree for a small scattering angle of 13°.

$$S(\mathbf{k}, \omega) = \frac{\varepsilon_0 \hbar k^2}{\pi e^2 n_e} \frac{\mathrm{Im}(\varepsilon^{-1}(\mathbf{k}, \omega))}{1 - \exp(\hbar\omega / k_B T_e)}. \tag{11}$$

Independent of evaluating the dielectric function, the shape of the structure factor, and modeling collisional effects, the dielectric function fulfills the requirement

$$\frac{S(\mathbf{k}, \omega)}{S(-\mathbf{k}, -\omega)} = e^{-\hbar\omega / k_B T_e}. \tag{12}$$

This relation is referred to as the detailed balance relation. As a general consequence, the structure factor shows an asymmetry with respect to **k** and ω. When applied to plasmons, it provides the electron temperature independent of a detailed formulation of dissipative processes in the plasma. In past laser experiments, due to small signal-to-noise ratios limited information was obtained on T_e [36, 97], making a future systematic studies with small noise amplitudes highly interesting.

3.2.5. Collisions

The width of the plasmon feature is sensitive to collisions, thus providing a measurement of the e–i collision frequency and consequently a novel method for measuring conductivity and equilibration. Figure 17 shows the comparison between measured and calculated collision frequency ν_{ei} and sensitivity of the plasmon width to ν_{ei}. From the e–i collision rate the conductivity can be inferred through the Ziman formalism [98, 99]. This can readily be seen from the dependence of the x-ray scattering spectrum on the dielectric function $\varepsilon(\mathbf{k}, \omega)$. The generalized expression for the dielectric function in the Mermin approximation employs the standard collision-less value obtained in the RPA, $\varepsilon^{RPA}(\mathbf{k}, \omega)$, and the collision frequency $\nu(\omega)$

$$\varepsilon^{M}(\mathbf{k}, \omega) = f(\varepsilon^{RPA}(\mathbf{k}, \omega + i\nu(\omega)), \varepsilon^{RPA}(\mathbf{k}, \omega), \nu(\omega)). \quad (13)$$

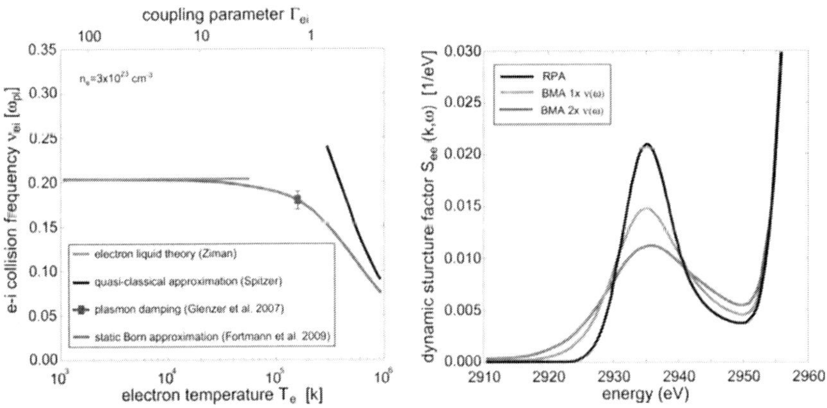

Figure 17. (Left) Electron–ion collision frequency is shown as function of the electron temperature. Previous data inferred from the plasmon measurement [36] compare well with the Born approximation [101]. To validate the theory, future experiments at LCLS will require an independent temperature measurement. (Right) The plasmon width increases significantly with increasing collision frequency.

The collision frequency obtained in this way is related to the conductivity through the plasma coupling parameter Γ [100]

$$\sigma(\omega) = \lambda(\Gamma)\frac{e^2 n_e}{m_e \nu(\omega)} \text{ and } \Gamma = \frac{e^2}{kT_e}(4\pi n_i/3)^{1/3}. \quad (14)$$

Thus, measurements of the plasmon spectrum have the potential to provide critically important information on equilibration and hydrodynamics but must use high spectral resolution and high signal-to-noise detection unique to LCLS.

3.2.6. The ion feature

LCLS high photon flux of narrowband and coherent x-rays are now available to study the dynamics of the low-frequency correlations (the ion-acoustic waves). The dispersion curve of and the details of ion correlations have long remained elusive in the HED physics regime [82, 88], with almost no comparison with experimental data except for the case of liquid metals at room temperature [102]. New studies at LCLS will be able to investigate contributions of LFC and the relevance of hydrodynamics and generalized hydrodynamics formulations on plasma transport properties [103, 104]. Further, predictions for additional dispersion at large wave numbers can be tested [63, 105].

Of particular interest are pump–probe experiments where unequal electron and ion temperatures will be produced. Thus, the measurement of the ion acoustic resonances will provide a direct estimate of the ion temperature, while the measurement of plasmon resonances will give the electron temperature. This platform can provide new insights in the e–i coupling in HED experiments at MEC.

3.2.7. Ion–ion correlations

Unique to LCLS MEC is the fact that one can further develop x-ray scattering diagnostic to simultaneously resolve both the spectral and angular (**k**-vector) components of the XRTS signal during a dynamic laser and x-ray heating experiment. In particular, when significant compression occurs, for example due to strong shock waves, accurate measurements can be achieved. The momentum-resolution of such methods has not previously been achieved in WDM research and is critically needed to measure the material structure in conditions with simultaneous spectrally-resolved scattering measurements. Such experiments will take advantage of the strong correlation peaks in the scattering intensity of the elastic scattering component (1st term of equation (8)).

The ionic structure factor peak has been observed recently for the first time in shock-compressed aluminum. In these experiments, aluminum was shocked from one side [46] and data were accumulated over 3 separate shot days at the Omega laser [45]. The absolute strength of the scattering amplitude and overall predicted shape have shown that HNC calculations with repulsive potentials and SOCP models provide an improved description while Debye–Hückl screening models show weaker correlations and smaller scattering amplitudes than the experimental data.

For matter with strong ion–ion correlations the frequency integral of equation (8), $S^{total}(\mathbf{k})$, is given by the Fourier transform of the pair–correlation function $g(\mathbf{r})$:

$$S^{total}(\mathbf{k}) = 1 - n \int d^3\mathbf{r} \exp(-i\mathbf{kr})[g(\mathbf{r}) - 1] \qquad (15)$$

with

$$g(\mathbf{r}) = \frac{1}{Nn} \int d^3\mathbf{r}' \, n(\mathbf{r} + \mathbf{r}')n(\mathbf{r}'), \qquad (16)$$

where n is the average density, N is the number of particles in the scattering volume. Hence, $S^{total}(k)$ is a direct measure of pair correlations in dense matter. Equations (15) and (16) predict a shift of the calculated maximum elastic x-ray scattering amplitudes to higher wave number with increased density. For higher densities, the mean separation between ions decreases providing scattering peaks at larger scattering angles and thus larger wave number **k**. Here, the small **k**-vector spread of the incoming LCLS beam is required to resolve the shift of the structure peak. Importantly, different theoretical models of the ion–ion correlation predict differences in the widths, which have been resolved in experiments at MEC and which are discussed further below.

3.3. DFT-MD simulations

DFT-MD simulations of this study were performed for a range of temperature and density conditions [61–63] spanning the experimental parameters accessible by the MEC laser capabilities. The primary goals were to calculate the ion–ion correlation feature and to determine the material physical properties. Here, the core radius is $r_C = 1.7a_B$ with $a_B = 0.53$ Å. The Mermin functional accounts for thermal excitations. The electronic cut off for the plane wave representation of the wave function is set to 550 eV as tested to provide the free energy and pressure with about 0.7% accuracy.

The simulations have been performed in a super-cell with constant number of particles, volume and temperature with periodic boundary conditions. A total of $N = 256$ ions have been used for the fluid. Initial runs have been performed at lower temperatures starting with the ions in an fcc lattice at the appropriate densities of the experiment. The resulting ionic configurations were then used as starting configurations for runs at the temperatures of interest. The simulations run for 2000–23 000 steps with a 0.2 fs time step. An initial relaxation time of varying length was excluded from the analysis of the EoS.

Figure 18 shows results from DFT-MD simulations of the total x-ray scattering amplitude versus scattering vector **k** for WDM conditions. These simulations show a well distinguished peak and determine temperature and density sensitivity of the angularly-resolved scattering measurements.

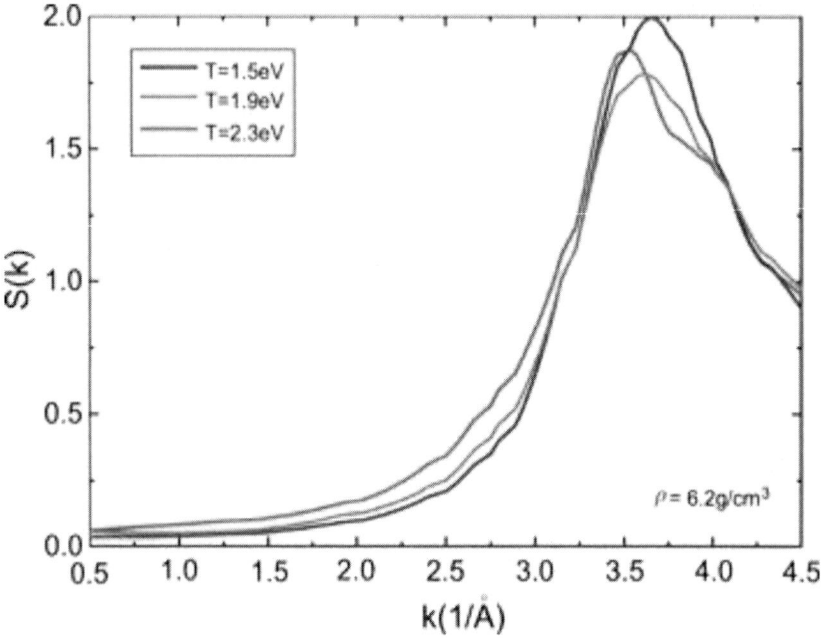

Figure 18. The total x-ray scattering amplitude is shown as function of the wave number **k** for compressed matter states. Calculations for various temperatures ($T_e = T_i$) show that the width of the correlation peak increases with increasing temperatures. Also, the absolute peak amplitude is sensitive to temperature.

4. RESULTS AND DISCUSSIONS

4.1. Shock compressed matter

4.1.1. Experimental results

Figure 19 shows the schematic of dynamically compressed matter experiments at MEC. Also shown are data records from the x-ray scattering detectors. The spectrometers have been fielded for density and temperature measurements from plasmons and angular resolved scattering provides diffraction and structure factor data. Here, two 4.5 J laser beams irradiate 50 μm thick Al foils (initial density of $\rho_0 = 2.7$ g cm^{-3}) that are coated with 2 μm thick parylene. The laser beams are absorbed in the parylene, heat the material and launch two counter-propagating multi-megabar shock waves into the solid aluminium by ablation pressure. For the conditions achieved in this experiment, we find that the shocks propagate inward at 14 km s^{-1}, each shock wave compressing the aluminium target to about 5 g cm^{-3}.

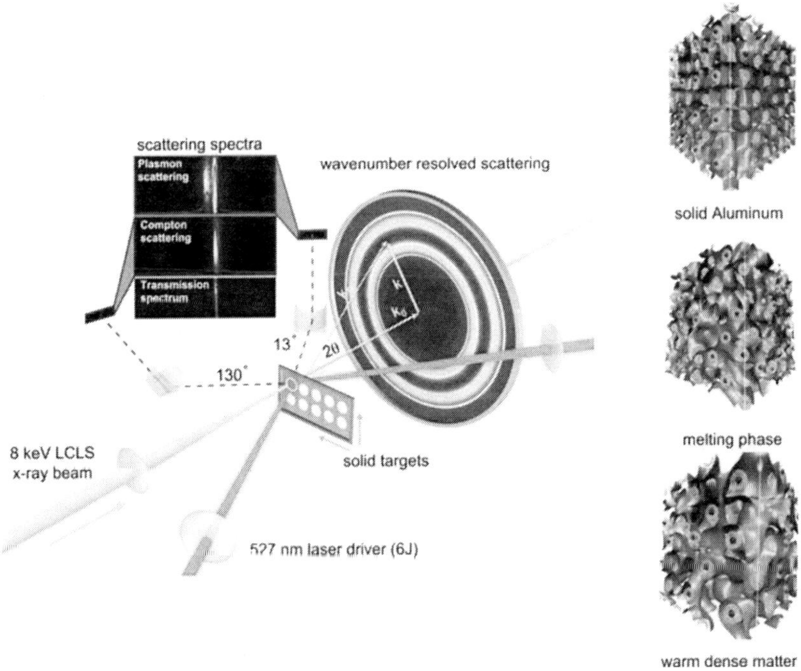

Figure 19. (Left) A schematic of experimental setup for compressed matter studies at MEC is shown. The LCLS x-ray beam is focused into the dense laser-compressed aluminium target. The 13° forward x-ray scattering spectrum shows inelastic plasmon scattering together with the elastic scattering feature at the incident x-ray energy. The 130° backscattering shows no scattering feature in this energy range validating the identification of the plasmon. A CSPAD area detector observes the total wavenumber-resolved x-ray scattered intensity indicating the transition from Debye–Scherrer diffraction rings (white) to an intense ion–ion correlation scattering feature (red). (Right) DFT-MD simulations of the formation of WDM indicate that the ions (blue) abandon their lattice positions. Whilst core electrons (gray) remain mostly unchanged the delocalized conduction electrons (represented by orange iso-surfaces) are disturbed from the very regular structure in the lattice.

The shock speed and planarity over the central 10 μm diameter region were verified by shock breakout measurements from 25 μm thick foils irradiated by a single beam. Thus, when adding the second beam on the opposite side of the target, we doubled the foil thickness to 50 μm to mirror the shock propagation. Figure 20shows that the averaged shock velocity varies by <2% over the region probed by the x-ray beam. Also, the shock velocity is seen to scale with laser intensity for two different focal spot sizes. When the shocks coalesce in the center of the foil, the density is expected to significantly increase to about 7.5 cc^{-1} reaching peak pressures of 5 Mbar.

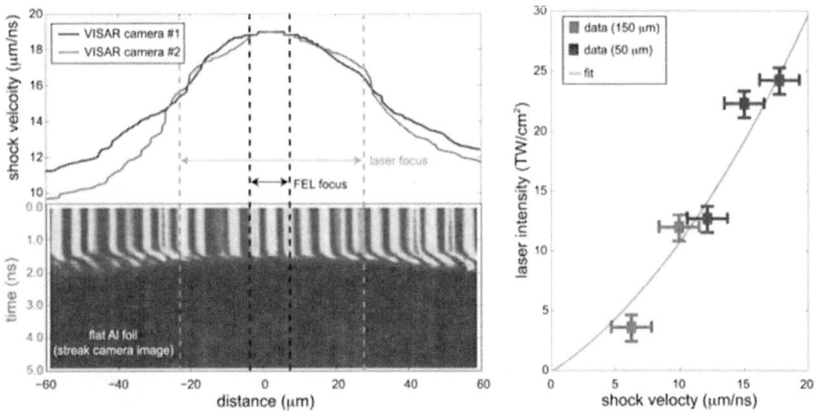

Figure 20. (Left) Results from shock breakout measurement from flat aluminium foils. The averaged shock velocity planarity is <2% in the central region probed by the LCLS x-ray beam. (Middle) An example of a VISAR streak camera record is shown from laser-driven aluminium. (Right) The averaged shock velocity is plotted for various laser intensities using two different phase plates with 50 μm and 150 μm spot size.

In this study, accurate measurements of compression have been achieved by measurements of the electron plasma (Langmuir) oscillations. These measurements test the predictions of compression based on equations (1)–(6) by applying the theory of equations (8) and (10). Figure 21 shows the plasmon scattering spectra and theoretical fits of the dynamic structure factor $S(\mathbf{k}, \omega)$ near peak-compression at $t = 1.9$ ns. These data are compared with measurements from uncompressed aluminium. The plasmon feature is downshifted from the incident 8 keV x rays as determined by the generalized Bohm–Gross dispersion relation, see equation (10), and its resonance frequency is used as a sensitive marker of the electron density, see figure 16. While the data from uncompressed aluminium yield a total plasmon shift of 19 eV providing a free electron density of

1.8×10^{23} cm^{-3} the total shift increases to 29 eV at shock coalescence. For the example shown in figure 21, a density of 4.1×10^{23} cm^{-3} has been measured and corresponding to 2.3× compression. The error in electron density is ±5% determined by noise in the data and the fit of the theoretical dynamic structure factor.

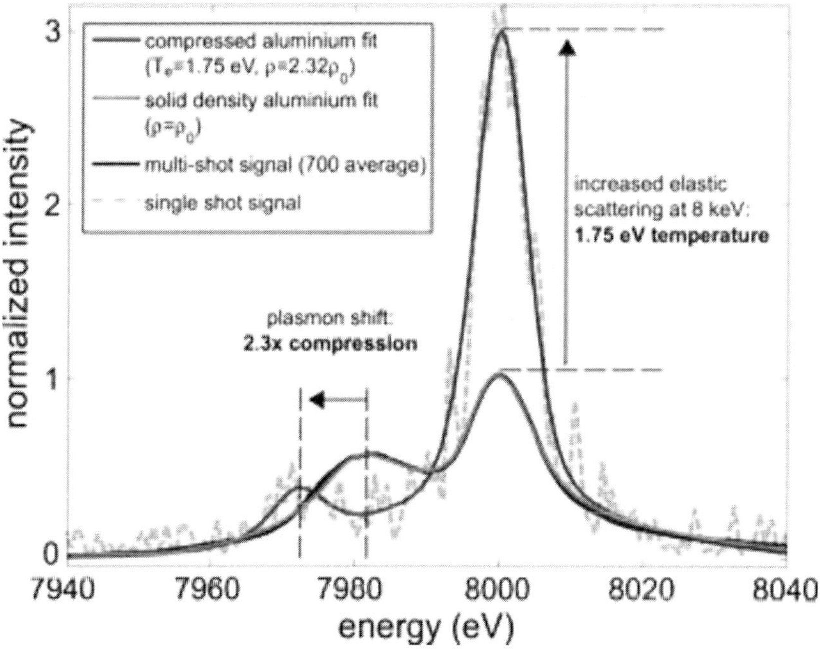

Figure 21. X-ray scattering spectra from compressed and solid-density aluminium are shown. A significant increase in plasmon frequency shift and elastic Rayleigh scattering amplitude occurs with compression. The experimental data (black and gray dashed curves) are fitted with theoretical dynamic structure factor (equation (8)) calculations (light straight curves). The elastic scattering amplitude provides the temperature of 1.75 eV. The plasmon shift $\hbar\omega$ is primarily determined by the increase in plasma frequency measuring the electron density. In this example, the plasmon shift indicates material compression of 2.3 times solid density with ρ = 6.3 g cc^{-1}.

The maximum density measured at shock coalescence is 7 g cc^{-1} slightly less than expected from estimates discussed above. Specifically, we find that compression in the first shock is slightly lower than predicted due to varying total peak powers and probably also due to the slow rise time of the laser pulse. Typical values are $1.2 < \rho_1/\rho_0 < 1.9$, which when combined with the compression due to coalescence of $\rho_2/\rho_1 \sim 5/3$ results in final compression values of $\rho_2/\rho_0 > 2$.

The scattering spectra further provide a measure of the temperature by analyzing the intensity of the elastic scattering features. For this purpose, structure factors are determined from wavenumber resolved scattering while the absolute intensity is determined by the intensity ratio with the plasmon feature; the plasmon intensity is determined by the *f*-sum rule (particle conservation)

$$\int_{-\infty}^{\infty} \left(\frac{d^2\sigma}{d\Omega d\omega} \right)_{\text{Plasmon}} \omega d\omega = \frac{Z\hbar k^2}{2m_e}. \tag{17}$$

where $(d^2\sigma/d\Omega\ d\omega)_{\text{Plasmon}}$ is the measured plasmon intensity from the free electron Langmuir oscillations. Here, $Z = 3$ is the ion charge state, $\hbar$ is Planck's constant, k is the wavenumber and m_e the electron mass. At peak compression, the intensity of the elastic scattering feature increases by a factor of 2.8 over the cold scattering amplitude. We use a model [84] to fit the measured intensities; the potential and structure factors are shown in section 4.1.2; they are in excellent agreement with the experimental data and DFT-MD simulations, see figure 18. For the example shown in figure 21 the temperature is $T = 20\ 000$ K, i.e., 1.75 eV ± 0.5 eV. For our conditions, the ultrafast x-ray pulse deposits 3500 K or 0.3 eV per electron into the target. A fraction of the energy can be expected to heat the electron fluid, but the total energy is small compared to shock heating and will not affect the plasmon shift. In addition, within the duration of the x-ray pulse, the ionic correlations have no time to respond to changes of the screening properties as demonstrated by the observation of solid and solid compressed states of aluminium.

Figure 22 shows the wavenumber resolved scattering data from which we obtain the high-pressure ion–ion structure factors. At $t = 0$, before the rise of the optical laser beam power, the data show peaks from Debye–Scherrer rings indicating the ionic lattice in the solid. The peaks at 38°, 45°, and 65° correspond to (111), (200), and (220), respectively. When the shocks are launched and compress the solid, the lattice spacing d is reduced and we observe that the peaks shift to larger scattering angles as determined by the Bragg scattering equation $n\lambda_{\text{x-ray}} = 2d \sin\theta$, here $\lambda_{\text{x-ray}} = 1.55$ Å the wavelength of the incident x-ray laser, and 2θ the scattering angle.

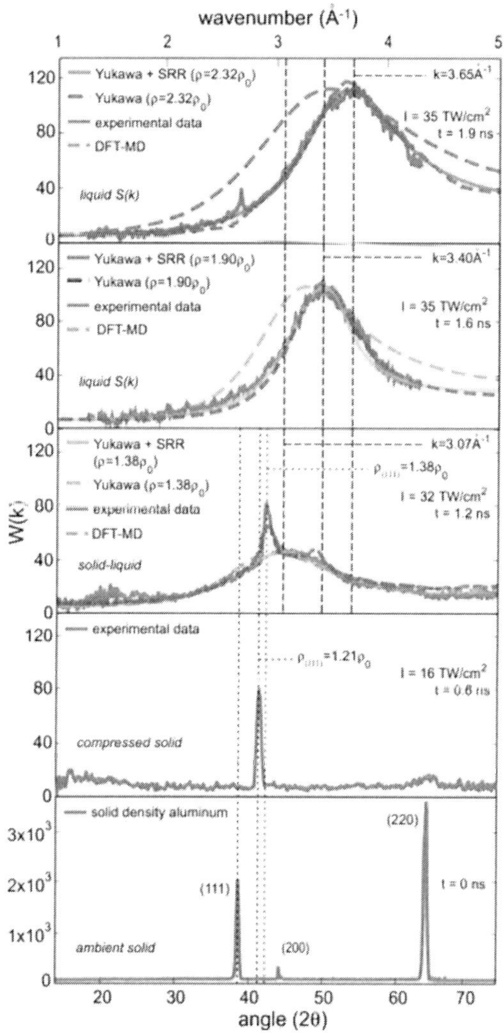

Figure 22. X-ray scattering data $W(k)$ are shown as function of scattering angle. (Bottom) Bragg peaks from Debye–Scherrer rings are shown at $t = 0$ ns. (2nd from bottom) Data from compressed solid aluminium show a shift of the (111) Bragg peak by 3° and compression of 1.21 ρ_0. (Middle) Appearance of a broad ion–ion correlation peak is observed together with a Bragg peak shifted further to larger angles. (2nd from top) At higher compression, aluminium melts, Bragg peaks disappear, and the ion–ion correlation peak shifts to 50°. (Top) The angle of the correlation peak increases further to 56° when higher densities are reached after coalescence. The data (black curves) show excellent agreement with DFT-MD simulations (top three panels). The data can also be described by a model that accounts for both screening and short-range repulsion (red, blue and green solid curves), but a screened potential alone is not sufficient (dashed curves).

With increasing pressures, the aluminium melts and transitions into the WDM state. We first observe a comparatively broad ion–ion correlation peak at an angle of 45° at 1.38× compression together with shifted Bragg scattering of solid compressed aluminium. For these conditions the comparison with simulations shows that the wave number of both the ion–ion correlation peak and the shifted Bragg peak provide the same density indicating a co-existence regime. With higher pressure accessed later in time and with higher laser intensity, we compress aluminium up to 1.9× the initial density and the strong ion–ion correlation peak shifts to $2\theta = 51°$. At shock coalescence, the peak shifts further to $2\theta = 55°$ and the data indicate 2.3× compressed aluminium consistent with densities inferred from the simultaneous measurements of the plasmon shift.

The measured ion–ion correlation feature is sensitive to the ion–ion structure factor, see equation (8). For nearly elastic scattering the scattering angle 2θ determines the wavenumber of the correlations. With $\boldsymbol{k} = 2\boldsymbol{k}_{x\text{-ray}} \sin\theta$ and $k = 2\pi/\lambda$ determining the scale length of the density fluctuations measured in this scattering experiment, we find 1.38 Å < λ_{SL} < 5.7 Å. On these atomic scale lengths, the scattering amplitude is probing important details of the ion–ion interaction potential V_{ii}.

4.1.2. Physical properties

For the analysis of the measured scattering data and the ion–ion correlation feature various theoretical approximations have been considered. The data depend on both material compression, temperature, and on the choice of the many particle theory to describe the interaction physics. Here, densities and temperatures are inferred from the plasmon scattering spectra providing an excellent experimental test of theory and simulations by comparing predictions with the measured wavenumber-resolved scattering data. In previous studies, the measurements of the elastic scattering amplitude alone was often not sufficient to provide information about the state of the dense plasma. This is due to the fact that calculations of $S_{ii}(k)$, e.g., using the HNC approximation, must assume an effective interaction potential. In this study, $Sii(k)$ is directly obtained from the measured wavenumber resolved scattering data using $W(k) = S_{ii}(k)(f(k) + q(k))^2$ with the atomic form factor $f(k)$ and the screening function $q(k)$ calculated from the number of bound core electrons, $Z_C = 10$ for aluminium.

Figure 23 shows various screening functions $q(k)$ and atomic form factors $f(k)$. Combining these results with our measurements of the wavenumber resolved scattering data, $W(k)$, demonstrates the sensitivity of the choice of the function $q(k)$. HNC calculations with a screened Coulomb potential using a Yukawa screening term (Y) together with a SRR term provide excellent agreement with the measured data. This is illustrated as the full curves in figure 22 that agree with both experiment and the results from the DFT-MD simulations. Similar results can be obtained with a hard sphere model. However, it is apparent that a Coulomb potential or a screened Coulomb potential cannot account for the experimental observations, see dashed curves in figure 22. At small wavenumbers, we note that these models show different sensitivity to the temperature. This is due to the fact that the repulsive core part of the potential is temperature independent; the remaining sensitivity is due to the temperature dependence of the screening cloud. Here, DFT-MD simulations shown in figure 18 provide the temperature dependence that is applied for the interpretation of the elastic scattreing amplitude in the plasmon spectra, see figure 21. The error in temperature can be estimated to be in the range of 0.5 eV–0.7 eV including contributions from both uncertainties in the models and noise of the data.

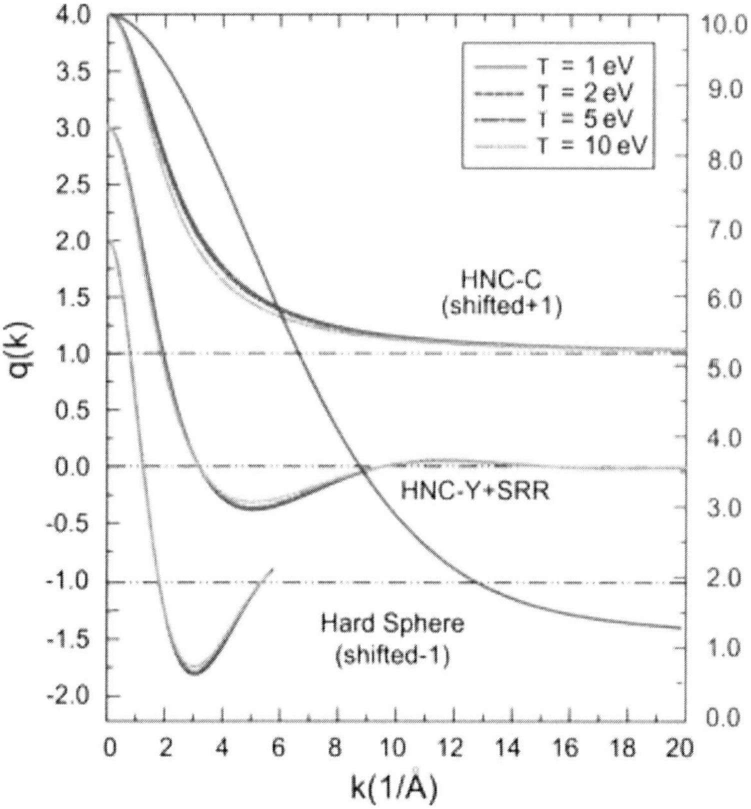

Figure 23. Three screening functions $q(k)$ are shown for temperatures ranging from 1 eV to 10 eV along with the atomic form factor $f(k)$. While HNC-Y + SRR and hard sphere models agree well with the data, models that only include screening provide a poor match with the data.

The analysis presented in figures 22 and 23 provide the following physical picture for the ion–ion structure factor in WDM: by matching the pair correlation functions $g(r)$, the effective ion–ion pair potentials can be directly extracted from the DFT-MD simulations that show the importance of including both SSR and screening. The interaction potential can be described as

$$\beta V_{\mathrm{ii}}(r) = \frac{Z^2 e^2}{r k_B T} e^{-\kappa r} + \frac{\sigma^4}{r^4}, \qquad (18)$$

with r being the distance between ions, k_B is Boltzmann's constant, e the electron charge number. The exponent $\kappa = \kappa\,(n_e, T_e)$ is the Yukawa screening function. For our conditions, κ approaches the Thomas–Fermi screening length with accuracy of better than 1%. The last term in equation (19) accounts for the additional repulsion from overlapping bound electron wave functions, with the parameter σ fitted to match DFT-MD simulations [59]. Figure 22 (top three panels) shows that the DFT-MD simulations, results based on equation (18), and the wavenumber resolved scattering data agree well with each other for densities obtained independently from the plasmon or Bragg scattering data. Importantly, linearly screened Coulomb potentials, denoted here by Yukawa in figure 22, show no agreement with the experimental data.

The data and simulations can further be used to describe the angle of the maximum scattering amplitude; the latter is sensitive to the density. Figure 24 compares the angle of maximum scattering amplitude with the density from the plasmon x-ray scattering data. Also shown are various theoretical approximations. The result show that using equation (18) for the calculation of the dynamic structure factor $S(k, \omega)$ provides a model to determine density from wavenumber resolved scattering.

The therory and simulations can now be applied to analyze the scattering data for potential gradients in the x-ray probing region. Potential non-uniform shock conditions could give rise to different conditions either transversly over the 10 μm x-ray probe cross section or along the shock propagation direction. Clearly, the latter will always include the low density blow-off plasma and if probed early in time the unshocked material ahead of the shock. However, the x-ray scattering amplitude is weighted towards the highest density and the broad ion–ion correlation feature can only be observed after shock melting. To quantitativly estimate the effect of gradients the wavenumber resolved scattering can be compared with calculations for the ion–ion structure factor for a range of density conditions.

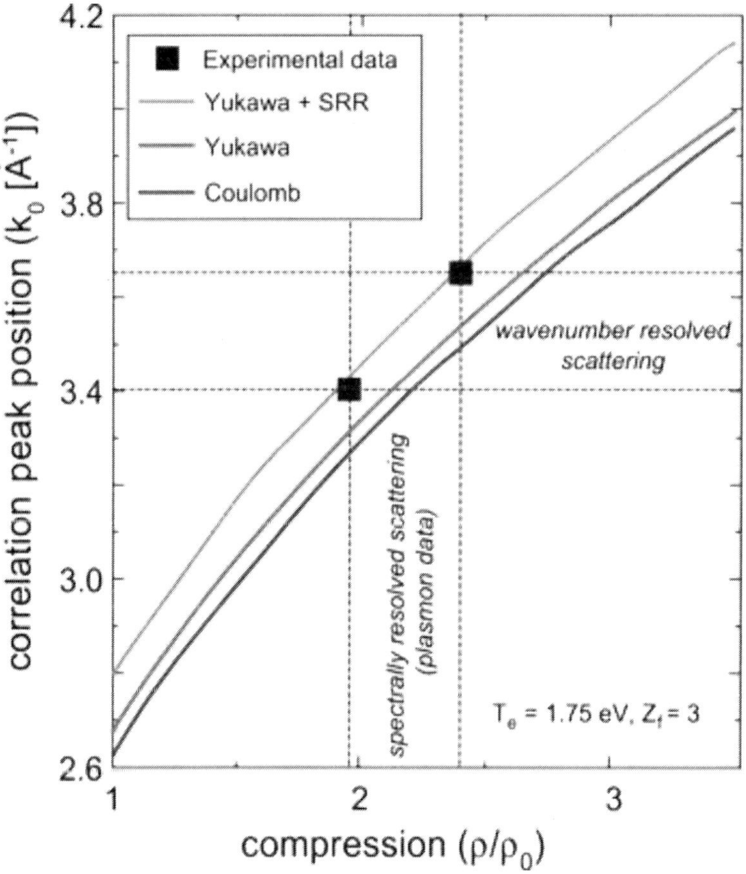

Figure 24. The measured wavenumber of the ion–ion correlation peak that is reflected by the maximum x-ray scattering amplitude is shown versus compression from plasmons (error bars are of the size of the symbols). The comparison with theory shows that including short-range repulsion correctly predicts the scattering angle.

Figure 25 shows the x-ray scattering data together with calculations for a density of $\rho = 2.32\ \rho_0 \pm \Delta\rho$. The density range is assumed to be linear over the ranges indicated in the figure. The results indicate that density gradients are negligibly small for the present experimental conditions and are less than 5%.

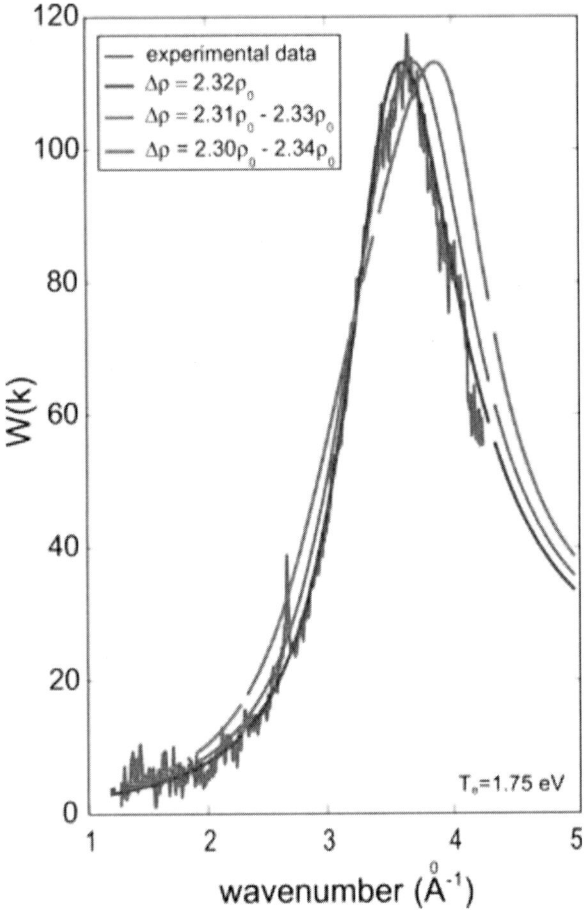

Figure 25. The wavenumber resolved scattering data $W(k)$ are compared with calculations of the ion–ion correlation functions. The latter represents the sum of scattering results calculated for a range of density values and combined to an effective synthetic scattering result. This analysis indicates that density gradients are less than 5%; larger values result in a width of the correlation peak not consistent with the experimental results.

Finally, the experimental and theoretical structure factors further provide a new method for determining the material pressures. In particular, the measured wavenumber resolved scattering data $W(\mathbf{k})$ yield the structure factor $S(\mathbf{k})$ using $W(k) = S(k)[f(k) + q(k)]^2$ with the atomic form factor $f(k)$ and the screening function $q(k)$ that can be integrated and which provides a novel method for determining the pressure [106]. The total pressure consists of ionic and electronic contributions with $P_{\text{tot}} = P_{\text{i}} + P_{\text{e}}$. In nearly

Fermi degenerate conditions, P_e includes well-known density-sensitive contributions accounting for Fermi pressure, degeneracy, Coulomb, and exchange terms. On the other hand, the ion pressure, when divided into an ideal gas part $P_G = n_i k_B T$ and the correlation part $P_i = px + P_G$, is dominated by the (negative) ion excess pressure

$$p^x = \frac{n_i U}{3N} - \frac{n_i (Ze)^2}{12\pi^2} \int_0^\infty S(\mathbf{k}) \frac{\kappa^4}{(k^2 + \kappa^2)^2} \mathrm{d}\mathbf{k} \qquad (19)$$

with U/N being the internal energy. In equation (19), the structure factor directly obtained from the measured wavenumber resolved scattering data can be used.

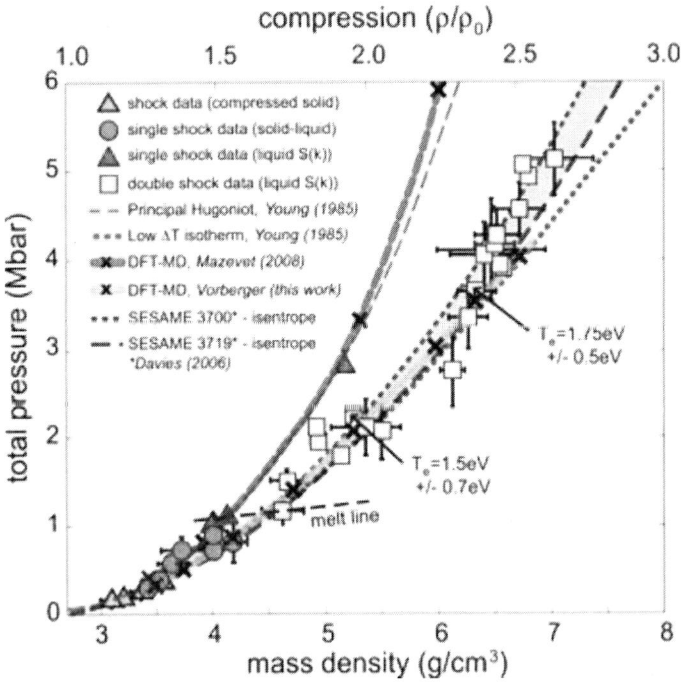

Figure 26. The comparison of pressure data of compressed dense aluminium measured for varying laser intensity is shown from single shocks and during shock coalescence of two counter-propagating shock waves. The single shock data follow the shock Hugoniot and DFT-MD simulations from $\rho = 3.5$ g cm^{-3} to about $\rho = 4$ g cm^{-3}. In addition, a high-density Hugoniot data point at > 5.2 g cm^{-3} was obtained in single shot targets. When the shock waves coalesce the increase in pressure is nearly adiabatic and is approaching an isentrope to $\rho = 7$ g cm^{-3}. The melt line (dashed) indicates the complete disappearance of shifted Bragg peaks. The DFT-MD simulations show excellent agreement when using the measured density and temperature data.

Figure 26 compares the experimental pressure data [22] with available empirical data [107] and state-of-the-art simulations. In particular, we show previous DFT-MD simulations for temperatures calculated for the shock Hugoniot [62] together with our DFT-MD simulations performed for temperatures and densities of the present experiment [61]; the simulations provide excellent agreement with the data from the experiment. The data further show no indication of Bragg peaks for compressed aluminium above $P = 1.2$ Mbar consistent with previous melt line measurements on the Hugoniot [108]. The present data set extends further showing the disappearance of Bragg scattering data above 1.2 Mbar and up to densities of 4.5 g cc^{-1}. At higher pressures, the shock coalescence data are on the isentrope [109] slightly above the isotherm [107] validating our understanding of dense aluminium utilizing pressure, temperatures and densities solely based on measurements.

These new experimental and theoretical findings demonstrate that spectrally and wavenumber resolved x-ray scattering is applicable for thorough testing of radiation-hydrodynamic calculations and EoS models. The method presents unique highly resolved data for dynamic high-pressure material science studies that require accurate knowledge of material properties and the thermodynamic state at high densities and are applicable for future studies aimed at observing effects of ionization on the EoS at high compression.

5. SUMMARY AND OUTLOOK

The first experiments at the MEC instruments have utilized record peak brightness x-ray laser radiation to probe high-power laser-driven targets demonstrating the unique capability to accurately measure the thermodynamic state of dense dynamically compressed matter. In particular, the forward x-ray scattering measurements have resolved the plasmon scattering spectrum providing a highly accurate tool to provide *in situ* measurements of the density and temperature evolution. This capability is accurate enough to test dense plasma theory used in radiation-hydrodynamic modeling and predictions of the physical properties of dense matter. Comparisons of the wavenumber resolved scattering data with DFT and reduced models have revealed the importance of including ion–ion interaction potentials that account for SRR. Differences can be seen between the data and results using orbital free Thomas–Fermi modeling or one component plasma theory. This success indicates that future applications of this technique may allow ultrafast pump–probe studies for measurements of e–i equilibration, heat and radiation transport and the thermodynamic equilibration of matter. Dense plasma effects can be studied with precision measurements of the plasmon dispersion, amplitude and plasmon broadening through Landau and collisional damping processes. Further, future x-ray scattering experiments in the Compton regime may reveal Pauli blocking and test models of continuum lowering.

For these studies, it is important to combine spectrally resolved scattering with angularly resolved scattering and optical velocity interferometer measurements. Here, dynamic x-ray diffraction data have visualized the transitions from a solid, compressed solid, into the co-existence regime, and to a WDM state. The measurements provide structural information of the compressed lattice and further allow inferring pressures in the multi-Mbar regime. We find excellent agreement with DFT-MD simulations on the shock Hugoniot or at shock coalescence using the temperatures and densities of the experiment. These data demonstrate measurements of structural phase transitions, melting curves, shock Hugoniot, and isentropic shock coalescence states. Importantly, larger explorations of the phase space of matter at high pressures will require future combinations of free electron x-ray lasers with higher-energy optical drive lasers. Our studies show that it will be possible to characterize conditions [110–113] found in white dwarfs to help determine the age of galaxies, provide insights into new high-pressure materials [114], and to test theoretical models that are applied for inertial confinement fusion experiments and calculations of the structure and magnetic fields of extrasolar planets.

This rich area of physics will be explored at multiple free electron laser facilities. Currently two hard x-ray free electron lasers are in operation worldwide, LCLS at SLAC in Stanford running since 2009 for users with the MEC instrument available since 2012 and which is the subject of this review. In addition, SACLA is in operation since 2012 at Spring8 in Harima. HED science experiments have recently begun, for example saturable absorption has recently been measured in iron targets using ultrahigh x-ray intensities of 10^{20} W cm^{-2} [115]. Current capabilities already include high-power drive lasers for pump–probe experiments. Two additional free electron lasers are in operation in the extreme ultraviolet and soft x-ray spectral ranges. The first facility to become available was FLASH at DESY in Hamburg, which started operation in 2005 [29–31] where pump–probe capabilities were recently added for HED experiments [32]. FERMI at ELETTRA in Trieste is running since 2012 [116]. MEC is an active area of research at all four facilities soon to be joined by a fifth instrument at the European XFEL.

In the near future, these facilities will begin exploring a new area for research of MEC by combining short pulse high-power lasers with the high brightness x-ray laser beam. Accurate measurements of density fluctuations using x-ray scattering and phase contrast imaging have the potential to reveal laser-induced collision-less shocks and characterize the nonlinear evolution of Weibel instabilities with the goal to demonstrate the physics of cosmic rays. Further, these techniques are important for measuring density fluctuations in laser-proton acceleration or laser-produced anti-matter studies. Finally, short pulse lasers are itself an excellent source of bright and ultrafast sources of x-rays and energetic particles allowing future ultrafast pump–probe studies with complementary sources to measure the fundamental process in photon–matter interactions.

ACKNOWLEDGMENTS

This work was supported by DOE Office of Science, Fusion Energy Science under FWP 100182. The experiments were performed at the Matter at Extreme Conditions (MEC) instrument of LCLS, supported by the DOE Office of Science, Fusion Energy Science under contract No. SF00515. This work was further supported by Laboratory Directed Research and Development grant, the Peter-Paul-Ewald Fellowship of the VolkswagenStiftung, and the Theodor-von-Lynen Fellowship of the Alexander-von-Humboldt foundation.

REFERENCES

1. Motz H 1979 The Physics of Laser Fusion (London: Academic)
2. Lindl J D, Amendt P, Berger R L, Glendinning S G, Glenzer S H, Haan S W, Kauffman R L, Landen O L and Suter L J 2004 The physics basis for ignition using indirect-drive targets on the National Ignition Facility Phys. Plasmas **11** 339
3. Lind J D et al 2014 Review of the National Ignition Campaign 2009–2012 Phys. Plasmas **21** 020501
4. Glenzer S H et al 2012 First implosion experiments with cryogenic thermoniclear fuel on the National Ignition Facility Plasmas Phys. Control. Fusion **54** 04013
5. Atzeni S and Meyerter-Vehn J 2004 The Physics of Inertial Fusion (Oxford: Clarendon)
6. Clayton C E et al 2010 Self-guided laser wakefield acceleration beyond 1 GeV using ionization-induced injection Phys. Rev. Lett. **105** 105003
7. Zepf M et al 2003 Proton acceleration from high-intensity laser interactions with thin foil targets Phys. Rev. Lett. **90** 064801
8. Kemp A J and Divol L 2012 Interaction physics of multipicosecond petawatt laser pulses with overdense plasma Phys. Rev. Lett. **109** 195005
9. Sentoku Y, d'Humières E, Romangnani L, Audebert P and Fuchs J 2011 Dynamic control over mega-ampere electron currents in metals using ionization-driven resistive magnetic fields Phys. Rev. Lett.**107** 135005
10. Krushelnick K et al 2008 Effect of relativistic plasma on extreme-ultraviolet harmonic emission from intense laser-matter interactions Phys. Rev. Lett. **100** 125005
11. Hegelich B M et al 2006 Laser acceleration of quasi-monoenergetic MeV ion beams Nature **439** 441–4
12. Blandford R and Eichler D 1987 Particle acceleration at astrophysical shocks: a theory of cosmic ray origin Phys. Rep. **154** 1–75

13. Kugland N L et al 2012 Self-organized electromagnetic field structures in laser-produced counter-streaming plasmas Nat. Phys. **8** 809–12
14. Guillot T 1999 Interiors of giant planets inside end outside the solar system Science **286** 72
15. Nettelmann N, Holst B, Kietzmann A, French M, Redmer R and Blaschke D 2008 Ab initio equation of state data for hydrogen, helium, and water and the internal structure of Jupiter Astrophys. J. **683**1217
16. Ross M 1981 The ice layer in uranus and neptune—diamonds in the sky? Nature **292** 435
17. Coppari F et al 2013 Experimental evidence for a phase transition in magnesium oxide at exoplanet pressures Nat. Geosci. **6** 926–9
18. Fiuza F, Fonseca R A, Tonge J, Mori W B and Silva L O 2012 Weibel-instability-mediated collisionless shocks in the laboratory with ultraintense lasers Phys. Rev. Lett. **108** 235004
19. Emma P et al 2010 First lasing and operation of an ångstrom-wavelength free-electron laser Nat. Photon. **4** 641–7
20. Glenzer S H and Redmer R 2009 X-ray Thomson scattering in high energy density plasmas Rev. Mod. Phys. **81** 1625
21. Sperling P et al 2015 Phys. Rev. Lett. **115** 115001
22. Fletcher L B et al 2015 Ultrabright x-ray laser scattering for dynamic warm dense matter physics Nat. Photon. **9** 274
23. Nagler B et al 2015 The matter in extreme conditions instrument at the linac coherent light Source J. Synchrotron Radiat. **22** 520–5
24. Gleason A et al 2015 Ultrafast visualization of crystallization and grain growth in shock-compressed SiO2 Nat. Commun. **6** 8191
25. Gorman M G et al 2015 Direct observation of melting in shock-compressed bismuth with femtosecond x-ray diffraction Phys. Rev. Lett. **115** 095701
26. Amann J et al 2012 Demonstration of self-seeding in a hard-x-ray free electron laser Nat. Photon. **6**693–8
27. Ciricosta O et al 2012 Direct measurements of the ionization potential depression in a dense plasma Phys. Rev. Lett. **109** 065002
28. Albert F et al 2013 Phys. Rev. Lett. **111** 235004
29. Zastrau U et al 2008 Bremsstrahlung and line spectroscopy of warm dense aluminum plasma heated by xuv free-electron-laser radiation Phys. Rev. E **78** 066406
30. Nagler B et al 2009 Turni aren ng solid aluminium transpt by intense soft x-ray photoionization Nat. Phys. **5** 693
31. Faeustlin R R et al 2010 Observation of ultrafast nonequilibrium collective dynamics in warm dense hydrogen Phys. Rev. Lett. **104** 125002
32. Zastrau U et al 2014 Phys. Rev. Lett. **112** 105002
33. Patel P K, Mackinnon A J, Key M H, Cowan T E, Foord M E, Allen M, Price D F, Ruhl H, Springer P T and Stephens R Isochoric heating of solid-density matter with an ultrafast proton beam Phys. Rev. Lett. **91**125004
34. Pelka A et al 2010 Ultrafast melting of carbon induced by intense proton beams Phys. Rev. Lett. **105**265701
35. Glenzer S H, Gregori G, Lee R W, Rogers F J, Pollaine S W and Landen O L 2003 Demonstration of spectrally resolved x-ray scattering in dense plasmas Phys. Rev. Lett. **90** 175002
36. Glenzer S H et al 2007 Observations of plasmons in warm dense matter Phys. Rev. Lett. **98** 065002
37. Gregori G et al 2008 X-Ray scattering measurements of radiative heating and cooling dynamics Phys. Rev. Lett. **101** 045003
38. Visco A J et al 2012 Measurement of radiative shock properties by x-ray Thomson scattering Phys. Rev. Lett. **108** 145001
39. Kritcher A L et al 2008 Ultrafast x-ray Thomson scattering of shock-compressed matter Science **322**69

40. Lee H J et al 2009 X-ray Thomson-scattering measurements of density and temperature in shock-compressed beryllium Phys. Rev. Lett. **102** 115001
41. Kritcher A L et al 2009 Measurements of ionic structure in shock compressed lithium hydride from ultrafast x-ray Thomson scattering Phys. Rev. Lett. **103** 245004
42. Neumayer P et al 2010 Plasmons in strongly coupled shock-compressed matter Phys. Rev. Lett. **105**075003
43. Kugland N L et al 2009 Evolution of elastic x-ray scattering in laser-shocked warm dense lithium Phys. Rev. E **80** 066406
44. Fortmann C, Lee H J, Döppner T, Falcone R W, Kritcher A L, Landen O L and Glenzer S H 2012 Measurement of the adiabatic index in be compressed by counterpropagating shocks Phys. Rev. Lett.**108** 175006
45. Regan S P et al 2012 Inelastic x-ray scattering from shocked liquid deuterium Phys. Rev. Lett. **109**265003
46. Ma T et al 2013 X-ray scattering measurements of strong ion–ion correlations in shock-compressed aluminum Phys. Rev. Lett. **110** 065001
47. Zel'Dovich Y and Raizer Y 2002 Physics of Shock Waves and High-Temperature Hydrodynamic Phenomena ed W D Hayes and R F Probstein (New York: Dover)
48. Drake R P 2006 High-Energy-Density Physics: Fundamentals, Inertial Fusion, and Experimental Astrophysics (Berlin: Springer)
49. Frontiers in High-Energy-Density Physics: The X Games of Contemporary Science 2003 (http://nap.edu/catalog.php?record_id=10544)
50. Betti R et al 2010 Thermonuclear ignition in inertial confinement fusion and comparison with magnetic confinement Phys. Plasmas **17** 058102
51. Marinak M M, Kerbel G D, Gentile N A, Jones O, Munro D, Pollaine S, Dittrich T R and Haan S W 2001 Phys. Plasmas **8** 2275
52. Fortov V E and Iakubov I T 2000 The Physics of Non-Ideal Plasmas (Singapore: World Scientific)
53. Ichimaru S 1991 Statistical Plasma Physics (Reading, MA: Addison-Wesley)
54. Hansen J P and McDonald I R 2000 Theory of Simple Liquids (London: Academic)
55. Baker L M and Hollenbach R E 1972 Laser interferometer for measuring high velocities of any reflecting surface J. Appl. Phys. **43** 4669
56. Geindre J P, Audebert P, Rousse A, Falliès F, Gauthier J C, Mysyrowicz A, Dos Santos A, Hamoniaux G and Antonetti A 1994 Frequency-domain interferometer for measuring the phase and amplitude of a femtosecond pulse probing a laser-produced plasma Opt. Lett. **23** 1997–9
57. Fletcher L B, Galtier E, Heimann P, Lee H J, Nagler B, Welch J, Zastrau U, Hastings J B and Glenzer S H 2013 Plasmon measurements with a seeded x-ray laser J. Instrum. **8** C11014
58. Gericke D O, Vorberger J, Gregori G and Wünsch K 2010 Screening of ionic cores in partially ionized plasmas within linear response Phys. Rev. E **81** 065401(R)
59. Wünsch K, Vorberger J and Gericke D O 2009 Ion structure in warm dense matter: benchmarking hypernetted-chain equations by first-principle simulations Phys. Rev. E **79** 010201(R)
60. Chapman D A et al 2015 Observation of finite-wavelength screening in high-energy-density matter Nat. Commun. **6** 6839
61. Mazevet S and Zerah G 2008 Ab initio simulations of the K-edge shift along the aluminum hugoniot Phys. Rev. Lett. **101** 155001
62. Vorberger J, Donko Z, Tkachenko I M and Gericke D O 2012 Dynamic ion structure factor of warm dense matter Phys. Rev. Lett. **109** 065002
63. Rüter H R and Redmer R 2014 Ab initio simulations for the ion–ion structure factor of warm dense aluminum Phys. Rev. Lett. **112** 145007
64. Clérouin J, Grégory R, Arnault P, Ticknor C, Kress J D and Collins L A 2015 Evidence for out-of-equilibrium states in warm dense matter probed by x-ray thomson scattering Phys. Rev. E **91** 011101
65. Starrett C E, Daligault J and Saumon D 2015 Phys. Rev. E **91** 013104

66. Souza A N, Perkins D J, Starrett C E, Saumon D and Hansen S B 2014 Predictions of x-ray scattering spectra for warm dense matter Phys. Rev. E **89** 023108
67. Starrett C E and Saumon D 2015 Models of the elastic x-ray scattering feature for warm dense aluminum Phys. Rev. E **92** 033101
68. White W E, Robert A and Dunne M 2015 The linac coherent light source J. Synchrotron Radiat. **22**472–6
69. Tonks L and Langmuir I 1929 Oscillations in ionized gases Phys. Rev. **33** 195–210
70. Marinelli A et al 2015 High-intensity double-pulse x-ray free-electron laser Nat. Commun. **6** 6369
71. Minitti M P et al 2015 Optical laser systems at the linac coherent light source J. Synchrotron Radiat.**22** 526–31
72. Gamboa E J, Montgomery D S, Hall I M and Drake R P 2011 Imaging x-ray crystal spectrometer for laser-produced plasmas J. Instrum. **6** P04004
73. Pak A et al 2004 X-ray line measurements with high efficiency Bragg crystals Rev. Sci. Instrum. **75**3747
74. Urry M K, Gregori G, Landen O L, Pak A and Glenzer S H 2006 J Quant. Spectrosc. Radiat. Transfer. **99**636
75. Zastrau U et al 2013 Characterization of strongly bent HAPG crystals for von-Hamos x-ray spectrographs J. Instrum. **8** P10006
76. Legall H et al 2006 New generation of x-ray optics based on pyrolytic graphite Proc. FEL (BESSY, Berlin, Germany,)
77. Zastrau U et al 2012 Focal aberrations of large-aperture HOPG von-Hamos x-ray spectrometers J. Instrum. **7** P09015
78. Dixit S N, Lawson J K, Manes K R, Powell H T and Nugent K A 1994 Kinoform phase plates for focal plane irradiance profile control Opt. Lett. **19** 417–9
79. MacFarlane J J, Golovkin I E and Woodruff P R 2006 HELIOS-CR A 1D radiation-magnetohydrodynamics code with inline atomic kinetics modeling J. Quant. Spectrosc. Radiat. Transfer **99** 381–97
80. Asay J R and Shahinpoor M 1993 High-Pressure Shock Compression of Solids (Berlin: Springer)
81. Neal T 1979 High-pressure science and technology AIRAP Conf. Proc. vol 1 and 2 ed K D Timmerhaus and M S Barber (New York: Plenum)
82. Froula D H, Glenzer S H, Luhmann N C and Sheffield J 2010 Plasma Scattering of Electromagnetic Radiation by Sheffield 2nd edn (New York: Academic)
83. Gregori G, Glenzer S H, Rozmus W, Lee R W and Landen O L 2003 Theoretical model of x-ray scattering as a dense matter probe Phys. Rev. E **67** 026412
84. Gregori G, Glenzer S H and Landen O L 2006 Generalized x-ray scattering cross section from nonequilibrium plasmas Phys. Rev. E **74** 026402
85. Chihara J 1987 Difference in x-ray scattering between metallic and non-metallic liquids due to conduction electrons J. Phys. F: Met. Phys. **17** 295
86. Johnson W R, Nilsen J and Cheng K T 2013 Resonant bound-free contributions to Thomson scattering of x-rays by warm dense matter High Energy Density Phys. **9** 407
87. Fletcher L B et al 2014 Observations of continuum depression in warm dense matter with x-ray Thomson scattering Phys. Rev. Lett. **112** 145004
88. Ichimaru S 1982 Strongly coupled plasmas—high-density classical plasmas and degenerate electron fluids Rev. Mod. Phys. **54** 1017
89. Mermin N D 1970 Phys. Rev. B **1** 2362–3
90. Redmer R R, Reinholz H, Röpke G, Thiele R and Höll A 2005 Theory of x-ray Thomson scattering in dense plasmas IEEE Trans. Plasma Sci. **33** 77
91. Hoell A et al 2007 Thomson scattering from near-solid density plasmas using soft x-ray free electron lasers High Energy Density Phys. **3** 120
92. Gregori G, Glenzer S H and Landen O L 2003 Strong coupling corrections in the analysis of x-ray Thomson scattering measurements J. Phys. A: Math. Gen. **36** 5971

93. Fortmann C, Wierling A and Röpke G 2010 Influence of local-field corrections on Thomson scattering in collision dominated two-component plasmas Phys. Rev. E **81** 026405
94. Bohm D and Pines D 1953 A collective description of electron interactions: III. Coulomb interactions in a degenerate electron gas Phys. Rev. **92** 609–25
95. Zimmerman R 1987 Many-Particle Theory of Highly Excited Semiconductors (Leipzig: Teubner)
96. Thiele R et al 2008 Plasmon resonance in warm dense matter Phys. Rev. E **78** 026411
97. Döppner T et al 2009 Temperature measurement through detailed balance in x-ray Thomson scattering High Energy Density Phys. **5** 182–6
98. Ziman J M 1961 A theory of the electrical properties of liquid metals: I. The monovalent metals Phil. Mag. **68** 1013–34
99. Reinholz H, Redmer R, Röpke G and Wierling A 2000 Long-wavelength limit of the dynamical local-field factor and dynamical conductivity of a two-component plasma Phys. Rev. E **62** 5648
100. Lee Y T and Moore R 1984 Phys. Fluids **27** 1273
101. Fortmann C 2009 Single-particle spectral function for the classical one-component plasmas Phys. Rev. E **79** 16404
102. Scopigno T, Ruocco G and Sette F 2005 Microscopic dynamics in liquid metals: the experimental point of view Rev. Mod. Phys. **77** 881
103. Hansen J-P and McDonald I R 2000 Theory of Simple Liquids (New York: Academic)
104. Mithen J, Daligaut J and Gregori G 2011 Extent of validity of the hydrodynamic description of ions in dense plasmas Phys. Rev. E **83** 015401(R)
105. Gregori G and Gericke D O 2009 Low frequency structural dynamics of warm dense matter Phys. Plasmas **16** 056306
106. Salpeter E E 1961 Energy and pressure of a zero-temperature plasma Astrophys. J. **134** 669
107. Young D A, Wolford J K, Rogers F J and Holian K S 1985 Theory of the aluminum shock equation of state to 104 Mbar Phys. Lett. **108** 157–60
108. Ross M, Yang L H and Boehler R 2004 Melting of aluminum, molybdenum, and the light actinides Phys. Rev. B **70** 184112
109. Davis J-P 2006 Experimental measurement of the principal isentrope for aluminum 6061-T6 to 240 GPa J. Appl. Phys. **99** 103512
110. Hau-Riege S et al 2012 Ultrafast transitions from solid to liquid and plasma states of graphite induced by x-ray free-electron laser pulses Phys. Rev. Lett. **108** 217402
111. Fortmann C, Niemann C and Glenzer S H 2012 Theory of x-ray scattering in high-pressure electrides Phys. Rev. B **86** 174116
112. Hammel B A, Haan S W, Clark D S, Edwards M J, Langer S H, Marinak M M, Patel M V, Salmonson J D and Scott H A 2010 High-mode Rayleigh–Taylor growth in NIF ignition capsules High Energy Density Phys. **6** 171–8
113. Collins G W et al 1998 Measurements of the equation of state of deuterium at the fluid insulator-metal transition Science **281** 1178
114. Gamboa E J et al 2015 Single-shot measurements of plasmons in compressed diamond with an x-ray laser Phys. Plasmas **22** 056319
115. Yoneda H et al 2014 Saturable absorption of intense hard x-rays in iron Nat. Commun. **5** 5080
116. Mincigrucci R et al 2015 Role of the ionization potential in nonequilibrium metals driven to absorption saturation Phys. Rev. E **92** 011101

CHAPTER 2

Structure, Equation of State, Diffusion and Viscosity of Warm Dense Fe Under The Conditions of a Giant Planet Core

Jiayu Dai[1,3], Yong Hou[1], Dongdong Kang[1], Huayang Sun[1], Jianhua Wu[1] and Jianmin Yuan[1,2,3]

[1] *Department of Physics, College of Science, National University of Defense Technology, Changsha 410073, People's Republic of China*

[2] *State Key Laboratory of High Performance Computing, National University of Defense Technology, Changsha 410073, People's Republic of China*

[3] *Department of Physics, College of Science, National University of Defense Technology, Changsha 410073, People's Republic of China*

ABSTRACT

Fe exists abundantly in the universe. In particular, the dynamic structures and transport properties of warm dense Fe are crucial for understanding the evolution and structures of giant planets. In this paper, we present the ionic structures, equation of states, diffusion and viscosity of Fe at two typical densities of 33.385 and 45 g cm^{-3} in the temperature range of 1–10 eV, giving the data by the first principles calculations using quantum Langevin molecular dynamics. Furthermore, the validation of the Stokes–Einstein (SE) relation in this regime is discussed, showing the importance of choosing the effective atomic diameter. The results remind us of the careful usage of the SE relation under extreme conditions.

1. INTRODUCTION

The properties of complex materials such as Fe under extreme conditions are crucial for understanding the evolution of planets (Earth, giant planets and extro-solar giant planets) [1] and stars such as the Sun [2, 3]. More interestingly, the laser-driven and Z-pinch experiments can lead to very high densities up to hundreds of g cm^{-3} and temperatures up to thousands of eV [4, 5]. In particular, temperatures of 1–10 eV with associated pressures of 10–1000 TPa are likely to exist in the interiors of massive planets [6, 7]. Recent studies have shown the possible stable structures of Fe at these pressures [6, 7], where face-centered cubic (fcc) structures are stable in the range of 7–21 TPa. However, the effect of temperature on these structures is still not known, since the dynamic structures such as melting, diffusion and viscosity can be induced by temperatures; the phases of Fe in Earth-like exoplanets are likely to be liquid. In order to understand these behaviors, since the experimental determination currently seems to be extremely expensive and difficult, accurate simulations are required for the determination of the structures and dynamics. The dynamic structures of Fe and its compounds from first principles molecular dynamics under the physical conditions of the Earth's core have been studied widely [8–10], showing the complexity of the new phenomena in high energy density physics. In particular, it is worth verifying whether the Stokes–Einstein (SE) relation at such high densities is valid, since the SE relation is one important bridge between dynamics and statistics; it has usually been adopted in previous calculations of viscosities from diffusion coefficients [8–12]. Up to now, the diffusion and viscosity can be obtained directly by molecular dynamics simulations with the Kubo–Green relation [10–13]. Moreover, we can construct statistical models according to the simulation results by introducing easy applications in hydrodynamics [14, 15]. However, molecular dynamics simulations are strongly dependent on the accuracy of many-body interatomic potentials; there are usually some limitations to statistical methods.

A cold stable fcc structure, with its inducing liquid at a high temperature, is considered an important phase in giant planets such as Jupiter and exoplanets [6, 7]. In this regime, the densities are from 33.9 to 66.7 g cm^{-3} and the temperatures are from 1 to 10 eV. For these states, the melting behaviors and the transport properties are interesting and deserve to be studied carefully. Besides, a high density will introduce many different chemical bonds since a pressure greater than 1 Mbar will change the traditional chemistry dramatically [16–20]. Furthermore, when the pressure increases to 100 Mbar, the core electron charge density can be changed significantly [16–18]. Therefore, the transport properties such as equation of states (EOS), diffusion and viscosity will change a lot since these properties are related to the electronic structures. With respect to the methods of calculating these properties accurately, first principles calculations are thus required because we did not understand their behaviors from semiclassical models, such as the average atom (AA) model [21, 22] and the Thomas–Fermi model or the orbital free method [23–25], in which the orbital behaviors or chemical bonds information cannot be described well.

Molecular dynamics combining the finite-temperature density functional theory (DFT) [26], called quantum molecular dynamics (QMD), and quantum Langevin molecular dynamics (QLMD) have been successfully used in warm and hot dense matter, especially under very high pressures [8, 10, 18–20, 27, 28]. The validations of QMD and QLMD at high densities and temperatures have been verified many times by comparing with experiments and interestingly reasonable results are found. In particular, QMD is considered as one of the most promising methods and has been successfully used to predict the transport properties of warm dense matter (WDM) [29, 31]. The biggest argument against QMD applications is that it limits the number of particles because of its computational cost at high temperature. However, QMD and QLMD can be efficient for the calculations of WDM under the conditions of a giant planet core; they can give relatively accurate transport properties using a few hundred particles [29–32]. In this paper, we adopted QLMD to simulate the transport properties of dense Fe at 45 g cm^{-3} from 1 to 10 eV. Except for considering the density of 33.385 g cm^{-3} at 100 eV along the Hugoniot curve [18], we also calculate the dynamical properties at 33.385 g cm^{-3} up to 10 eV. Furthermore, the approximation of the SE relation is discussed, showing the debatable validation under such dense regimes.

2. METHODS

For the complexity of Fe at high density, both QMD with the velocity rescaling method and QLMD simulations are tested first. In fact, at low temperatures, the electron–ion collision induced friction is small since the friction is proportional to the temperature. Therefore, QMD and QLMD are equivalent for the equilibrium properties at low temperatures. Eventually, the advantage of QLMD is its higher computational efficiency compared to traditional QMD [33]. The electronic structures are solved based on finite temperature DFT implemented in the Quantum ESPRESSO package [34]. Considering the small atomic sizes at high density, the time step of 0.5 fs is used in order to keep the correct trajectories of ions. The pseudopotential (PP) is one of the key points in QMD and QLMD simulations. Here, we construct a new PP with 16 electrons in the valence and 0.9 atomic units for the radius cutoff within the generalized-gradient approximation [35], which can promise correctness with respect to the conditions we are studying. Using this PP, we can reproduce the bulk modulus properties of Fe as previous results calculated by full-electron calculations and experiments. More importantly [18, 36], at high pressure, the pressure and band structures at the density of 33.9 and 48.23 g cm^{-3} are in good agreement with the pressure in [7] using the same PP [18]. Furthermore, the same PP has been successfully used to calculate the EOS of Fe in a hot dense regime within a wide range of densities and temperatures [18], indicating the validation of PP in this work. Besides, 2000 time steps with a large convergent tolerance of 1.0×10^{-4} in electronic structure calculations are adopted in order to reach the thermalization; 10 ps time lengths with a small convergent tolerance of 1.0×10^{-6} [33] are simulated to obtain the transport properties and thermal average. In order to accelerate the calculations, the Gamma point is only used for the representation of the Brillouin zone. To compare with the results from semiclassical methods, averaged atom molecular dynamics (AAMD) [22, 37] is performed for all cases. When two atoms are close enough, their electronic distributions overlap, inducing interactions between them. In order to describe these interactions, we should establish the interatomic potential. In the AAMD method, the temperature-dependent pair potential, which can describe the contributions of overlapping electronic distributions, is constructed from the AA electronic calculations, including the ionic correlations within the framework of the AA model. Within this framework, the ions are moving on this pair potential, giving the ionic trajectories and ionic configurations at different times. From AAMD, we can basically use a large number of particles and give the convergent results corresponding to the size effects. The comparison between QLMD and AAMD has been reported before [19, 33], showing their agreement at high temperatures. The details of AAMD can be found in [37].

The self-diffusion coefficient D can be obtained from two methods. One is from the trajectories by the mean-square displacement DR

$$D^R = \frac{1}{6t}\langle |\mathbf{R}_i(t) - \mathbf{R}_i(0)|^2 \rangle, \tag{1}$$

where the $\mathbf{R}i$ is the position of the ith particle. The other method is from the velocity autocorrelation function DV

$$D^V = \frac{1}{3}\int_0^{\infty} \langle \mathbf{V}_i(t)\,\mathbf{V}_i(0) \rangle \, \mathrm{d}t, \tag{2}$$

where $\mathbf{V}i$ is the velocity of the ith particle. Besides, the viscosity η can be obtained from the autocorrelation function of the off-diagonal component of the stress tensor, the so-called Green–Kubo equation [10, 38]

$$\eta = \lim_{t\to\infty} \overline{\eta}(t) = \lim_{t\to\infty} \frac{V}{k_B T} \int_0^t \langle P_{12}(0)\, P_{12}(t') \rangle \, dt', \tag{3}$$

where P_{12} represents the averaged result for the five independent off-diagonal components of the stress tensor *Pxy*, *Pyz*, *Pzx*, $(Pxx - Pyy)/2$ and $(Pyy - Pzz)/2$. Using this method, the self-diffusion coefficient and viscosity of warm and hot dense matter have been reported widely [8–12, 29, 39–42] within the framework of QMD, showing its validation in these extreme conditions.

On the other side, the viscosity can also be calculated directly from the SE relation

$$D\eta = k_B T/(2\pi a), \tag{4}$$

where a is an effective atomic diameter. This relation is statistically obtained from the Brownian motion of a macroscopic particle in liquid, but it is only an approximation for the atoms. If the validation of the SE relation can be verified, the size of particles and the transport behaviors can be understood well [43]. In fact, for a Brownian particle, the SE relation is equivalent to equations (2) and (3). In particular, for dense matter, the diameters of atoms are very small so the validation of the SE relation should be examined very carefully.

3. CONVERGENCE TESTS

For calculating the transport properties from first principles a number of atoms should be tested, since diffusion and viscosity are strongly dependent on the system sizes. For this purpose, in QLMD simulations, we tested face-centered cubic (fcc) structures with 32, 108 and 256 atoms and a body-centered cubic (bcc) structure with 54 atoms in a supercell at 45 g cm^{-3}. The pressure and diffusion versus temperature relations with different numbers of atoms are shown in figure 1. It is concluded that the pressure is not incredibly sensitive to the system sizes. In particular, the pressures are closer for different sizes when the temperatures are higher. The small difference at low temperatures may be caused by the long ranged orders with crystal structures. It is worth noting that the initial configurations of bcc and fcc structures are important for calculating diffusion in the solid phases, but the differences from different initial configurations disappear gradually with the increasing temperatures in the liquid phases (here above 5 eV).

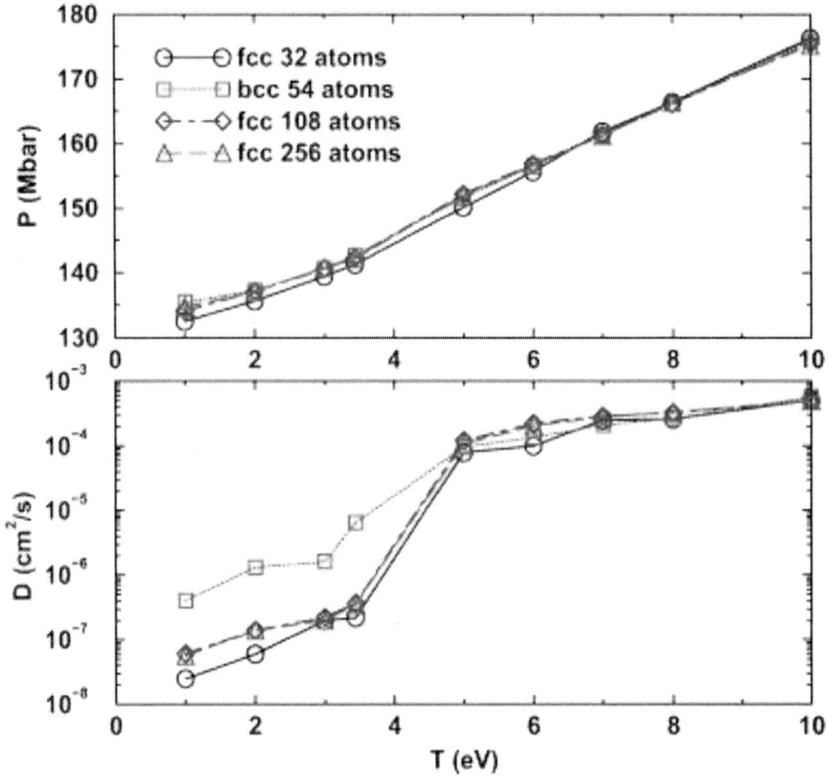

FIGURE 1. Size effect on the pressure (upper panel) and diffusion coefficient (lower panel) of Fe at 45 g cm^{-3}.

According to the tests, it can be stated that our systems are in fact convergent for the calculation of pressure and diffusion coefficients when we use 108 atoms with initial fcc structures. It should be noted that when the system becomes liquid (after the jump in the diffusion coefficient), the results of 54 atoms, 108 atoms and 256 atoms are consistent. The difference at low temperatures would be from the different initial structures of bcc or fcc crystals. In fact, the small differences from different sizes can also come from the statistical errors for small sizes. Basically, this error can be overcome by longer time simulations.

For the viscosity, the results of the convergent tests are shown in figure 2. It is only useful for the liquids here since the viscosity is very large for solids. Here, the liquid phases are shown above 5 eV in our simulations. By comparison, we can know that the viscosity is incredibly sensitive to the sizes. For a small size of 32 atoms, the statistical errors are really large, which is not likely to be compensated by increasing the simulation time.

When the number of atoms is increased up to 108 and 256, the viscosities at different temperatures seem convergent. Therefore, we can safely use the 108 atoms for the calculations of pressure, diffusion and viscosity in our cases. In fact, about 100 or 200 atoms are usually used in QMD simulations for the transport properties [8, 10, 12, 29–31], which has been shown to be convergent within reasonable errors.

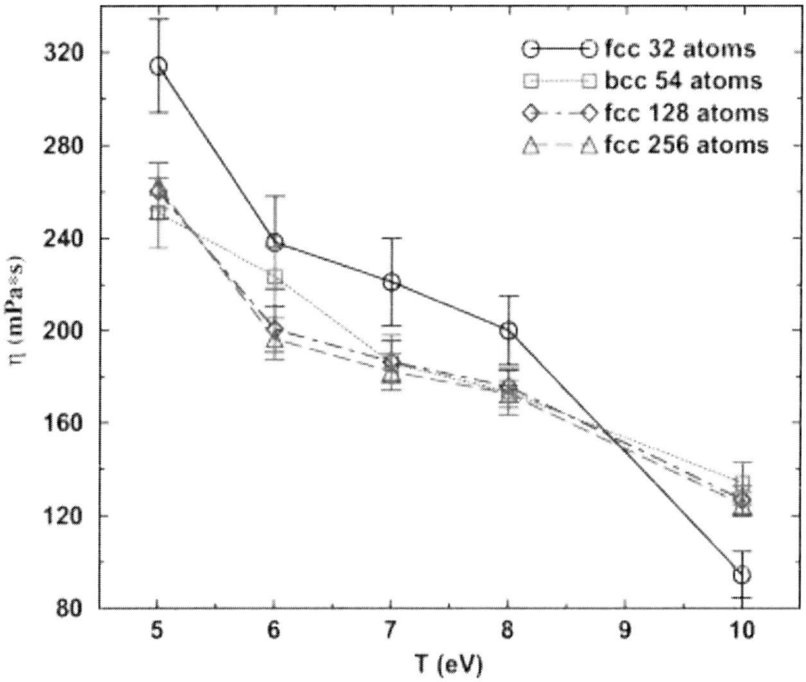

FIGURE 2. Size effect on the viscosity of Fe at 45 g cm^{-3} from QLMD simulations.

For the effect of the particle number, we use a large number of particles up to 5324 in AAMD calculations to verify the convergence, as shown in figure 3. The simulation time is up to 100 ps in each case in order to obtain the correct correlations of particles in these semiclassical calculations. It can be shown that although the number of particles can affect the convergence of diffusion and viscosity, all the differences are within 10% or smaller. When the number reaches 4000, the diffusion and viscosity are almost convergent. In order to compare the results, we use 4000 particles in AAMD to calculate the transport properties below. It is worth noting that AAMD adopted pair potentials, which should be more sensitive to the size and debatable for the viscosity since the unsymmetrical shear of the system can not be described well by pair potentials.

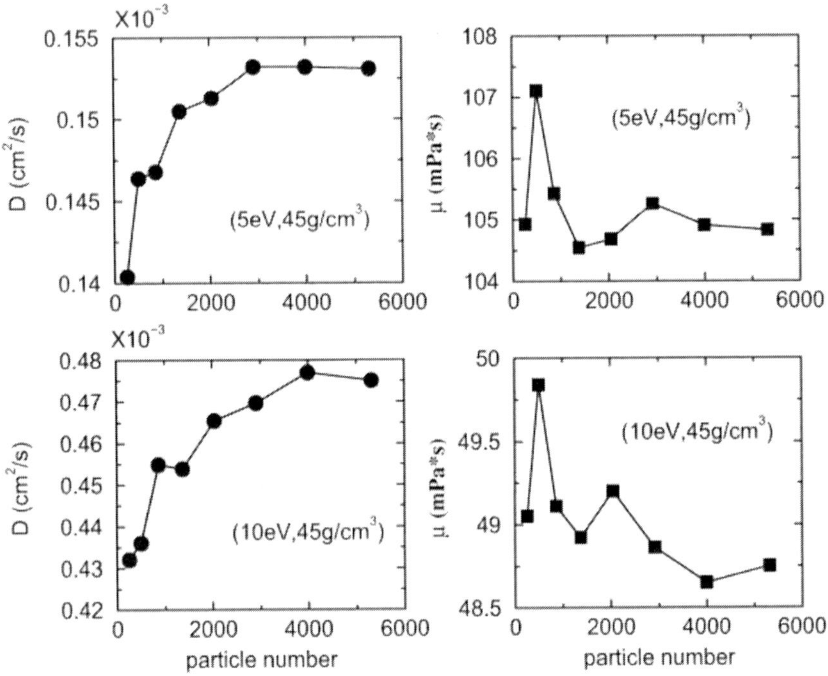

FIGURE 3. Size effect on the viscosity of Fe at 45 g cm^{-3} from AAMD simulations.

For the density of 33.385 g cm^{-3}, we used 54 atoms with initial bcc structures to calculate their transport properties. The simulation details are also the same as those of 45 g cm^{-3}. Basically, the solid phase of Fe at 33.385 g cm^{-3} may be hexagonal close-packed (hcp). However, determining the lattice constants of hcp at high temperatures is very difficult; the dynamic properties such as the structure and melting behavior are very similar to the bcc structures [7, 45]. Therefore, we used bcc structures as the initial configurations. In fact, from the tests of Fe at 45 g cm^{-3}, it is now known that the transport properties are independent of the initial configurations when Fe melts. Furthermore, the effects of larger system sizes on calculated viscosities have been previously studied for models such as hard-sphere and Lennard–Jones liquids [44], showing meaningful values for μ even with only 32 atoms. Thus, the initial configurations and sizes in this work can promise the reasonable accuracy of our simulations.

4. STRUCTURAL DYNAMICS

The dynamic structures for dense matter are basic when attempting to understand the melting behaviors and the dynamics in a planet's core. However, dense liquid Fe has rarely been studied before due to the lack of effective methods. Dense Fe can hold the solid phase or ordered structures at very high temperatures, form clusters and chemical bonds assisted by core electrons [18] and keep a high melting temperature [45–47]. Here, the structures are investigated by looking at the radial distribution functions (RDF) $g(r)$, as shown in figures 4 and 5.

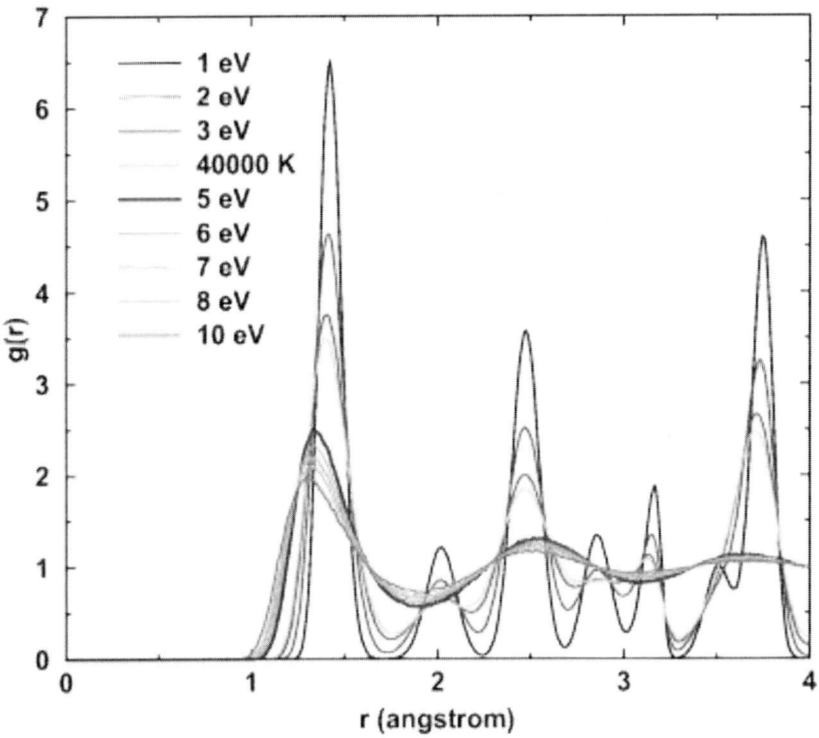

FIGURE 4. RDF versus temperature for Fe at 45 g cm^{-3}.

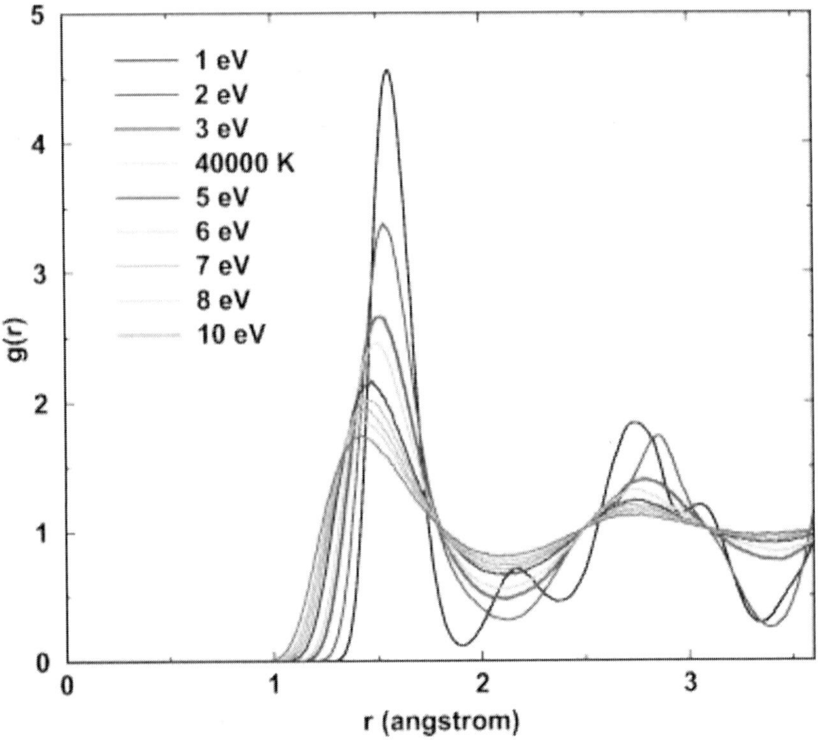

FIGURE 5. RDF versus temperature for Fe at 33.385 g cm^{-3}.

From the RDF, we can know that when the temperature increases up to 5 eV, Fe exhibits liquid behaviors at 45 g cm^{-3} and above 3 eV at 33.385 g cm^{-3}. We calculated the details of the dynamic structures near 5 and 3 eV for the two densities respectively, using the two phase approach [45] with 216 atoms and 108 atoms to obtain the melting temperature. The melting temperatures are found to be around 57 000 and 30 000 K, within the errors of 200 K in our simulations. The same behaviors can also be found in the data of diffusion coefficients and viscosities, which are shown in figures 1and 2 and will be discussed in detail later. In this stage, the Fe at 45 g cm^{-3} and 40 000 K is also in the solid phase in the Fe phase diagram [7].

We should note that the first peaks of the RDF move toward to the zero point, indicating that the effective distance between ions decreases with the increasing temperatures. Therefore, we may use the positions of the first peaks in the RDF as the effective atomic diameters when discussing the SE relation later. If we use only the averaged ionic radius, the temperature effect cannot be included since the radius is only dependent on the type of elements and system densities.

5. TRANSPORT PROPERTIES

After the convergence tests, we can obtain the transport properties including EOS, diffusion coefficients and viscosities for dense Fe. The summaries of the data are shown in tables 1 and 2. In the tables, pressures and diffusion coefficients from AAMD are also shown. When the solid phases are not broken, the diffusion coefficients are very small and the viscosities are very large (therefore they are not convergent in the Green–Kubo equation).

Table 1. Summary of the results of dense Fe at different temperatures at 45 g cm^{-3}. *DR* and *DV*are respectively defined in equations (1) and (2). D_{AAMD} and μ_{AAMD} are the diffusion coefficient and viscosity from AAMD calculations, respectively.

T(eV)	*P*(Mbar)	*DR*(cm^2 s^{-1})	*DV*(cm^2 s^{-1})	μ(mPa s)	P_{AAMD}(Mbar)	D_{AAMD}(cm^2 s^{-1})	$\mu AAMD$(mPa s)
1	134.128	0.61×10^{-7}	0.60×10^{-7}	–	245.13	2.39×10^{-8}	–
2	137.148	1.40×10^{-7}	1.61×10^{-7}	–	250.25	9.55×10^{-8}	–
3	140.594	2.24×10^{-7}	2.68×10^{-7}	–	255.73	2.07×10^{-7}	–
3.447	142.295	3.72×10^{-7}	3.82×10^{-7}	–	258.11	2.91×10^{-7}	–
5	152.218	1.21×10^{-4}	1.23×10^{-4}	62.07±6	270.77	1.54×10^{-4}	104.29
6	156.825	2.22×10^{-4}	2.11×10^{-4}	44.11±5	275.11	2.11×10^{-4}	83.38
7	161.296	3.15×10^{-4}	3.05×10^{-4}	41.29±4	279.43	2.73×10^{-4}	69.47

T(eV)	P(Mbar)	DR(cm^2 s^{-1})	DV(cm^2 s^{-1})	μ(mPa s)	P_{AAMD}(Mbar)	D_{AAMD}(cm^2 s^{-1})	$\mu AAMD$(mPa s)
8	166.081	3.26×10^{-4}	3.37×10^{-4}	38.13±4	283.65	3.39×10^{-4}	60.34
10	175.421	4.86×10^{-4}	4.83×10^{-4}	27.72±3	292.46	4.77×10^{-4}	48.64

Table 2. Summary of the results of dense Fe at different temperatures at 33.385 g cm^{-3}.

T(eV)	P(Mbar)	DR(cm^2 s^{-1})	DV(cm^2 s^{-1})	μ (mPa s)	P_{AAMD}(Mbar)	D_{AAMD}(cm^2 s^{-1})	$\mu AAMD$(mPa s)
1	60.547	3.50×10^{-7}	3.21×10^{-7}	-	79.52	2.11×10^{-6}	-
2	63.471	2.45×10^{-6}	2.19×10^{-6}	-	84.05	2.69×10^{-5}	-
3	67.739	6.70×10^{-5}	6.71×10^{-5}	47.1996±5	86.99	1.49×10^{-4}	29.43
3.447	69.786	1.16×10^{-4}	1.14×10^{-4}	40.3466±5	88.36	1.84×10^{-4}	27.32
5	75.282	2.98×10^{-4}	2.85×10^{-4}	24.217±3	92.57	3.28×10^{-4}	22.39
6	79.727	3.67×10^{-4}	3.74×10^{-4}	19.8484±2	95.33	4.15×10^{-4}	20.41
7	82.248	5.99×10^{-4}	5.91×10^{-4}	18.1716±2	98.30	5.08×10^{-4}	19.00
8	85.825	6.08×10^{-4}	5.98×10^{-4}	17.1372±2	101.11	6.13×10^{-4}	17.83
10	93.098	7.51×10^{-4}	7.69×10^{-4}	15.7396±2	107.30	8.01×10^{-4}	16.28

For a density of 45 g cm^{-3}, from table 1, the diffusion coefficients change rapidly between the temperatures of 40 000 K (3.447 eV) and 5 eV, and reach the orders of liquids diffusion at 5 eV. Also, the viscosities above 5 eV go to convergence when we integrate the pressure autocorrelations in equation (3), indicating the existence of liquid phases. However, AAMD gives the liquid Fe at 4 eV (not shown here), showing its limitations from pair potentials in very strongly coupling regimes. Furthermore, we can see that the diffusion coefficients from equations (1) and (2) are equal, which shows the validation and convergence of our simulations. For the pressure, AAMD gives much larger pressures than those of QLMD simulations, which is caused by the electronic structure calculations from Thomas–Fermi methods. It is very interesting that when the temperatures are above 5 eV, the diffusion coefficients from QLMD and AAMD are very close. One reason might be that the diffusion coefficients are strongly dependent on the nearest neighbor distributions. Pair potentials can deal with the nearest neighbor interactions between ions with good accuracy. This can be verified using the comparisons of RDF from QLMD and AAMD, as shown in figure 6, where the positions of the first peaks of RDF are almost overlapped. Therefore, the nearest neighbor interactions should be reasonable in AAMD. However, AAMD cannot deal with many-body interactions, suggesting that the shear viscosities from AAMD are incorrect here, since dense Fe liquid holds a lot of medium ordered structures; therefore, the melting behavior is not correct within the AAMD framework.

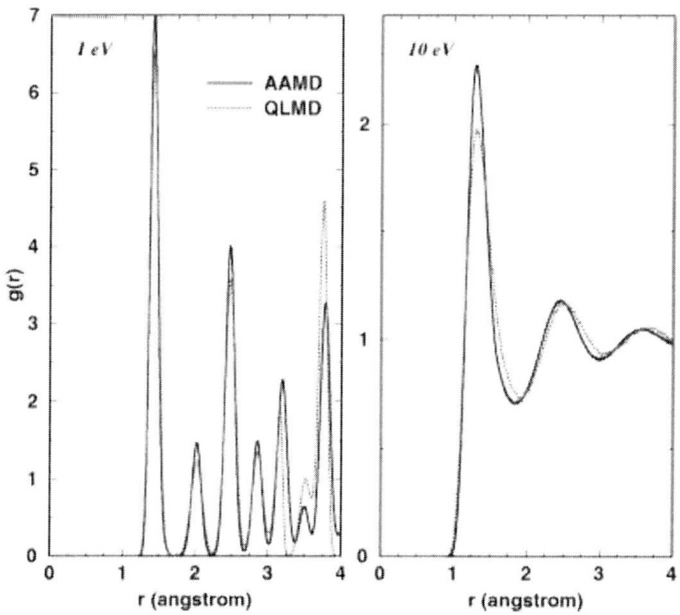

FIGURE 6. RDF comparison between AAMD and QLMD for Fe at 45 g cm^{-3}.

For a density of 33.385 g cm^{-3}, the behaviors are almost the same. The melting temperature locates below 3 eV, according to the RDF and diffusion coefficients. It is noted that the pressures from AAMD are much closer to those from QLMD compared with the results of 45 g cm^{-3}. This means that the electronic structures from the AA model are reasonable here but give more free electrons at 45 g cm^{-3}. Furthermore, the melting temperature from AAMD is also lower than that from QLMD; the diffusion coefficients are in the same order when Fe melts above 3 eV.

The small sizes in the calculations can introduce statistical errors. Especially for the viscosities, the errors are within 10%. Therefore, calculations with more than 10 000 atoms at least and accurate many-body potentials are needed basically. We need to construct more accurate semiclassical potentials for obtaining more accurate diffusion coefficients and viscosities.

6. STOKES–EINSTEIN RELATION

With respect to the simulation time and system sizes, we find that the diffusion coefficients can reach convergence much more easily than the viscosities. Therefore, if we can obtain viscosities from calculated diffusions, a lot of computational resources can be saved. Therefore, the validation of the SE relation is important and meaningful. In order to test the validation, we calculate viscosities from different definitions of effective atomic diameters: averaged ionic distance at a specific density $r_i = V^{(1/3)}$ (where V is the averaged volume of one atom at a specific density), the position of the first peaks in RDF (r_f) and the effective radius (r_E) from effective coordination number (ECN) definition. ECN can reveal the topology of the structures partly and can also give the local information around one atom. The definition of ECN is as follows:

$$\mathrm{ECN} = \frac{1}{N}\sum_{i=1}^{N}\mathrm{ECN}_i = \frac{1}{N}\sum_{i=1}^{N}\sum_{j\neq i}\exp\left[1-\left(\frac{d_{ij}}{d_{\mathrm{av}}^{i}}\right)^{6}\right], \quad (5)$$

where dij is the distance between atoms i and j; N is the total number of atoms in the system; di_{av} and its average value d_{av} are defined as

$$d_{\mathrm{av}}^{i} = \frac{\sum_j d_{ij}\exp\left[1-\left(\frac{d_{ij}}{d_{\mathrm{av}}^{i}}\right)^{6}\right]}{\sum_j \exp\left[1-\left(\frac{d_{ij}}{d_{\mathrm{av}}^{i}}\right)^{6}\right]}, \quad d_{\mathrm{av}} = \frac{1}{N}\sum_{i=1}^{N} d_{\mathrm{av}}^{i}. \quad (6)$$

With this definition, the effective diameters $r_E = d_{\mathrm{av}}$, which includes the temperature and density effects.

The results of different effective atomic diameters and the calculated viscosities from the SE relation are shown in tables 3 and 4. Comparing the results with the calculated viscosities from the Green–Kubo equation, we find that the viscosities are very different at low temperatures. With the increasing temperature, the SE relation seems to be more and more validated. Moreover, the choice of the effective atomic diameters plays an important role in the validation of the SE relation, where definitions of the positions of the first peaks of RDF and diameters from ECN calculations are more appropriate. Considering the calculation uncertainties and statistical errors, the SE relations can be considered validated only when the choice of the effective diameters are reasonable.

Table 3. Summary of the effective diameters and their corresponding viscosities for dense Fe at different temperatures at 45 g cm^{-3}. $\bar{Z}$ is the average ionization degree from AA calculations. μ_i, μ_f and μ_E are the viscosities corresponding to different definitions of effective atomic diameters of ions r_i, r_f and r_E, respectively. μ_{YVM} is the viscosity from the Yukawa viscosity model (YVM) (see details in [14]).

T (eV)	r_i (Å)	μ_i (mPa s)	r_f (Å)	μ_f (mPa s)	r_E (Å)	μ_E (mPa s)	$\bar{Z}$	μ_{YVM} (mPa s)
5	1.275	83.361	1.335	79.612	1.338	79.428	13.011	43.99
6	1.275	54.002	1.330	51.767	1.323	52.052	13.009	37.97
7	1.275	46.527	1.325	44.770	1.309	45.302	13.009	33.77
8	1.275	49.051	1.320	47.377	1.297	48.219	13.002	30.71
10	1.275	41.166	1.295	40.529	1.276	41.127	12.993	26.64

Table 4. Summary of the effective diameters and their corresponding viscosities for dense Fe at different temperatures at 33.385 g cm^{-3}.

T (eV)	r_i (Å)	μ_i (mPa s)	r_f (Å)	μ_f (mPa s)	r_E (Å)	μ_E (mPa s)	$\overline{Z}$	μ_{YVM} (mPa s)
3	1.408	81.002	1.515	75.301	1.488	76.667	9.344	28.65
3.447	1.408	53.831	1.505	50.375	1.473	51.459	9.343	25.81
5	1.408	30.300	1.475	28.931	1.437	29.687	9.332	20.23
6	1.408	29.633	1.465	28.488	1.419	29.410	9.326	18.41
7	1.408	21.147	1.455	20.469	1.404	21.214	9.323	17.28
8	1.408	23.847	1.450	23.162	1.388	24.197	9.312	16.56
10	1.408	24.122	1.425	23.840	1.364	24.912	9.296	15.91

In order to avoid using the SE relation, some models have been constructed. For example, Murrilo reported the construction of the YVM based on molecular dynamics simulations [14], which can be used in a warm dense regime. We verify the validation of the YVM model here. First of all, we give the average ionization degree ($\overline{Z}$) from the AA model with electronic energy-level broadening [22] and then show the viscosities (μ_{YVM}) from the YVM model, as shown in tables 3 and 4. It can be shown that the YVM model improves the accuracy of the SE relation, especially for cases at high temperatures. This suggests that the Yukawa potential might be reasonable for states at such high densities and at relatively high temperatures.

What is the possible reason for the debatable validation of the SE relation in this regime? The basic physics is whether the ions exhibit as Brownian particles. For the specific case in this work, heavy Fe ions move in the liquid-like electron sea and randomly collide with many free-like electrons. However, the other dominant factor for ionic motions is the interactions between ions, i.e. the strong coupling of ions. Every ion moves in a specific force field determined by both electrons and ions; this force field is not constant. The ions in dense matter have a memory effect, i.e. the ionic positions at this time play an important role in the ionic movements the next time. In other words, if the correlation time of ions is larger than the observable time, the motion of the ions should not have a Markov process. In order to analyze the correlation time of the systems, the velocity autocorrelation functions (VAF) of ions at 5 and 10 eV are shown in figure 7. Here, we adopt the simplest definition of the correlation time (τ_c) to compare the correlation time at different temperatures. Here, τ_c is defined as the time for which the VAF(t) decreases to $1/e$ [48]. With this definition, τ_c is about 4.1 and 2.7 fs for the temperatures of 5 and 10 eV, respectively. Therefore, with the increasing temperature, the behaviors of ions have a more and more Brownian-like motion. In fact, if the system is ideal gas, the correlation time will be zero, indicating uncorrelated behaviors between two near time steps. That is to say, when the temperature is high enough, the random collisions between particles are dominant and the statistical behaviors can work well. If the correlation time of the system is zero, we can recover the SE relation directly from equations (2) and (3) [49]. We can then conclude that the SE relation is strongly dependent on the coupling parameters of ions $\Gamma = Z^{*2}/(k_B T a)$, with T the system temperature, k_B the Boltzmann constant, a the mean ionic sphere radius defined as $a = (3/(4\pi n_i))^{1/3}$, Z^* the average ionization degree and n_i the ionic number density. If Γ is small enough, the SE relation can be valid. Furthermore, one possible reason that QMD or QLMD can obtain reasonable transport properties within a relatively short time (within 10 ps) is the short correlation time of particles, as shown in figure 7.

7. CONCLUSION

In conclusion, the dynamic structures and transport properties including EOS, diffusion and viscosities are calculated using the QLMD method within the framework of first principles. The dynamic structures provide information about RDF, melting behaviors, diffusions and viscosities. The validation of the SE relation is also discussed, showing the importance of the methods of choosing the effective atomic diameters. This work studies the dense liquid Fe existing in giant planets such as Jupiter and exoplanets, indicating the necessity of first principles calculations or constructing accurate many-body interactions. Furthermore, the results are crucial for understanding laser-shock induced compression experiments, reminding us of the complexity of dense matter. In this field, a great deal of new physics needs to be studied to a greater depth.

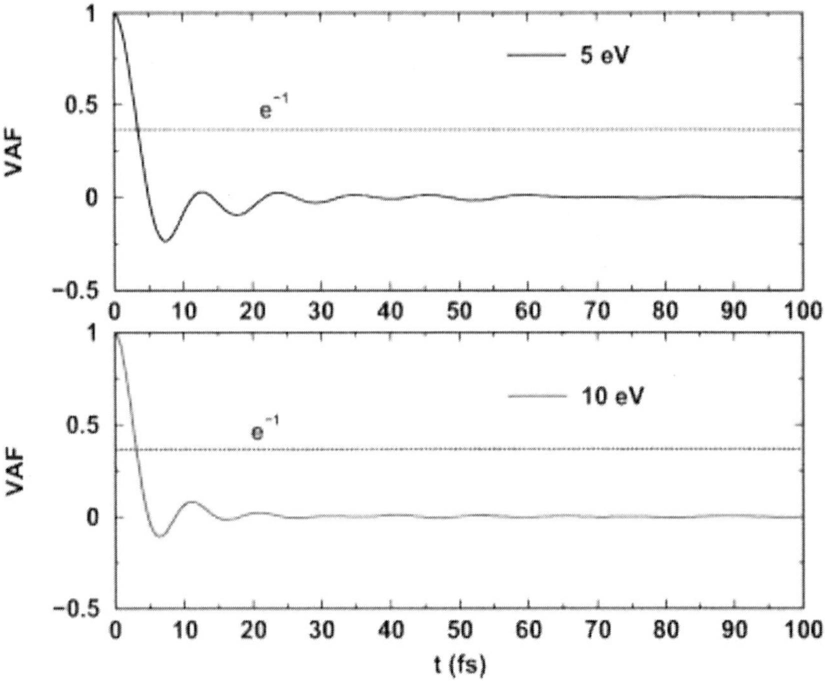

FIGURE 7. Velocity autocorrelation function (VAF) of Fe ions at the temperatures of 5 eV (upper panel) and 10 eV (lower panel) from QLMD calculations for Fe at 45 g cm^{-3}.

ACKNOWLEDGMENTS

This work was supported by the National NSFC under grant numbers 11104351, 11274383, 60921062, 11005153 and 11104350. Calculations are carried out at the Research Center of Supercomputing Application, NUDT.

REFERENCES

1. *Swift D C et al 2012 Mass-radius relationships for exoplanets Astrophys. J.* ***744*** *59*
2. *Jin F T et al Radiative opacity of iron studied using a detailed level accounting model Astrophys. J.* ***693****597–609*
3. *Gao C and Zeng J L 2008 Spectrally resolved and Rosseland and Planck mean opacities of iron plasmas at temperatures above 100 eV: a systematic study Phys. Rev. E* ***78*** *046407*
4. *Moses E I et al 2009 The National Ignition Facility: Ushering In a new age for high energy density science Phys. Plasmas* ***16*** *041006*
5. *Jeanloz R et al 2007 Achieving high-density states through shock-wave loading of precompressed samples Proc. Natl Acad. Sci. USA* ***104*** *9172–7*
6. *Pickard C J and Needs R J 2009 Stable phases of iron at terapascal pressures J. Phys.: Condens. Matter* ***21*** *452205*
7. *Stixrude L 2012 Structure of iron to 1 Gbar and 40 000 K Phys. Rev. Lett.* ***108*** *055505*
8. *Wijs Gilles A et al 1998 The viscosity of liquid iron at the physical conditions of the Earth's core Nature****392*** *805*
9. *Alfè D, Gillan M J and Price D G 1999 The melting curve of iron at the pressures of the Earth's core from ab initio calculations Nature* ***401*** *462*
10. *Alfè D and Gillan M J 1998 First-principles calculation of transport coefficients Phys. Rev. Lett.* ***81****5161*
11. *Alfè D and Gillan M J 1998 First-principles simulations of liquid Fe–S under Earth's core conditions Phys. Rev. B* ***58*** *8248*
12. *Vočadlo L, Alfè D, Price G D and Gillan M J 2000 Transport coefficients of liquids from first principles Phys. Earth Planet. Inter.* ***120*** *145*
13. *Saigo T and Hamaguchi S 2002 Shear viscosity of strongly coupled Yukawa systems Phys. Plasmas* ***9****1210–6*
14. *Murillo M S 2008 Viscosity estimates of liquid metals and warm dense matter using the Yukawa reference system High Energy Density Phys.* ***4*** *49–57*
15. *Daligault J 2012 Diffusion in ionic mixtures across coupling regimes Phys. Rev. Lett.* ***108*** *225004*
16. *Ma Y M et al 2009 Transparent dense sodium Nature* ***458*** *182*
17. *Pickard C J and Needs R J 2010 Aluminium at terapascal pressures. Nature Mater.* ***9*** *624–27*
18. *Dai J Y et al 2011 Dynamic ionic clusters with flowing electron bubbles from warm to hot dense iron along the Hugoniot curve Phys. Rev. Lett.* ***109*** *175701*
19. *Dai J Y, Hou Y and Yuan J M 2010 Unified first principles description from warm dense matter to ideal ionized gas plasma: electron–ion induced friction Phys. Rev. Lett.* ***104*** *245001*
20. *Dai J Y, Hou Y and Yuan J M 2010 Quantum Langevin molecular dynamics determination of the solar-interior equation of state Astrophys. J.* ***721*** *1158–63*
21. *Yuan J M 2002 Self-consistent average-atom scheme for electronic structure of hot and dense plasmas of mixture Phys. Rev. E* ***66*** *047401*
22. *Hou Y, Jin F T and Yuan J M 2006 Influence of the electronic energy level broadening on the ionization of atoms in hot and dense plasmas: an average atom model demonstration Phys. Plasmas* ***13****093301 and references therein*
23. *Mazevet S, Lambert F, Bottin F, Zérah G and Clérouin J 2007 Ab initio molecular dynamics simulations of dense boron plasmas up to the semiclassical Thomas–Fermi regime Phys. Rev. E* ***75****056404*
24. *Lambert F, Clérouin J and Zérah G 2006 Very-high-temperature molecular dynamics Phys. Rev. E* ***73****016403*
25. *Danel J-F, Kazandjian L and Zérah G 2012 Numerical convergence of the self-diffusion coefficient and viscosity obtained with Thomas–Fermi–Dirac molecular dynamics Phys. Rev. E* ***85*** *066701*

26. *Mermin N D 1965 Thermal properties of the inhomogeneous electron gas Phys. Rev.* **137** *A1441–3*
27. *Graziani F R et al 2012 Large-scale molecular dynamics simulations of dense plasmas: the Cimarron project High Energy Density Phys.* **8** *105–31*
28. *Dai J Y, Hou Y and Yuan J M 2011 Influence of ordered structures on electrical conductivity and XANES from warm to hot dense matter High Energy Density Phys.* **7** *84–7*
29. *Kress J D, Cohen James S, Horner D A, Lambert F and Collins L A 2010 Viscosity and mutual diffusion of deuterium–tritium mixtures in the warm-dense-matter regime Phys. Rev. E* **82** *036404*
30. *Horner D A, Lambert F, Kress J D and Collins L A 2009 Transport properties of lithium hydride from quantum molecular dynamics and orbital-free molecular dynamics Phys. Rev. B* **80** *024305*
31. *Lambert F and Recoules V 2012 Plastic ablator and hydrodynamic instabilities: a first-principles set of microscopic coefficients Phys. Rev. E* **86** *026405*
32. *Kress J D et al 2011 Quantum molecular dynamics simulations of transport properties in liquid and dense-plasma plutonium Phys. Rev. E* **83** *026404*
33. *Dai J Y and Yuan J M 2009 Large scale efficient Langevin dynamics and why it works Europhys. Lett.***88** *20001*
34. *Giannozzi P et al 2009 Quantum ESPRESSO: a modular and open-source software project for quantum simulations of materials J. Phys.: Condens. Matter* **21** *395502*
35. *Perdew J P, Burke K and Ernzerhof M 1996 Generalized gradient approximation made simple Phys. Rev. Lett.* **77** *3865–8*
36. *Zhang H L et al 2010 Static equation of state of bcc iron Phys. Rev. B* **82** *132409*
37. *Hou Y and Yuan J M 2009 Alternative ion–ion pair-potential model applied to molecular dynamics simulations of hot and dense plasmas: Al and Fe as examples Phys. Rev. E* **79** *016402*
38. *Allen M P and Tildesley T J 1987 Computer Simulation of Liquids (Oxford: Clarendon)*
39. *Collins L, Kwon I, Kress J, Troullier N and Lynch D 1995 Quantum molecular dynamics simulations of hot, dense hydrogen Phys. Rev. E* **52** *6202–19*
40. *Mithen J P, Daligault J and Gregori G 2012 Molecular dynamics simulations for the shear viscosity of the one-component plasma Contrib. Plasma Phys.* **52** *58–61*
41. *Lorenzen W, Holst B and Redmer R 2009 Demixing of hydrogen and helium at megabar pressures Phys. Rev. Lett.* **102** *115701*
42. *Plagemann K U et al 2012 Dynamic structure factor in warm dense beryllium New J. Phys.* **14** *055020*
43. *Cappelezzo M, Capellari C A, Pezzin S H and Coelho L A F 2007 Stokes–Einstein relation for pure simple fluids J. Chem. Phys.* **126** *224516*
44. *Schoen M and Hoheisel C 1985 The shear viscosity of a Lennard–Jones fluid calculated by equilibrium molecular dynamics Mol. Phys.* **56** *653–72*
45. *Morard G, Bouchet J, Valencia D, Mazevet S and Guyot F 2011 The melting curve of iron at extreme pressures: implications for planetary cores High Energy Density Phys.* **7** *141–4*
46. *Nguyen Jeffrey H and Holmes Neil C 2004 Melting of iron at the physical conditions of the Earth's core Nature* **427** *339–42*
47. *Alfè D 2009 Temperature of the inner-core boundary of the earth: melting of iron at high pressure from first-principles coexistence simulations Phys. Rev. B* **79** *060101*
48. *Kesselring T A, Franzese G, Buldyrev S V, Herrmann H J and Stanley H E 2012 Nanoscale dynamics of phase flipping in water near its hypothesized liquid–liquid critical point Sci. Rep.* **2** *474*
49. *Frenkel D and Smit B 1996 Understanding Molecular Simulation (San Diego, CA: Academic)*

CITATIONS

1. *Average-atom model for two-temperature states and ionic transport properties of aluminum in the warm dense matter regime Yong Hou et al 2017 High Energy Density Physics* ***22*** *21*
2. *Melting properties of Pt and its transport coefficients in liquid states under high pressures Pan-Pan Wang et al 2016 International Journal of Modern Physics B* ***30*** *1550250*
3. *Equations of state and transport properties of mixtures in the warm dense regime Yong Hou et al 2015 Physics of Plasmas* ***22*** *022711*
4. *Properties of Liquid Nickel along Melting Lines under High Pressure Cao Qi-Long et al 2015 Chinese Physics Letters* ***32*** *086201*
5. *Quantum molecular dynamics study on the structures and dc conductivity of warm dense silane Huayang Sun et al 2014 Physical Review E* ***89***
6. *Nuclear quantum dynamics in dense hydrogen Dongdong Kang et al 2014 Scientific Reports* ***4***
7. *Transport coefficients and entropy-scaling law in liquid iron up to Earth-core pressures Qi-Long Cao et al 2014 The Journal of Chemical Physics* ***140*** *114505*
8. *Focus on high energy density physics R Paul Drake and Peter Norreys 2014 New Journal of Physics* ***16*** *065007*

CHAPTER 3

Phenomenological Review on Quark–Gluon Plasma: Concepts vs. Observations

Roman Pasechnik [1,*] and Michal Šumbera [2]

[1]*Department of Astronomy and Theoretical Physics, Lund University, SE-223 62 Lund, Sweden*

[2]*Nuclear Physics Institute ASCR, 250 68 Řez/Prague, Czech Republic*

ABSTRACT

In this review, we present an up-to-date phenomenological summary of research developments in the physics of the Quark–Gluon Plasma (QGP). A short historical perspective and theoretical motivation for this rapidly developing field of contemporary particle physics is provided. In addition, we introduce and discuss the role of the quantum chromodynamics (QCD) ground state, non-perturbative and lattice QCD results on the QGP properties, as well as the transport models used to make a connection between theory and experiment. The experimental part presents the selected results on bulk observables, hard and penetrating probes obtained in the ultra-relativistic heavy-ion experiments carried out at the Brookhaven National Laboratory Relativistic Heavy Ion Collider (BNL RHIC) and CERN Super Proton Synchrotron (SPS) and Large Hadron Collider (LHC) accelerators. We also give a brief overview of new developments related to the ongoing searches of the QCD critical point and to the collectivity in small ($p + p$ and p + A) systems.

Keywords: extreme states of matter; heavy ion collisions; QCD critical point; quark–gluon plasma; saturation phenomena; QCD vacuum

1. INTRODUCTION

Quark–gluon plasma (QGP) is a new state of nuclear matter existing at extremely high temperatures and densities when composite states called hadrons (protons, neutrons, pions, etc.) lose their identity and dissolve into a soup of their constituents—quarks and gluons. The existence of this novel phase of matter was proposed in the mid-seventies [1,2], just ten years after the birth of the *Quark Model* of hadrons [3,4], and two years after it was realised that the candidate non-Abelian field theory of inter-quark forces—*quantum chromodynamics* (QCD) [5]—predicts their weakening at short distances, the so-called *asymptotic freedom* [6,7].

Contrary to atoms and molecules which can be ionized to reveal their constituents, quarks and gluons are never found free, but are confined inside the hadrons. This situation is quite similar to decomposing the magnet into two when trying to isolate its north pole from its south pole. Even deeper goes the analogy [8] between field lines confining quarks inside the hadrons and the magnetic field in the vicinity of a superconductor which expels the magnetic flux lines (*Meissner effect*). If two magnetic poles are surrounded by a superconducting medium, the field is confined into a thin tube. A hadronic string with quark and antiquark sitting at its end points has a similar one-dimensional field confined not by a superconducting medium, but by the vacuum1.

Experimental attack on producing QGP under laboratory conditions started in the world-leading particle physics facilities at CERN (Conseil Européen pour la Recherche Nucléaire, Geneva, Switzerland) and Brookhaven National Laboratory (BNL, New York, NY, USA) in the late 1980s [10,11,12]. In the year 2000, after finishing the main part of its heavy-ion program at the Super Proton Synchrotron (SPS) accelerator, CERN announced circumstantial evidence for the creation of a new state of matter in Pb + Pb collisions [13]. The real discovery of QGP took place in 2005, when four international collaborations studying Au + Au collisions at the Relativistic Heavy Ion Collider (RHIC) at BNL announced the results of their first five years of measurements [14,15,16,17]. Surprisingly, the properties of the new state of matter [18] differed markedly from predictions made over many years before its discovery2.

This review aims at giving a short historic introduction into the vast research field of QGP physics and the underlined phenomenological aspects with a comprehensive list of corresponding references. Effects of the hot/dense medium such as the nuclear suppression, initial-state interactions, in-medium energy loss, color screening and saturation are important for a proper understanding of the collective phenomena in heavy-ion collisions, and are included in the scope of this review.

Besides, we have qualitatively overviewed and confronted with existing observations such fundamental theoretical concepts as the QCD vacuum and phase diagram, equation of state (EoS) of deconfined QCD matter, initial-state effects, collectivity, flow, hydrodynamic properties of the QGP, and associated electromagnetic effects.

The paper is organized as follows. Section 2 presents a short history of the theoretical understanding of extreme states matter. The phase diagram of QCD is discussed in Section 3. Section 4 gives a historical perspective on experiments operating with with collisions of heavy ions. Section 5 describes the basic signatures of QGP production, while current developments in QGP research are provided in Section 6. The concluding remarks are given in Section 7. For further reading on fundamental concepts and the latest studies of QGP dynamics, we recommend several textbooks [9,20,21,22,23,24] and reviews [25,26,27,28,29,30] published in recent years.

2. MATTER UNDER EXTREME CONDITIONS

The properties of matter under extreme conditions at high values of state variables have always attracted the curiosity of scientists, owing to the possibility of advancing to new domains of the phase diagram and producing the exotic states of matter in laboratory [31,32]. The first attempts to discuss the properties of matter at densities well above the normal nuclear density $\rho 0 = 2 \times 1014$ g·cm−3(≈0.16 GeV·fm−3) date back to the seminal Oppenheimer–Volkoff paper from 1939 [33]. A study of the gravitational stability of a new phase of neutron matter suggested a few years earlier by Landau [34] led them to carry out their computations to several tens of $\rho 0$before smoothly extrapolating the results to the black-hole singularity. In 1962, when discussing relativistic limitations of the equation of state (EoS) of matter made of baryons interacting via a massive vector field, Zeldovich used the density twenty times exceeding $\rho 0$[35]. In 1976, the same value of density was shown by Baym and Chin [36] to be energetically favourable for the neutron matter–quark phase transition.

In cosmology, a very dense matter with $\rho \propto 106$ g·cm−3 (≈1 eV·fm−3) was first studied in 1946 by Gamow [37] when discussing the relative abundances of elements in the universe. The key discovery of the cosmic microwave background radiation by Penzias and Wilson in 1965 [38] not only provided a strong basis for the hot universe scenario which was used by Gamow, but also motivated Sakharov [39] to push it ad extremum. Considering the properties of hot matter at densities when gravitational interaction between photons becomes significant, he established that the absolute maximum of the temperature of any substance in equilibrium with radiation is on the order of Planck temperature TP=ℏc5Gk2−−−√∝ 1032 K (≈1022 MeV). Unfortunately, the theoretical apparatus of that period was completely inadequate to deal with the thermal history of the universe from TP downwards, but even at temperatures twenty orders of magnitude lower [40].

The problem was due to two successful but mutually conflicting contemporary models of hadrons: the Bootstrap model [41,42] based on the hypothesis that all hadrons are composite of one another, and—at that time not fully developed composite model of hadrons—the Quark model [3,4]. The Bootstrap model predicted that after reaching some limiting value of temperature—the so-called *Hagedorn temperature* (TH = 170–180 MeV) that can be estimated from the spectrum of hadronic masses [42,43]—the subsequent heating of strongly interacting matter will lead to the creation of more and more massive hadron species, but not to an increase of its temperature. The quark model, on the other hand, predicted relic cosmological quarks [44] roaming free through our universe. The leftover quarks were predicted to be as abundant as gold atoms [45].

The conflict was finally resolved in 1973 when Gross, Wilzek [6], and Politzer [7] discovered the asymptotic freedom in non-Abelian gauge theories of elementary particle interactions. Shortly after the idea of asymptotic freedom was introduced, two groups—Collins and Perry [1], and Cabibbo and Parisi [2]—independently realized its fascinating consequence for the properties of hot and dense matter. The first group argued that since the interaction between quarks weakens as quarks get closer at sufficiently high density, these quarks are no longer confined inside the hadrons and become free. The superdense matter at densities higher than the nuclear one consists of a quark soup. The other group re-interpreted the existence of the Hagedorn limiting temperature TH as a signal of a second-order phase transition between the hadronic and quark–gluon phases of matter.

The discovery of asymptotic freedom also paved the way to our current understanding of the evolution of the early universe. The commonly-accepted scenario of the subsequent cooling of the universe assumes a series of first- or second-order phase transitions associated with the various spontaneous symmetry-breakings of the basic non-Abelian gauge fields [46,47,48]. The Standard Model (SM) of elementary particles predicts two such transitions [48]. One taking place at temperatures of a few hundred GeV is responsible for the spontaneous breaking of the electroweak (EW) symmetry, providing masses to elementary particles. It is also related to the EW baryon-number-violating processes, which had a major influence on the observed baryon-asymmetry of the universe

[49]. The lattice simulations have shown that the EW transition in the SM is an analytic crossover [50].

The second—the transition of QGP to hadronic matter—happens at T<200 MeV, and is related to the spontaneous breaking of the chiral symmetry of the non-Abelian theory of strong interactions, which is based on the SU(3)c color group: the QCD. The nature of this phase transition affects to a great extent our understanding of the evolution of the early universe [48]. For instance, in a strong first-order phase transition, the QGP supercools before bubbles of hadron gas are formed. Since the hadronic phase is the initial condition for nucleosynthesis, the inhomogeneities in this phase could have a strong effect on the nucleosynthesis epoch [48]. Here, the lattice non-perturbative QCD calculations developed since the late 1970s [51] (for a recent review, see Ref. [52]) can be of great help. Knowing that the typical baryon chemical potentials μB are much smaller than the typical hadron masses (μB≈45 MeV at sNN−−−−√ = 200 GeV [16] and negligible in the early universe), we can use the lattice QCD calculations performed at μB=0. The results [53] not only confirm the previous finding [51] that confinement of quarks into hadrons is strictly a low-temperature phenomenon, but provide strong evidence that the QCD transition is also a crossover, and thus the above mentioned scenarios—and many others—are ruled out. The same conclusion was made in Ref. [54], where the first lattice analysis of the QCD phase transition in a model with chiral quarks having physical masses was performed. Numerical simulations on the lattice also indicate that at vanishing μB≈0 MeV, the two phase transitions which are possible in the QCD—de-confining and chiral symmetry restoring—occur at essentially the same point [55].

The situation at large μB and *T* is more complicated (see the left panel of Figure 1). Here, the wealth of novel QCD phases is predicted to exist [56], including the so-called quarkyonic phase [57]. At T≈0 MeV and μB≥1 GeV, a variety of color superconducting phases occur [56,58]. Somewhere on the phase boundary at μB≈400 MeV, a critical point separating the first- and second-order phase transitions is predicted [56]. The search for this point is now underway at RHIC [59], and some of the recent results will be discussed in Section 6.

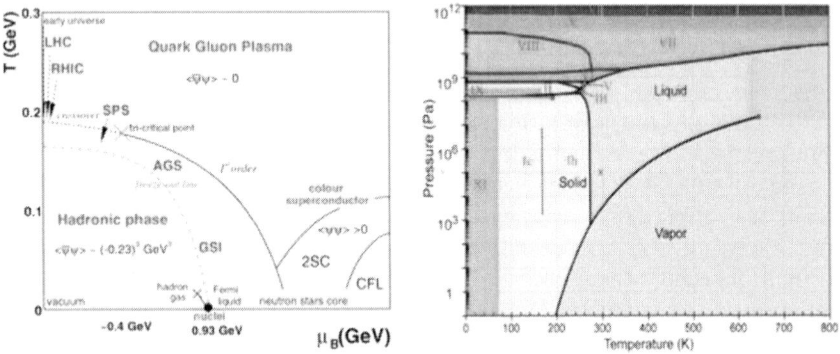

Figure 1. Left: The schematic phase diagram of quantum chromodynamics (QCD) in terms of *T* and μB state variables adapted from Ref. [71]; **Right**: The phase diagram of water illustrates a similar complexity, and is taken from Ref. [72]. AGS: Alternating Gradient Synchrotron; CFL: color–flavor locking; LHC: Large Hadron Collider; RHIC: Relativistic Heavy Ion Collider; SPS: Super Proton Synchrotron.

3. PHASES OF QCD MATTER

Although the quark matter was mentioned as early as 1970 by Itoh [60] in the context of neutron stars, the term "hadronic plasma" was first introduced in 1977 by Shuryak [61] to describe a new state of matter existing at temperatures above 1 GeV. This makes a good analogy with a classical gaseous plasma case, when electrically neutral gas at high enough temperatures turns into a statistical system of mobile charged particles [62]. While in the latter case their interactions obey the U(1)em gauge symmetry of Quantum Electrodynamics (QED), in the QCD case, the interactions between plasma constituents is driven by their SU(3)c color charges. For this reason, the SU(3)c plasma is now called the quark–gluon plasma (QGP). For an exhaustive collection of papers tracing the development of theoretical ideas on the topic of QGP up to 1990, see Ref. [19]. For a summary of later developments, see recent reviews [26,63].

Let us note that, contrary to the first oversimplified expectations [19], strongly interacting multi-particle systems feature numerous emergent phenomena that are difficult to predict from the underlying QCD theory, just like in condensed matter and atomic systems where the interactions are controlled by the QED theory. In addition to the hot QGP phase, several additional phases of QCD matter were predicted to exist [64,65]. In particular, the long-range attraction between quarks in the color antitriplet ($\bar{3}$) channel was predicted to lead to the color superconductivity (CSC) with the condensation of 1S0 Cooper pairs [66]. The analysis of CSC two-flavor deconfined quark matter at moderate densities [67] has revealed quite spectacular properties of this novel phase of matter, such as the spontaneous breakdown of rotation invariance manifested in the form of the quasi-fermion dispersion law. At high baryon density, an interesting symmetry breaking pattern $SU(3)_c \times SU(3)_R \times SU(3)_L \times U(1)_B \rightarrow SU(3)_{c+L+R} \times Z(2)$ leading to the formation of quark Cooper pairs was found in QCD with three massless quark flavours (i.e., mu = md = ms = 0) [65,68]. This breaking of color and flavor symmetries down to the diagonal subgroup SU(3)c+L+R implies a simultaneous rotation of color and flavor degrees of freedom, called the color–flavor locking (CFL). Let us note that the CSC and CFL phases of deconfined QCD matter might play an important role when studying the EoS of neutron stars [69]. Another interesting phase is the matter–pion condensate studied by Migdal [70].

Current knowledge on the QCD phase diagram is summarized on the left panel of Figure 1. The arrows indicate the expected crossing through the deconfinement transition during the expansion phase in heavy-ion collisions at different accelerators. The red and black full circles denote the critical endpoints of the chiral and nuclear liquid–gas phase transitions, respectively. The (dashed) freeze-out curve indicates where hadro-chemical equilibrium is attained at the final stage of the collision. The nuclear matter ground-state at T=0 and μB = 0.93 GeV, and the approximate position of the QCD critical point at μB~0.4 GeV are also indicated. The dashed line is the chiral pseudo-critical line associated with the crossover transition at low temperatures. Comparing this diagram to the phase diagram of water shown on the right panel, one notices that (at least theoretically) the complexity of the former approaches the latter.

3.1. The Role of the QCD Ground State

The quantum ground state of QCD plays an immensely important role in both particle physics and cosmology. In particular, the quark–gluon condensate is responsible for spontaneous chiral symmetry breaking, color confinement, and hadron mass generation (for a comprehensive review on the QCD vacuum, see Refs. [73,74,75,76,77] and references therein). It determines the properties (and possibly the generation mechanism) of the quark–gluon plasma, and dynamics of the phase transitions and hadronization. The latter phenomena are the most critical QCD phenomena taking place beyond the Perturbation Theory (PT), and thus are very difficult to explore by means of the well-known approaches. This strongly motivates further even deeper studies in this direction.

Let us start with the classical Yang–Mills (YM) gauge theory in the SU(Nc) (Nc=3 for QCD) determined by the gauge-invariant Lagrangian

$$\mathcal{L}_{\rm cl} = -\frac{1}{4} F^a_{\mu\nu} F^{\mu\nu}_a \,, \tag{1}$$

where

$$F^a_{\mu\nu} = \partial_\mu A^a_\nu - \partial_\nu A^a_\mu + g_s\, e^{abc} A^b_\mu A^c_\nu$$

is the gluon field stress tensor with SU(Nc) adjoint a,b,c=1,⋯N2c−1 and Lorentz μ,ν=0,1,2,3 indices, and with the strong coupling constant gs. The generating functional of such a classical theory is given by the Euclidean functional integral

$$Z \propto \int [DA]\, e^{-S_{\rm cl}[A] + \int J^a_\mu A^a_\mu d^4x} \,, \qquad S_{\rm cl}[A] = \int \mathcal{L}_{\rm cl} d^4x \,, \tag{2}$$

which is dominated by minima of the classical action corresponding to the classical vacuum state with unaltered by quantum corrections. The field excitations about the classical YM vacuum are referred to as instantons [78,79].

In fact, the classical YM equations of motion corresponding to Equation (1) are form-noninvariant with respect to small quantum fluctuations which break the conformal invariance of the gauge theory [80]—the effect known as *the conformal (or trace) anomaly*. Indeed, there is no threshold for the vacuum polarisation of a massless quantum gluon field by its classical component such that the solutions of the classical YM equations are unstable w.r.t the radiative corrections and cannot be used in physical applications. The conformal anomaly in QCD has notable implications; for example, in Cosmology, leading to an appearance of the Lorentz-invariant negative-valued contribution to the cosmological constant,

$$\epsilon^{\rm QCD} = \frac{\beta(g_s^2)}{8} \langle 0| : F^a_{\mu\nu} F^{\mu\nu}_a : |0\rangle + \frac{1}{4} \sum_{q=u,d,s} \langle 0| : m_q \bar{q} q : |0\rangle \simeq -(5 \pm 1) \times 10^{-3}\, {\rm GeV}^4 \,, \tag{3}$$

where the one-loop expression for the QCD β-function , accounting for three light flavours (for pure gluodynamics,) is typically used. Besides the wrong sign, the QCD vacuum density is over forty orders of magnitude larger in absolute value than the positive cosmological constant observed in astrophysical measurements,

$$\epsilon_{\rm CC} > 0\,, \qquad \left|\frac{\epsilon_{\rm CC}}{\epsilon^{\rm QCD}}\right| \simeq 10^{-44}\,, \tag{4}$$

The nonperturbative QCD vacuum effect is expected to be dynamically cancelled at macroscopically large distances in the course of cosmological expansion (see Ref. [81] and references therein). A dynamical mechanism of such a cancellation of vacua terms is yet unknown (for the existing scenarios discussed in the literature, see Refs. [81,82,83,84].

Consider a consistent effective Lagrangian formulation of the YM theory incorporating the conformal anomaly. In the corresponding variational technique, the strong coupling is treated as an operator depending on operators of quantum gluon fields by means of the RG equations in the operator form. Namely, the gauge field operator is considered as a variational variable which—together with the corresponding stress tensor operator—are related to those in the standard normalisation as follows

$$\mathcal{A}_\mu^a \equiv g_s A_\mu^a\,, \qquad \mathcal{F}_{\mu\nu}^a \equiv g_s F_{\mu\nu}^a = \partial_\mu \mathcal{A}_\nu^a - \partial_\nu \mathcal{A}_\mu^a + f^{abc}\mathcal{A}_\mu^b \mathcal{A}_\nu^c\,. \tag{5}$$

The effective action and Lagrangian operators of the quantum gauge theory is given in terms of the gauge-invariant operator of the least dimension J by [85]

$$S_{\rm eff}[\mathcal{A}] = \int \mathcal{L}_{\rm eff} d^4x\,, \qquad \mathcal{L}_{\rm eff} = -\frac{J}{4g_s^2(J)}\,, \qquad J = \mathcal{F}^2 \equiv \mathcal{F}_{\mu\nu}^a \mathcal{F}_a^{\mu\nu} = 2g_{s,*}^2(B^2 - E^2)\,, \tag{6}$$

respectively, whose variation w.r.t leads to the energy–momentum tensor of the gauge theory

$$T_\mu^{\nu,g} = \frac{1}{g_s^2}\left[1 - \frac{1}{2}\beta(g_s^2)\right]\left(-\mathcal{F}_{\mu\lambda}^a \mathcal{F}_a^{\nu\lambda} - \frac{1}{4}\delta_\mu^\nu J\right) - \frac{\delta_\mu^\nu \beta(g_s^2)}{8g_s^2} J\,, \qquad g_s^2 = g_s^2(J)\,. \tag{7}$$

In Equation (6), as a normalisation point, one can choose, for example, the strong coupling in the minimum of the effective action . One distinguishes chromomagnetic and chromoelectric condensates, such that one or both of the corresponding ground-state solutions (minima of the effective action) should be stable in order to contribute to the physical QCD vacuum.

The strong coupling dependence on J is determined by the RG evolution equation

$$2J\frac{dg_s^2}{dJ} = g_s^2\,\beta(g_s^2)\,, \qquad g_s^2 = g_s^2(J)\,, \tag{8}$$

The effective action (6) can be considered as an effective classical model [85] which possesses well-known properties of the full quantum theory such as (i) local gauge invariance; (ii) RG evolution and asymptotic freedom; (iii) correct quantum vacuum configurations; and (iv) trace anomaly.

In Ref. [84], it was noticed that the effective YM equation of motion in an expanding universe with the conformal metric ( )

$$\left(\frac{\delta^{ab}}{\sqrt{-g}}\partial_\nu\sqrt{-g} - f^{abc}\mathcal{A}_\nu^c\right)\left[\frac{\mathcal{F}_b^{\mu\nu}}{g_s^2\sqrt{-g}}\left(1-\frac{1}{2}\beta(g_s^2)\right)\right] = 0 \tag{9}$$

has a partial nonperturbative solution , where is the solution of the RG Equation (8) evaluated in the minimum of the effective action (6) . The corresponding value of the ground state density

$$T_0^{0,g} = -\frac{J^*}{4g_{s,*}^2} \equiv \mathcal{L}_{\text{eff}}\Big|_{J=J^*} \tag{10}$$

indicates that the QCD vacuum indeed has a chromomagnetic nature for in the deeply non-perturbative domain. This means that the corresponding solution is stable; namely, any small perturbation around the vacuum state effectively vanishes at the typical QCD time scale .

In asymptotically free gauge theories like QCD, the quantum vacuum configurations are controlled by the strong coupling regime. Performing an analysis in Euclidean spacetime, in Ref. [85] it was shown that the vacuum value of the gauge invariant in a strongly-coupled quantum gauge theory does not vanish as it does in the classical gauge theory, and the corresponding functional integral is not dominated by the minima of the classical action (2). Moreover, it was shown that there are no instanton solutions to the effective action (6), such that the ground state of the quantum YM theory does not contain the classical instanton configurations. Instead, the quantum vacuum can be understood as a state with ferromagnetic properties which undergoes spontaneous magnetisation, providing a consistent description of the nonperturbative

QCD vacuum and confinement alternative to the conventional instanton model.

How to understand the smallness of the observed cosmological constant within the effective QCD action approach? One way elaborated in Ref. [84] is to assume that such a compensation happens due to the presence of an additional QCD-like dynamics—Mirror QCD—with a confinement scale . The corresponding nonperturbative Mirror QCD vacuum contribution may have an opposite sign to that in QCD, and can therefore compensate the QCD one at a certain time scale in the course of cosmological expansion. Another interesting possibility explored in Ref. [82] is to assume that the QCD vacuum is degenerate itself, and at a given time scale consists of two opposite-sign (quantum-topological and quantum-wave) contributions. Then, as soon as such a compensation occurs, the observable small cosmological constant can be generated by means of weak gravitational interactions in the QCD vacuum. Both possibilities, however, require a fine tuning of vacuum parameters in order to provide an exact compensation of bare (zeroth-order in gravitational interactions) QCD contributions to the ground state density.

Can one avoid such a major fine tuning problem? The stable ground-state solution can actually be both chromomagnetic (when , , and) and chromoelectric (corresponding to , , and). Indeed, the standard argument in favor of the positive definiteness of is given in a classical YM theory, where

$$\mathcal{F}^2 \propto -\frac{\partial}{\partial t}A_i^a \frac{\partial}{\partial t}A_i^a \tag{11}$$

in Minkowski space, such that would lead to infinitely fast growth of the field and action would not have a minimum. In the quantum case, however, is a function of J, and can take negative values as long as the effective action has a minimum for . Besides, in close vicinity of the ground-state solution , the corresponding solution of the RG Equation (8) takes a linear behaviour

$$\frac{d\ln g_s^2}{d\ln J} \approx 1, \qquad g_s^2 = \pm |g_{s,*}^2| \frac{J}{J^*}, \qquad J,\ J^* > 0, \tag{12}$$

Adopting that the stable attractor solution is realised at macroscopically large time scales, the net QCD ground-state density would then asymptotically vanish [81]:

$$\epsilon^{\pm} \to \pm\frac{J^*}{4|g_{s,*}^2|}, \qquad \epsilon^-(T)+\epsilon^+(T) \to 0, \qquad T \ll T_{QCD} = \Lambda_{QCD} \sim 100\ \text{MeV}, \tag{13}$$

if both contributions coexist in the QCD vacuum, thus canceling each other beyond the confinement radius or after the QCD phase transition epoch in the cosmological history of the universe. The latter is an important example of conformal anomalies' cancelation in the classical limit of a YM theory without any fine tuning. In the deconfined (QGP) phase (i.e., at temperatures), the chromoelectric contribution should quickly vanish such that , providing a consistency with hadron physics phenomenology. This effect becomes plausible as long as is attributed to the ground state of hadronic degrees of freedom, which indeed becomes relevant only as soon as QGP is hadronised [82].

What is the possible role of the QCD vacuum in the generation of QGP? As was demonstrated in Ref. [86] within a semi-classical analysis, an interacting YM system of homogeneous gluon condensate and inhomogeneous wave modes evolves in real time in such a way that the amplitude of waves parametrically grows at the expense of decaying gluon condensate. The corresponding effect of the energy "swap" from the condensate to the gluon plasma waves is illustrated in Figure 2. Together with the growth of the plasma wave amplitudes, a vacuum average of their bi-linear products do not vanish and grow as well, inducing the positive-valued component of the ground-state density. This effect—if it holds in the full quantum formulation—can then be the basis for the QGP generation and reheating mechanism, both in heavy ion collisions and in Cosmology. A similar mechanism may be responsible for reheating of the cosmological plasma and particle production in the end of Cosmic Inflation due to decay of the dominant spatially-homogeneous chromoelectric condensate (inflaton) driving the inflationary epoch.

Can such parameteric growth of plasma modes due to a decay of the gluon condensate be studied in particle (e.g., heavy-ion) collisions? Can this effect be detectable, and if so, be used as a tool for the investigation of the dynamical evolution of the quantum ground state in QCD? Answers to these and other related questions are big unknowns, and very little has been done so far in this direction. An interesting insight into the problem of QCD ground state can be offered by the low- (<200 MeV) spectra of pions measured at the LHC, which show up to ~30%–50% enhancement compared to the hydrodynamic models (see Refs. [87,88,89]). A possible interpretation could be found in the framework of a hypothesis about hadronization and freeze-out processes in chemical non-equilibrium [90]. Among the possible reasons for the non-equilibrium dynamics are the QGP supercooling [91,92] and gluon condensation [93] phenomena. A particularly interesting possibility has been proposed in Refs. [94,95,96], where it was shown that the Bose–Einstein pion condensate at the level of 5% can account for the missing low- charged pion yields coming from a coherent source in Pb + Pb collisions (= 2.76 TeV) at various centralities. Moreover, if there is such a condensate, there must be large fluctuations of pions, which should be seen starting from the fourth moment of the multiplicity distribution [97]. Further studies of the non-equilibrium QCD dynamics accounting for the ground state are certainly required from both theoretical and experimental standpoints.

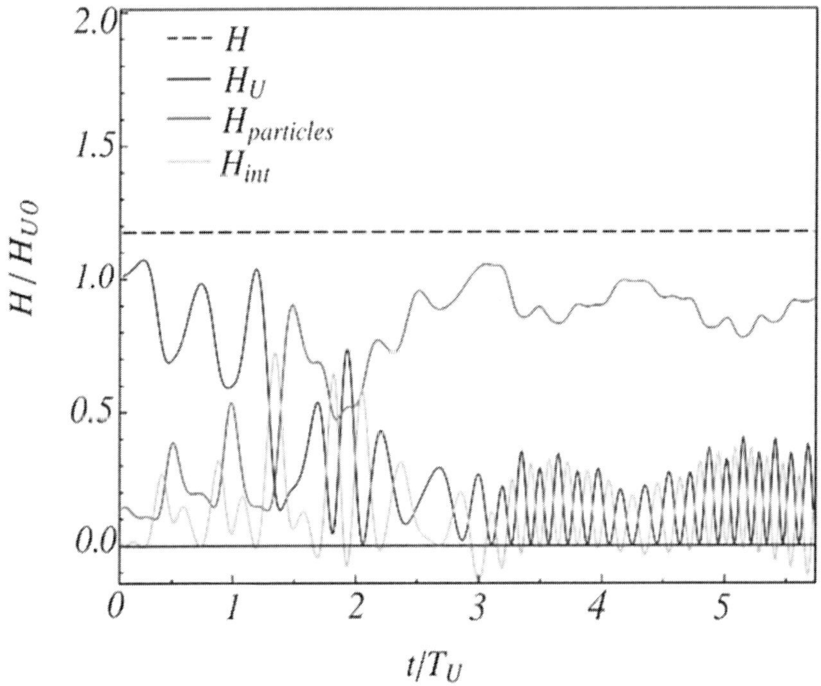

Figure 2. Real-time evolution of the energy density components of the interacting Yang–Mills (YM) system—homogeneous gluon condensate and inhomogeneous waves versus the interactions' contribution and the total Hamiltonian . Reproduced from Ref. [86].

3.2. Strongly Interacting Quark–Gluon Plasma

The surprising fact that the deconfined matter found at RHIC [14,15,16,17] does not behave as a gas of almost-free quarks and gluons, but as a strongly interacting liquid [18,26] was anticipated by only a few [98,99,100]. The fact that QGP close to the critical temperature is a strongly interacting system was used in Ref. [101,102] to exploit its analogy with strongly-coupled classical non-relativistic plasmas [62] in order to understand experimental observations and to interpret the lattice QCD results.

By definition, plasma is a state of matter in which charged particles interact via long-range (massless) gauge fields [26]. This distinguishes it from neutral gases, liquids, or solids in which the inter-particle interaction is of short range. So, plasmas themselves can be gases, liquids, or solids, depending on the value of the plasma parameter Γ, which is the ratio of interaction energy to kinetic energy of the particles forming the plasma [62].

A non-relativistic electromagnetic plasma is called strongly-coupled if the interaction energy (Coulomb energy) between the particles is larger than the thermal energy of the plasma particles; i.e., if the Coulomb coupling parameter = /( ) > 1, where q is the particle charge, a is the interparticle distance, and T is the plasma temperature (in the system of units where $\hbar = c = = 1$). Let us note that the strongly-coupled classical electromagnetic plasmas are not exotic objects at all [62]. For example, table salt (NaCl) can be considered as a crystalline plasma made of permanently charged ions Na and Cl [26]. At K (still too small to ionize non-valence electrons), it transforms into a molten salt, which is a liquid plasma with . An estimate of the plasma parameter for QGP was considered in Ref. [101] where it was found that = 2C /( ), where—depending on the type of plasma—$C = 4/3$ or $C = 3$ is the Casimir invariant for quarks or gluons and a is the interparton distance $a = 0.5$ fm. For QGP at temperature only slightly above the critical de-confining temperature (i.e., MeV), the corresponding coupling constant = 0.3–0.5, and = 1.5–6, the plasma can be considered as a strongly interacting one.

The strongly interacting plasmas are also a special case of strongly correlated systems where *correlated behavior* means a deviation from the trivial ideal gas behavior [103]. Prominent properties of all strongly correlated systems (see Figure 3) can be quantified by a few dimensionless parameters: the coupling parameter Γ, the degeneracy parameter , and the Brueckner parameter , where n is the number density of the particles, is the thermal de Broglie wavelength, a is the average interparticle distance, m is the particle mass, and is the Bohr radius.

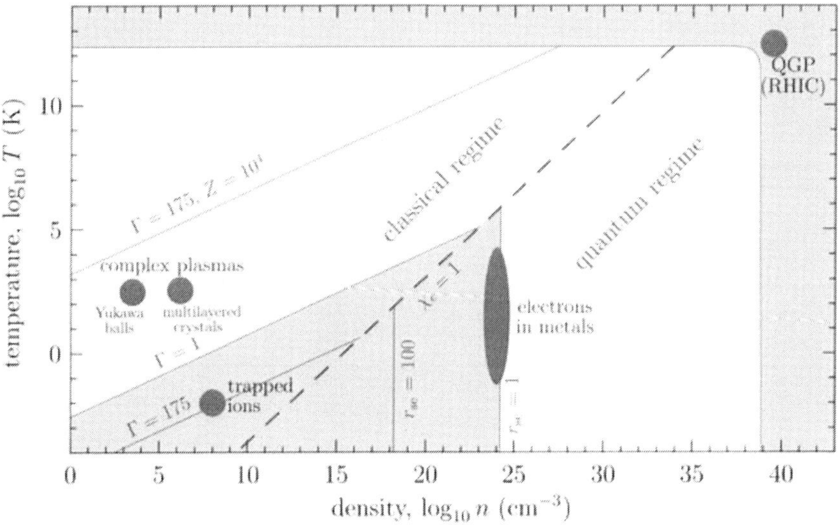

Figure 3. Examples of strongly correlated systems in thermodynamic equilibrium include complex plasmas, trapped ions, and the quark–gluon plasma (QGP) extending along the outer (**pink**) area, while dot shows the conditions at RHIC. The figure has been reproduced from Ref. [103].

Strongly interacting plasmas that can be studied in laboratory are ultracold atomic Fermi gases [104]; in particular, strongly-coupled Li atoms [105,106]. A distinctive property of these plasmas is that—similarly to the strongly- coupled QGP (sQGP)—their shear viscosity-to-entropy density ratio (see Section 4.3.1 for definition) characterizing how close the fluid is to a perfect liquid [107] is effectively negligible [26,105,108]. Cold atomic gases are produced in optical or magneto-optical traps containing typically - atoms [109]. The hydrodynamic behaviour is observed when the trapping potential is modified, or if the local density or energy density is modified using laser beams [105]. In this way, the scattering length *a* (and hence the interaction strength between the atoms) can be made almost infinite [26]. This is also the case of data point Li ( ) shown in Table 1, where the thermodynamical parameters for several other substances of interest are summarized. For H O and He, two points are displayed. First are the data at atmospheric pressure and temperatures just below the boiling point and the λ transition, respectively. These data points roughly correspond to the minimum of at atmospheric pressure. Second are the data near the critical point, which roughly corresponds to the global minimum of .

Table 1. The viscosity η, the viscosity over density ratio, and the viscosity over entropy density ratio for several fluids at particular values of pressure p and temperature T (from Ref. [108]).

Fluid	p [Pa]	T [K]	η [Pa·s]	η/n [$\hbar$]	η/s [$\hbar/k_B$]
H_2O	0.1×10^6	370	2.9×10^{-4}	85	8.2
^{4}He	0.1×10^6	2.0	1.2×10^{-6}	0.5	1.9
H_2O	22.6×10^6	650	6.0×10^{-5}	32	2.0
^{4}He	0.22×10^6	5.1	1.7×10^{-6}	1.7	0.7
^{6}Li ($a = \infty$)	12×10^{-9}	23×10^{-6}	$\leq 1.7 \times 10^{-15}$	≤ 1	≤ 0.5
QGP	88×10^{33}	2×10^{12}	$\leq 5 \times 10^{11}$		≤ 0.4

3.3. QCD at High Temperatures and Vanishing Chemical Potentials

The grand canonical partition function in gauge theory (such as QCD) with fermion flavours having a common chemical potential μreads

$$Z(T,\mu) = \int DU\, e^{-S_g(T)} \prod_{f=1}^{N_f} \text{Det}\, M(m_f,\mu,T), \tag{14}$$

where M is the Dirac operator, is the gauge part of the QCD action which depends on temperature T through boundary conditions. In the Hamiltonian formulation,

$$Z(T,\mu) = \text{Tr}\exp\left[-\frac{\hat{H}}{T} - \frac{\mu\hat{N}}{T}\right], \tag{15}$$

where is the number operator, and is the Hamiltonian. It is needless to mention that the analytic properties of the free energy

$$F(T,\mu) = -T\ln Z(T,\mu) \tag{16}$$

as a function of general complex μ are known to be useful for studying the phase structure of QCD on the lattice [110].

Let us consider QCD thermodynamics at high temperatures and zero chemical potentials—the region relevant for the LHC and also partly for RHIC—by recalling how the basic bulk thermodynamic observables can be obtained from the grand canonical partition function with vanishing quark chemical potentials, [55]. The grand canonical potential,, normalized in such a way that it vanishes at zero temperature,

$$\Omega(T,V) = T\ \ln Z(T,V) - \Omega_0\,, \qquad \Omega_0 = \lim_{T\to 0} T\ \ln Z(T,V)\,, \quad (17)$$

can be used to obtain the thermal part of the pressure (p) and energy density (ϵ)

$$p = \frac{1}{V}\Omega(T,V)\,, \qquad \epsilon = \frac{T^2}{V}\frac{\partial\Omega(T,V)/T}{\partial T}\,, \quad (18)$$

both vanishing at small temperature, by construction. Using these relations, one can express the difference between ϵ and ; i.e., the thermal contribution to the trace of the energy–momentum tensor (also called the trace anomaly or the interaction measure) in terms of a derivative of the pressure with respect to temperature:

$$\frac{\Theta^{\mu\mu}(T)}{T^4} \equiv \frac{\epsilon - 3p}{T^4} = T\frac{\partial}{\partial T}(p/T^4)\ . \quad (19)$$

In fact, it is which is the basic thermodynamic quantity conveniently calculated on the lattice as the total derivative of with respect to the lattice spacing a [111]:

$$\Theta^{\mu\mu} = \epsilon - 3p = -\frac{T}{V}\frac{d\ln Z}{d\ln a}\ . \quad (20)$$

Before moving to the results of lattice calculations, it is useful, for comparison, to recall a description of the strongly interacting matter below deconfinement temperature T. Here, all thermodynamic quantities are expected to be well-described by the hadron resonance gas (HRG) model consisting of non-interacting hadrons as proposed by Hagedorn in the mid 1960s [41] (see also Ref. [42]). The trace anomaly in the HRG model is given by

$$\left(\frac{\epsilon - 3p}{T^4}\right)^{HRG} = \sum_{m_i \le m_{max}} \frac{d_i}{2\pi^2} \sum_{k=1}^{\infty} \frac{(-\eta_i)^{k+1}}{k} \left(\frac{m_i}{T}\right)^3 K_1\left(\frac{km_i}{T}\right)\,, \quad (21)$$

where is a modified Bessel function, the different particle species of mass have degeneracy factors and for bosons (fermions), and the sum runs over all known hadrons up to the resonance mass of GeV.

The results on the temperature dependence of the trace anomaly and suitably normalized pressure, energy density, and entropy density from lattice calculations, together with the HRG predictions, are shown on the left and right panels of Figure 4, respectively. The vertical band in the right panel marks the crossover region, MeV. The horizontal line at ≈15.6 corresponds to the ideal Stefan–Boltzmann (SB) gas limit for the energy density of relativistic massless gas consisting of quark flavours and gluons with colors having altogether g degrees of freedom:

$$\frac{3p_{SB}}{T^4} = \frac{\epsilon_{SB}}{T^4} = g\frac{\pi^2}{30} , \qquad g = 2(N_c^2 - 1) + \frac{7}{2}N_c N_f = \frac{95}{2} . \tag{22}$$

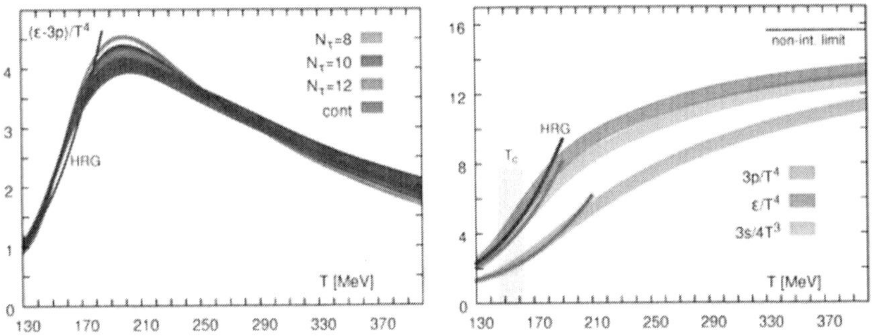

Figure 4. Left: The continuum-extrapolated trace anomaly for several values of the lattice spacing and its continuum extrapolation; **Right**: The continuum-extrapolated values of suitably normalized pressure, energy density, and entropy density as functions of the temperature. Both figures have been reproduced from Ref. [111]. The darker lines in both figures show the corresponding predictions of the hadron resonance gas (HRG) model.

The fact that even at 400 MeV, the pressure, energy density, and entropy density of the QGP are far from their ideal gas values indicates substantial remaining interactions among the quarks and gluons in the deconfined phase. It is interesting to compare (at least qualitatively) this behaviour with the gaseous two-component plasma of particles with charge . The pressure normalized to that of the ideal gas can be deduced from the standard textbook formula (e.g., Ref. [112])

$$\frac{p}{p_{id}} = 1 - \frac{\sqrt{\pi}}{3}\sqrt{\frac{nq^6}{T^3}} , \tag{23}$$

which is valid for . The non-ideal behaviour of gaseous two-component plasma thus increases very rapidly with charge q of the plasma particles, but much more slowly with their density n, and decreases rather quickly with the plasma temperature T.

3.3.1. Softest Point in the EoS of Deconfined QCD Matter

An important property of the QGP phase transition is the presence of a local minimum in the ratio of pressure to energy density as a function ϵ [113,114]. The possible existence of this *softest point* in the QCD EoS is distinguishable by a very small sound velocity of the deconfined medium, and has thus been suggested as a signal of the first-order phase transition. This also becomes evident in second-order derivatives of the QCD partition function with respect to temperature. The speed of sound, , is related to the inverse of the specific heat, ,

$$c_s^2 = \frac{\partial p}{\partial \epsilon} = \frac{\mathrm{d}p/\mathrm{d}T}{\mathrm{d}\epsilon/\mathrm{d}T} = \frac{s}{C_V}, \qquad \frac{C_V}{T^3} = \left.\frac{\partial \epsilon}{\partial T}\right|_V \equiv \left(4\frac{\epsilon}{T^4} + T\left.\frac{\partial(\epsilon/T^4)}{\partial T}\right|_V\right). \tag{24}$$

The quantity can be calculated directly from the trace anomaly and its derivative with respect to temperature,

$$T\frac{\mathrm{d}\epsilon/T^4}{\mathrm{d}T} = 3\frac{\Theta^{\mu\mu}}{T^4} + T\frac{\mathrm{d}\Theta^{\mu\mu}/T^4}{\mathrm{d}T}. \tag{25}$$

In Figure 5 (left panel), we show the speed of sound as a function of temperature. The softest point of the EoS predicted in Ref. [114] at (145–150) MeV (i.e., at the minimum of the speed of sound) lies on the low temperature side of the crossover region. At this point, the speed of sound is only slightly below the corresponding HRG value. Furthermore, the value is roughly half-way between zero, the value expected at a second-order phase transition with a divergent specific heat, and the value for an ideal massless gas, [111]. At the high temperature end, $T \sim$ 350 MeV, it reaches within of the ideal gas value.

The softest point of the EoS is of interest for the phenomenology of heavy ion collisions, as it characterizes the temperature and energy density range in which the expansion and cooling of matter slows down. The system spends a longer time in this temperature range, and one expects to observe characteristic signatures from this regime. The quantity as a function of the energy density is shown in Figure 5 (left). At the softest point, the energy density is only slightly above that of normal nuclear matter, . In the crossover region, MeV, the energy density varies from at the lower edge to at the upper edge—slightly above the energy density inside the proton .

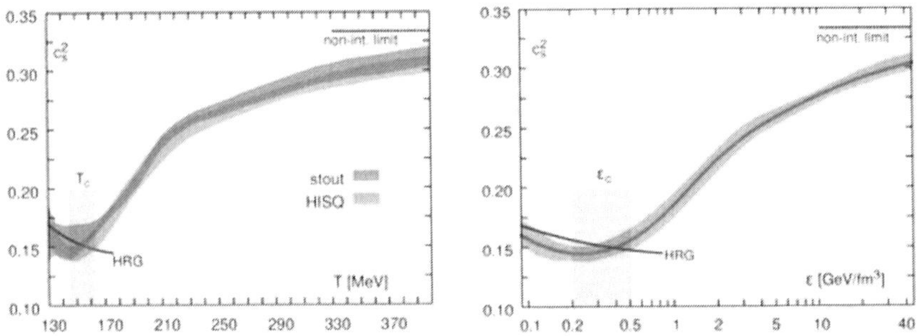

Figure 5. The speed of sound squared from lattice QCD and the HRG model versus (**Left**) temperature T and (**Right**) energy density ϵ. The figure has been reproduced from Ref. [111]. The vertical yellow bands mark the location of the crossover region MeV and the corresponding range in energy density, , respectively.

Thus, the QCD crossover region starts at or close to the softest point of the EoS, and the entire crossover region corresponds to relatively small values of the energy density, (1.2–3.1) . This value is about a factor of four smaller than that of an ideal quark–gluon gas in this temperature range.

3.3.2. Testing the Properties of the Medium with Infinitely Heavy Static Test Charges

An important property of the QGP medium is the color screening: the range of interaction between heavy quarks becomes inversely proportional to the temperature. This effect also forms the basis of the most common description of dynamics of quarkonia (mesons consisting of heavy), produced in heavy ion collisions—the potential between the heavy quarks or becomes screened by deconfined quarks and gluons, and the heavy quarks separate from each other, leading to a suppression of quarkonia yields [115].

On the lattice, this phenomenon is studied using (infinitely) heavy static test charges [116,117]. The color screening effect is estimated from the spatial correlation function of a static quark and anti-quark, which propagate in Euclidean time from to , where T is the temperature. The free energy of static quark pair is then calculated as the logarithm of the correlator [116,118]. In the zero temperature limit, the singlet free energy coincides with the zero temperature potential calculated on the lattice [119]. However, as argued in Ref. [117], using the free energies instead of potentials is preferable, since the latter are not gauge invariant. On the other hand, the gauge-invariant static quark–antiquark pair free energy is a non-perturbatively well-defined quantity that carries information about the deconfinement properties of the QGP.

The heavy-quark free energies for different temperatures *T* of the medium from two different lattice calculations are presented in Figure 6 (left). The solid black line on the upper plot is a parameterisation of the zero temperature potential. One can see that with increasing temperature the free energy—and hence, the spatial correlations between *Q* and —get more and more diluted.

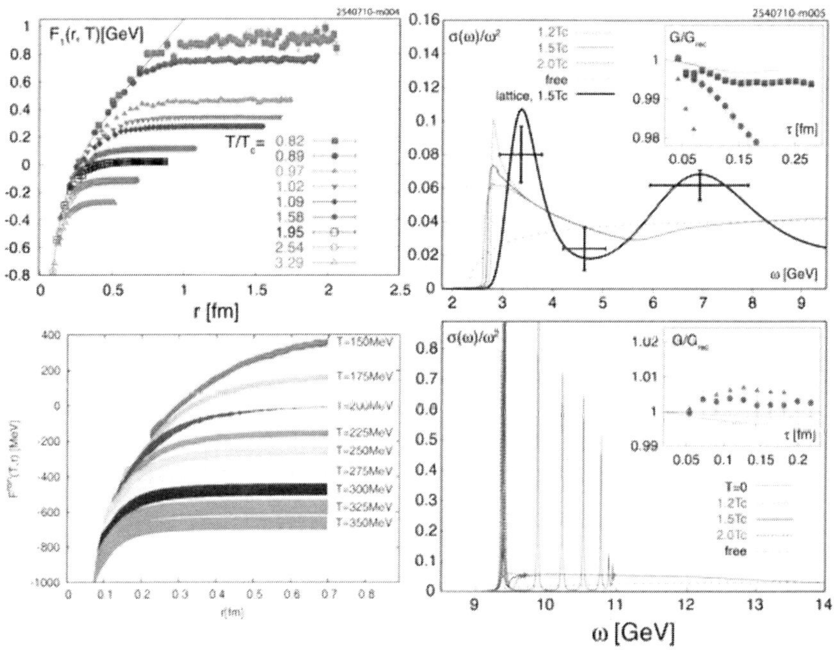

Figure 6. Left: Heavy-quark-singlet free energy versus quark separation calculated in 2 + 1 flavor QCD for (**top**) different values of [116] and (**bottom**) continuum-extrapolated values of the static free energy for different temperatures [117]. The solid black line on the right top plot is a parameterisation of the zero temperature potential; **Right**: The S-wave (**upper**) charmonium and (**lower**) bottomonium spectral functions calculated in potential models. Insets: correlators compared to lattice data. The dotted curves are the free spectral functions. Reproduced from Ref. [116].

The correlation functions in time variable are related to the spectral functions by

$$G(\tau,T) = \int_0^\infty d\omega\ \sigma(\omega,T)\frac{\cosh(\omega(\tau-1/(2T)))}{\sinh(\omega/2T)}\ . \quad (26)$$

While a stable quarkonium state in the vacuum contributes a δ-function-like peak at the value of its mass to the spectral function, in the medium, it gives a quasi-particle-like smeared peak with the width being the thermal width. As the temperature increases, the width increases, and at sufficiently high temperatures, the contribution from the meson state in the spectral function becomes sufficiently broad so that it is no longer meaningful to speak of it as a well-defined state (Figure 6, right). The effect is more prominent for the lighter mesons like charmonia, consisting of pairs, and much weaker for the bottomonia— mesons.

3.4. QCD at High Temperatures and Non-Zero Chemical Potentials

As direct lattice QCD calculations at non-zero are not yet possible, one has to analyze the EoS using Taylor expansion in quark chemical potentials , , and : [52,120]

$$\frac{p}{T^4} = \frac{1}{VT^3}\ln Z(T,\mu_u,\mu_d,\mu_s) = \sum_{ijk}\frac{1}{i!j!k!}\chi_{ijk}^{uds}\left(\frac{\mu_u}{T}\right)^i\left(\frac{\mu_d}{T}\right)^i\left(\frac{\mu_s}{T}\right)^j \quad (27)$$

$$\chi_{ijk}^{uds} = \frac{\partial^{i+j+k}p/T^4}{\partial(\mu_u/T)^i\partial(\mu_d/T)^j\partial(\mu_s/T)^k}\ , \quad (28)$$

where , , and are related to the chemical potentials corresponding to the baryon number B, electric charge Q, and strangeness S of hadrons, as follows:

$$\mu_u = \tfrac{1}{3}\mu_B + \tfrac{2}{3}\mu_Q\,,\quad \mu_d = \tfrac{1}{3}\mu_B - \tfrac{1}{3}\mu_Q\,,\quad \mu_s = \tfrac{1}{3}\mu_B - \tfrac{1}{3}\mu_Q - \mu_S\,. \quad (29)$$

The EoS at non-zero can thus be obtained from the coefficients of the Taylor expansion in hadronic chemical potentials expressed via[52]. Here, we report the result [120] for = = 0, which sufficiently illustrates the relative importance of higher-order corrections in different temperature and regions. The Taylor series for the pressure is given by

$$\frac{p(T,\mu_B)-p(T,0)}{T^4}=\frac{1}{2}\chi_2^B(T)\left(\frac{\mu_B}{T}\right)^2\left[1+\frac{1}{12}\frac{\chi_4^B(T)}{\chi_2^B(T)}\left(\frac{\mu_B}{T}\right)^2\right]+\mathcal{O}(\mu_B^6)\,. \tag{30}$$

The leading-order correction to the pressure at non-vanishing is proportional to the quadratic fluctuations of the net baryon number. The next-to-leading order corrections are proportional to the quartic fluctuations. In Figure 7, we show (left) and (right). With increasing temperature, the correction rapidly loses importance relative to the leading term. Moreover, the results for the -dependent contribution to the total pressure evaluated for different values of [120] suggest that the EoS given by Equation (30) works well for all values of the chemical potential below = 2, corresponding to the region of nuclear collisions at energies GeV.

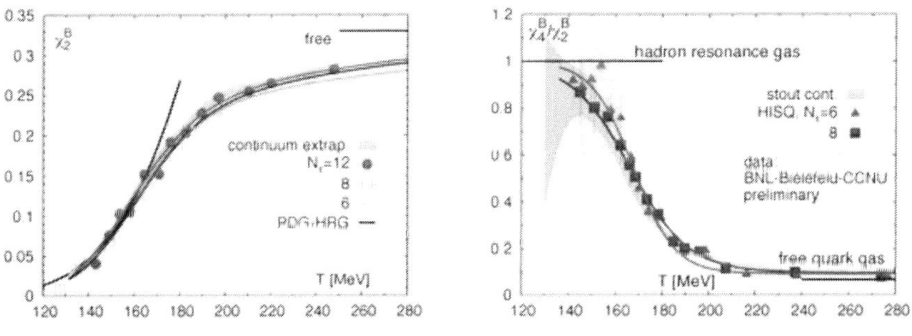

Figure 7. The expansion coefficients of the pressure at the non-zero baryon chemical potential adopted from Ref. [120]. **Left**: The leading-order correction; **Right**: The relative contribution of the next-to-leading order lcorrection. BNL: Brookhaven National Laboratory.

Let us note that are interesting in their own right, as they are related to the fluctuations and correlations of conserved charges. The latter are sensitive to the underlying degrees of freedom, which could be hadronic or partonic, and so they are used as sensitive probes of deconfinement. While the off-diagonal expansion coefficients are related to correlations among conserved charges (e.g.,), the diagonal ones describe their second- and higher order fluctuations

$$\chi_2^X=\frac{1}{VT^3}\langle N_X^2\rangle\,,\qquad \chi_4^X=\frac{1}{VT^3}\left(\langle N_X^4\rangle-3\langle N_X^2\rangle^2\right)\,,\quad \text{etc.} \tag{31}$$

4. STUDY OF HOT AND DENSE NUCLEAR MATTER USING NUCLEAR COLLISIONS

4.1. Heavy Ion Accelerators

The basic hopes and goals associated with investigations of very hot and dense nuclear matter in laboratory were first formulated in mid-seventies [121,122,123]. It was the experience with astrophysical objects like supernovae and neutron stars, and with thermonuclear ignition which led the authors to the idea that the nuclear matter shock compression [31] of about five-fold normal nuclear density should be accomplished in violent head-on collisions of heavy nuclei [11]. The goal was to find out the response of the nuclear medium under compression by pressure resisting that compression; i.e., to study the nuclear matter EoS. The original question was: is such a bulk nuclear matter EoS accessible within the dynamics of relativistic heavy ion collisions? [124,125]. The prospect of observing a phase transition in highly compressed nuclear matter [126] was lurking behind.

The interest in collisions of high-energy nuclei as a possible route to a new state of nuclear matter was substantially strengthened with the arrival of QCD as the microscopic theory of strong interactions. The particle physics community began to adapt existing high-energy proton accelerators to provide heavy-ion nuclear beams in the mid-1970s. The Berkeley Bevalac and JINR Synchrophasotron started to accelerate nuclei to kinetic energies from few hundreds of MeV to several GeV per nucleon [11,125]. By the mid-1980s, the first ultra-relativistic nuclear beams became available. Silicon and gold ions were accelerated to 10 GeV/nucleon at Brookhaven's Alternating Gradient Synchrotron (AGS) [10]. The first nuclear collisions took place at CERN in the early 1980s when alpha particles were accelerated to *center-of-mass energy per nucleon–nucleon pair* GeV at the ISR collider. The new era of research began at CERN in fall 1986, when oxygen—and later on in the summer of 1990, sulphur ions—were injected into the SPS and accelerated up to an energy of 200 GeV/nucleon (= 19.6 GeV) [10,11,12]. However, the genuine heavy ion program only started in 1994, after the CERN accelerator complex was upgraded with a new lead ion source which was linked to pre-existing interconnected accelerators, the Proton Synchrotron (PS) and the SPS. Seven large experiments involved (NA44, NA45/CERES, NA49, NA50, NA52, WA97/NA57, and WA98) studied different aspects of Pb + Pb and Pb+Au collisions at = 17.3 GeV and = 8.6 GeV [11,12].

In the meantime, at the Brookhaven National Laboratory (BNL), the Relativistic Heavy Ion Collider (RHIC) [127] rose up from the ashes of the ISABELLE/CBA collider project abandoned in 1983 by particle physicists. In 1984, the first proposal for a dedicated nucleus–nucleus machine accelerating gold nuclei up to = 200 GeV was submitted. Funding to proceed with the construction was received in 1991, and on 12 June 2000, the first Au + Au collisions at = 130 GeV were recorded by the BRAHMS, PHENIX, PHOBOS and STAR experiments [14,15,16,17].

The idea of the Large Hadron Collider (LHC) [128] dates even further back, to the early 1980s. Although CERN's Large Electron Positron Collider (LEP, which ran from 1989 to 2000) was not yet built, scientists considered re-using the 27-kilometer LEP ring for an even more powerful machine running at the highest possible collision energies (= 14 TeV) and intensities. The ion option (= 5.4 TeV per nucleon–nucleon pair for Pb + Pb collisions) was considered since the beginning. The LHC was approved in December 1994, and its official inauguration took place on 21 October 2008. The first proton–proton collisions occurred on 23 November 2009, and the first Pb + Pb collisions on 7 November 2010. The ALICE, ATLAS, and CMS experiments are currently involved in the heavy-ion program at the LHC [25,29].

Many years ago, American accelerator physicist M. Stanley Livingston noted that advances in accelerator technology increase the energy records achieved by new machines by a factor of ten every six years. This trend is illustrated on the left panel of Figure 8, summarising the worldwide advances in high-energy accelerators in the period of 1960–2008. One can see that the increase in the energy is even faster for the ion accelerators than for the proton accelerators. However, even in the most central nucleus–nucleus collisions, not all of the energy could be converted into a thermalised form of energy needed for the phase transition from hadronic into a QGP state of matter to occur. The experimentally accessible quantity measuring this transformation is the density of transverse energyper unit of pseudorapidity η. The latter can be used to estimate the initial energy density using the Bjorken formula [129]:

$$\epsilon_{BJ} = \frac{1}{S_{\perp}\tau}\frac{dE_T}{d\eta}\Big|_{\eta=0} , \tag{32}$$

where is the transverse overlap area of the nuclei and τ is the time scale for the evolution of the initial non-equilibrium state of matter into a (locally) thermalized system. The dependence of on for most central Au + Au and Pb + Pb collisions is presented on the right panel of Figure 8.

Even for a rather pessimistic value of the eqilibration time τ = 1 fm/c [17], the achieved energy density increases from 1.4 to 14 GeV/fm . It is thus not only much higher than than the normal nuclear density, but 3–30 times bigger than the energy density inside the nucleon = , and is definitely higher than the 1 GeV/fm scale required for the QCD deconfining transition from the lattice calculations that were discussed in Section 3.3.

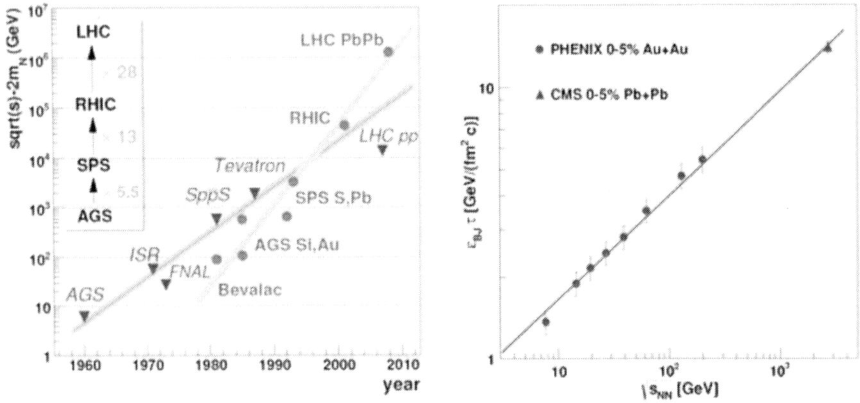

Figure 8. Left: Available center-of-mass energy -2m versus time for (anti)proton (blue triangles) and ion (magenta circles) accelerators, adapted from Ref. [71]; **Right**: The Bjorken estimate of the initial energy density (Equation (32)) multiplied by τ calculated from the data on transverse energy distributions in 5% most central Au + Au [130] and Pb + Pb [131] collisions as a function of c.m.s. energy per one nucleon–nucleon pair . The red line corresponds to a power law fit. From Ref. [132].

4.2. Heavy Ion Collisions as a Source of Strong Electromagnetic Fields

The important quantity determining the lifetime of heavy ion beams inside the accelerator tube (and hence their potential to produce an adequate number of nuclear collisions) is the loss of ions in the bunch. Its decay rate is given by the formula:

$$\lambda_T = -\frac{1}{N}\frac{dN}{dt} = \frac{N_{IR} L \sigma_T}{kN} , \qquad (33)$$

in which k is the number of bunches, N is the number of particles per bunch, is the number of interaction regions, is the total cross-section, and L is the available luminosity:

$$L = f\gamma \frac{N^2 k}{4\pi\epsilon\beta^*} F \quad , \tag{34}$$

which—in addition to k and N—also depends on the revolution frequency f, the Lorentz factor of the beam γ, the emittance ϵ, the beta function at the collision point , and the geometric luminosity reduction factor F due to the crossing angle at the interaction point [128]. Since the event rate for a certain process is given by its cross-section times the luminosity , the possibility of studying rare phenomena depends on the maximum luminosity accessible. At the same time, the rate of background processes (which in general have large cross-sections) will increase L, at some point reaching the maximum event rate that the experiment can handle. Secondary beams created by these background processes can limit the collider heavy ion luminosity, since they have a different charge-to-mass ratio than the primary beam, and can be lost in cryogenically cooled magnets.

Quite surprisingly, at RHIC and the LHC, these background processes are not part of the strong nuclear interaction cross-section determined primarily by the nuclear geometry , but are solely accounted for by the coherent action of all electric charges in colliding nuclei. The cross-section of electromagnetic processes—primarily due to the creation of pairs with subsequent atomic shell capture and electro-magnetic dissociation—becomes important for ions with . It is as large as hundreds of barns; i.e., about 30 (60) times larger than for Au + Au collisions at RHIC (Pb + Pb collisions at the LHC) [133].

A classical description of the electromagnetic action of a fast-moving charged particle on another one based on the equivalence between the perturbative action of its field and the flux of electromagnetic radiation dates back to Fermi [134], Weizsäcker [135], and Williams [136]. This equivalence is true as far as the effects caused by different spectral components add up incoherently (i.e., a perturbation caused by the fields is small enough). The solution for the time-dependent electromagnetic fields mutually seen by the two incident ions can be found, for example, in the textbook on "Classical Electrodynamics" [137]. The longitudinal ($\parallel$) and transversal ($\perp$) field components induced by a heavy ion I passing a target at distance b and with velocity β are given by the following formulas:

$$E_{\parallel}(t) = \frac{-Z_I e\gamma\beta t}{(b^2+\gamma^2\beta^2 t^2)^{3/2}}, \quad \vec{E}_{\perp}(t) = \frac{Z_I e\gamma\vec{b}}{(b^2+\gamma^2\beta^2 t^2)^{3/2}}, \quad B_{\parallel}(t) = 0, \quad \vec{B}_{\perp}(t) = \vec{\beta}\times\vec{E}_{\perp}(t). \tag{35}$$

Let us note that for , these fields act on a very short time scale of order . During this time, fields and are *equivalent* to a linearly polarized pulse of radiation incident on a target in the beam direction. Thus, according to the equivalent photon method, the strong and rapidly time-varying field of the point charge is seen by a passing charge as a flux of virtual (nearly real) photons with intensity

$$I(\omega, b) = \frac{1}{4\pi}|\vec{E}(\omega)\times\vec{B}(\omega)| \approx \frac{1}{2\pi}|E_{\perp}(\omega)|^2 \sim Z_I^2 \ , \tag{36}$$

where , , and are the Fourier components of the fields , and . The energy spectrum of these photons falls as , up to a maximum energy . The interaction between the colliding nuclei becomes dominantly electromagnetic for impact parameters b exceeding the size of the radii of colliding nuclei .

Interactions between ultra-relativistic nuclei taking place at are called the ultra-peripheral collisions. By taking advantage of the photon fields carried by relativistic nuclei, they are used to study photoproduction and two-photon physics at hadron colliders. This field of ultra-relativistic heavy ion collisions is sometimes called *"non-QGP" physics*, and is thus outside the scope of this article. We refer the interested reader to reviews [138,139,140,141] where more detailed information on these aspects can be found.

Quark–Gluon Plasma in a Strong Magnetic Field

More important from the point of view of QGP physics are the strong magnetic fields accompanying ultra-relativistic heavy ion collisions [142]. Consider the collision of two identical nuclei of radius R with electric charge , and use the Biot–Savart law to estimate the magnitude of the perpendicular magnetic field they create in the center-of-mass frame

$$B_{\perp} \sim \gamma\, Ze\, \frac{b}{R^3}. \tag{37}$$

Here, is the Lorentz factor. At RHIC, heavy ions are collided at GeV per nucleon, hence . Using for Gold and fm, we estimate G. At the LHC at = 5.02 TeV and , this value is even 30 times larger. To appreciate how strong this field is, compare it with the magnetic field of a neutron star – G [69], or that of its slowly rotating magnetic variant, the magnetar, G

[143]. It is very likely the strongest magnetic field in nature, though existing only for a minute period of time.

Calculation with the realistic distribution of protons in a nucleus shows that magnetic field rapidly decreases as a power of time, and after the first 3 fm/c drops from its maximal value (37) by more than three orders of magnitude [144]. However, different estimates to be discussed in the next sections indicate that a strongly interacting thermalised medium is formed as early as 0.5 fm/c. Therefore, a more realistic calculation going beyond the above field in the vacuum calculation has to include the response of the medium determined by its electrical conductivity. It has been found by lattice calculations [145] that the gluon contribution to the electrical conductivity of static quark–gluon plasma is

$$\sigma = (5.8 \pm 2.9)\,\frac{T}{T_c}\ \text{MeV}. \tag{38}$$

This result was confirmed and further extended by more elaborate lattice simulations with 2 + 1 dynamical flavors for temperatures T= (120–350) MeV [146,147]. The calculations have shown that already starts to deviate from zero for (i.e., in the confined phase), and increases towards the QGP value (38) and further on. The non-zero electrical conductivity in the QGP and (probably also) in the hadronic phase when taken at its face value would inevitably lead to a substantially prolonged lifetime of the magnetic field inside the medium, and might thus even influence the hadron decay widths [148].

A plethora of novel non-dissipative transport phenomena related to the interplay of quantum anomalies with the magnetic field and vorticity in systems with chiral fermions, including the QGP, is reviewed in Ref. [149]. The most direct effect of magnetic field on the QGP is the induction of electric currents carried by the charged quarks and antiquarks in the plasma, and later, by the charged hadrons. In Ref. [150], it was suggested that it may leave its imprint on the azimuthal distributions and correlations of the produced charged hadrons. Charged particles moving along the magnetic field direction y are not influenced by the magnetic Lorentz force, while those moving in the -plane (i.e., in the reaction plane) are affected the most. The result is an azimuthally anisotropic flow of expanding plasma in the -plane, even when the initial plasma geometry is completely spherically symmetric.

Another effect is related to the chiral symmetry restoration. In such a state, within a localized region of space-time, gluon fields can generate nontrivial topological charge configurations that lead to parity violation in strong interactions [126]. In ultra-relativistic heavy-ion collisions, interactions between quarks and these gluonic states can lead to an imbalance in left- and right-handed quarks which violates parity symmetry [151]. The presence of a strong magnetic field induced by the spectator protons transforms this chirality imbalance into an electromagnetic current perpendicular to the reaction plane. This interesting phenomenon stemming from the interplay of chirality, magnetic field, and the chiral anomaly is called the Chiral Magnetic Effect (CME) [149].

Several manifestations of the phenomena related to strong magnetic fields produced in Au + Au or Pb + Pb collisions have been reported by RHIC [152,153,154] and the LHC [155] experiments. Some doubts on the prevailing interpretation were cast by the recent observation of charge-dependent azimuthal correlations also in p + Pb collisions at the LHC [156]. Moreover, in the presence of elliptic flow (for its definition, see Section 5.1), practically all conventional two-particle correlations like the local charge conservation [157] may contribute to the reaction-plane-dependent correlation function used to quantify the CME [158]. Obviously, more investigations are needed. The program of varying the magnetic field by a controlled amount while keeping all else fixed by using nuclear isobars (pairs of nuclei with the same mass number A but different charge Z) is now under consideration at RHIC. The most attractive isobars are Zr+Zr and Ru+Ru, or some other combinations having charge differences of four like Sn /Xe , Te /Ba , and Xe /Ce [158].

An interesting suggestion addressing the simultaneous effects of the huge vorticity of nearly-perfect fluid and the strong magnetic field generated in non-central heavy-ion collisions was made in Ref. [159]. The authors suggest the measurement of a global polarization of the final hadrons in order to estimate the thermal vorticity due to the large orbital momentum of colliding nuclei as well as the electromagnetic field developed in the plasma stage of the collision.

4.3. Transport Models

One of the main tasks of the theory is to link experimental observables to different phases and manifestations of the QCD matter. To achieve this goal, a detailed understanding of the dynamics of heavy-ion reactions is essential. This is facilitated by transport theory, which helps to interpret or predict the quantitative features of heavy-ion reactions. It is particularly well-suited for a non-equilibrium situation, finite size effects, non-homogeneity, N-body phase space, particle/resonance production and freeze-out, as well as for collective dynamics. Microscopic [160,161,162,163,164,165,166,167,168], macroscopic (hydrodynamical) [169,170,171,172], or hybrid [173,174,175] transport models attempt to describe the full time evolution from the initial state of a heavy-ion reaction up to the freeze-out of all initial and produced particles after the reaction. This is illustrated in Figure 9, where a comparison of the data from heavy-ion collisions to the microscopic and hydrodynamical models is presented.

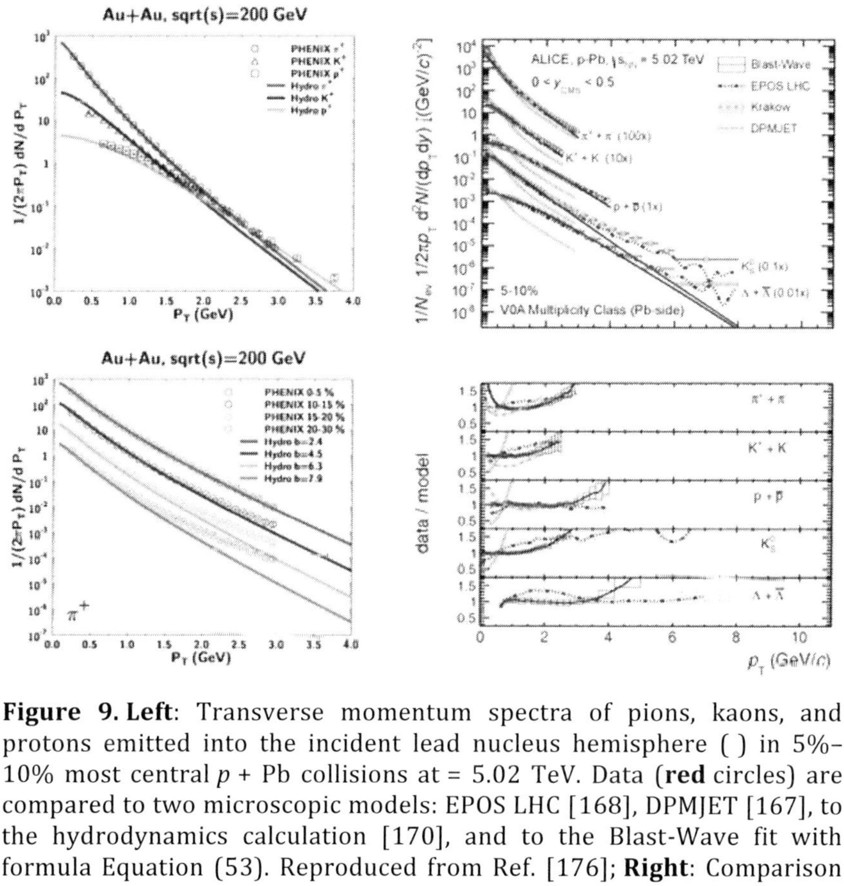

Figure 9. Left: Transverse momentum spectra of pions, kaons, and protons emitted into the incident lead nucleus hemisphere ( ) in 5%–10% most central p + Pb collisions at = 5.02 TeV. Data (**red** circles) are compared to two microscopic models: EPOS LHC [168], DPMJET [167], to the hydrodynamics calculation [170], and to the Blast-Wave fit with formula Equation (53). Reproduced from Ref. [176]; **Right**: Comparison of the experimental -spectra of , , and from Au + Au collisions at (**top**) GeV and (**bottom**) (for the case of) of their centrality dependence [177] with the hydrodynamical model calculations. Reproduced from Ref. [178].

The hadronic cascade models (some with mean-field interactions) have succeeded in reproducing the gross and many-detailed features of the nuclear reactions measured at SIS, AGS, and SPS [160,161,163,164]. They have become indispensable for experimentalists who wish to identify interesting features in their data or to make predictions to plan new experiments. The main strength of the models based on superposition of collisions, relativistic geometry, and final-state hadronic rescattering is not that it gives a precise agreement with experiment for individual observables in particular kinematic regions, but in its ability to give an overall qualitative description of a range of observables in a wide kinematic region. The price to be paid for this simplicity is to assume that either hadrons or hadron-like objects can exist at the earliest stage of the heavy-ion collision just after the two nuclei pass through each other; i.e., that the hadronization time in the frame of the particle is short and insensitive to the environment in which it finds itself. The general success of these models at lower energies can nonetheless easily lead to misconceptions at higher energies. The main concern is the relevance of these models at high particle densities, which are so characteristic for collisions of heavy systems. Here, all the models based on hadronic dynamics are fundamentally inconsistent [179]. Studying the size of the fraction of the energy contained in known hadrons and that temporarily stored in more elusive objects (such as pre-hadronized strings), it was found [163] that up to a time of 8 fm/c, most of the energy density resides in strings and other high-mass continuum states that have not fully decayed. The physical properties of these objects are poorly known, even when they occur in isolation [180]—not to speak about their interactions (or even their existence) in a dense environment. The application of these models to the early phase of collision of two ultra-relativistic heavy nuclei is therefore ill-founded [179].

A complementarity between the microscopic and macroscopic descriptions becomes obvious for the case of strongly-interacting plasmas. The fact that neither Boltzmann equation nor cascades can be used for liquids stems from the fact that particles are strongly correlated with several neighbours at all times. The very idea of "scattering" and cross-section involves particles coming from and going to infinity: it is appropriate for dilute gases, but not for condensed matter where interparticle distances do not exceed the range of the forces at any time [26].

The idea to use the laws of ideal hydrodynamics to describe the expansion of the strongly interacting matter formed in high energy hadronic collisions was first formulated by Landau in 1953 [181]. Later on, Bjorken [129] discovered a simple scaling solution that provides a natural starting point for more elaborate solutions in the ultra-relativistic domain. The phenomenological success of the Landau model was a big challenge for high energy physics for decades [182]. First, because hydrodynamics is a classical theory, and second because it assumes local equilibrium. Both of these assumptions imply many degrees of freedom, and it is by no means clear that the highly excited but still small systems produced in violent nuclear collisions satisfy the criteria justifying treatment in terms of a macroscopic theory [183]. Therefore, the Landau model (and other statistical models of strong interactions) were considered up to the mid-seventies as exotic approaches, lying outside of mainstream physics [182]. Then, the authors of Refs. [121,122] realized that the exploitation of hydrodynamics in an interpretation of data is the only chance of proving the existence of a new state of matter in laboratory. This is a trivial corollary of the well-known fact that a state of matter is defined by its EoS, and there is no other way to get information about the EoS than by using hydrodynamics [182,183,184].

4.3.1. Elements of Relativistic Hydrodynamics

Let us now briefly recall some of the basic results from relativistic hydrodynamics [9,20,22,185,186,187] upon which the contemporary models are based. The basic hydrodynamical equations describe the energy–momentum and the current conservation

$$\partial_\mu T^{\mu\nu} = 0, \quad \partial_\mu j_i^\mu = 0, \tag{39}$$

where is the conserved current. Both quantities can be decomposed into time-like and space-like components using natural projection operators, the local flow four-velocity , and the second-rank tensor perpendicular to it :

$$T^{\mu\nu} = \epsilon u^\mu u^\nu - p\Delta^{\mu\nu} + W^\mu u^\nu + W^\nu u^\mu + \pi^{\mu\nu}, \tag{40}$$

$$j_i^\mu = n_i u^\mu + V_i^\mu, \tag{41}$$

where is the energy density, is the hydrostatic + bulk pressure, is the energy (or heat) current, is the charge density, is the charge current, and is the shear stress tensor. The angular brackets in the definition of the shear stress tensor stand for the following operation:

$$\langle A^{\mu\nu}\rangle = \left[\frac{1}{2}(\Delta^\mu{}_\alpha\Delta^\nu{}_\beta + \Delta^\mu{}_\beta\Delta^\nu{}_\alpha) - \frac{1}{3}\Delta^{\mu\nu}\Delta_{\alpha\beta}\right] A^{\alpha\beta}. \tag{42}$$

To further simplify our discussion, we restrict ourselves in the following to only the one conserved charge, and denote the corresponding baryon current as . The various terms appearing in the decompositions (40) and (41) can then be grouped into ideal and dissipative parts

$$T^{\mu\nu} = T^{\mu\nu}_{id} + T^{\mu\nu}_{dis} = [\epsilon u^\mu u^\nu - p_s\Delta^{\mu\nu}]_{id} + [-\Pi\Delta^{\mu\nu} + W^\mu u^\nu + W^\nu u^\mu + \pi^{\mu\nu}]_{dis} \tag{43}$$

$$j^{\mu} = j^{\mu}_{id} + N^{\mu}_{dis} = [nu^{\mu}]_{id} + [V^{\mu}]_{dis} \,. \quad (44)$$

Neglecting the dissipative parts, the energy–momentum conservation and the current conservation (39) define *ideal hydrodynamics*. In this case (and for a single conserved charge), a solution of the hydrodynamical Equation (39) for a given initial condition describes the space-time evolution of the six variables—three state variables , , , and three space components of the flow velocity . However, since (39) constitute only five independent equations, the sixth equation relating p and ϵ—the EoS—has to be added by hand to solve them. For this, one can either use the relativistic non-interacting massless gas EoS or its generalization to the case of a non-zero interacting measure . In addition to many different phenomenological parameterizations of , one can exploit the relation (20) to obtain the EoS directly from the lattice QCD simulations. Examples of this approach were given in Section 3.3.1 (see particularly Equations (24) and (25)), and are illustrated in Figure 5.

Two definitions of flow can be found in the literature [9,20,185]; one related to the flow of energy (Landau) [181] reads

$$u^{\mu}_{L} = \frac{T^{\mu}{}_{\nu} u^{\nu}_{L}}{\sqrt{u^{\alpha}_{L} T_{\alpha}{}^{\beta} T_{\beta\gamma} u^{\gamma}_{L}}} = \frac{1}{e} T^{\mu}{}_{\nu} u^{\nu}_{L} \,, \quad (45)$$

while the other relating to the flow of conserved charge (Eckart) [188] as follows:

$$u^{\mu}_{E} = \frac{j^{\mu}}{\sqrt{j_{\nu} j^{\nu}}} \,. \quad (46)$$

Let us note that ( ) in the Landau (Eckart) frame. In the case of vanishing dissipative currents, both definitions represent a common flow. In ultra-relativistic heavy-ion collisions, the Landau definition is more suitable when describing the evolution of matter in the region with a small or zero baryon number deposition (i.e., when) like the mid-rapidity region at the LHC and at the top RHIC energy; see Figure 1. In this case, the heat conduction effects can be neglected.

In order to solve the hydrodynamic equations with the dissipative terms, it is customary to introduce the following two phenomenological definitions (so-called *constitutive equations*) for the shear stress tensor and the bulk pressure Π [129,

$$\pi^{\mu\nu} = 2\eta \left\langle \nabla^{\mu} u^{\nu} \right\rangle , \quad \Pi = -\zeta \partial_{\mu} u^{\mu} = -\zeta \nabla_{\mu} u^{\mu} , \tag{47}$$

where the coefficients η and ζ are called the *shear viscosity* and *bulk viscosity*, respectively.

For the boost-invariant Bjorken flow [185] which is also called the one-dimensional Hubble flow, since velocity in the z direction, , is proportional to z,

$$u^{\mu}_{\text{BJ}} = \frac{x^{\mu}}{\tau} = \frac{t}{\tau} \left(1, 0, 0, \frac{z}{t}\right) , \tag{48}$$

where is the proper time, one obtains the following equation of motion [185]:

$$\frac{d\epsilon}{d\tau} = -\frac{\epsilon + p_s}{\tau} \left(1 - \frac{4}{3\tau T} \frac{\eta}{s} - \frac{1}{\tau T} \frac{\zeta}{s}\right) . \tag{49}$$

Neglecting the last two terms in Equation (49), one obtains the famous Bjorken solution of ideal hydrodynamics [129]. The last two terms on the right-hand side in Equation (49) describe a compression of the energy density due to viscous corrections. The first one is due to the shear viscosity in compressible fluids, while the second comes from the bulk viscosity. Two dimensionless coefficients in the viscous correction, and , reflect the intrinsic properties of the fluids; see Table 1 and Figure 15 (left panel). The value has been obtained in the framework of SUSY Yang–Mills theory [107]. The conformal nature of this theory gives automatically. Moreover, for gluonic matter is obtained from the lattice calculations of pure gauge theory [189], while the bulk viscosity ζ has a prominent peak around resulting from the trace anomaly of QCD [190].

Hydrodynamics provides an effective description of a system that is in local thermal equilibrium, and can be derived from the underlying kinetic description through Taylor expansion of the entropy four-current = in gradients of the local thermodynamic variables. In zeroth order in gradients, one obtains ideal fluid dynamics. Then, the higher orders describe dissipative effects due to irreversible thermodynamic processes such as the frictional energy dissipation between the fluid elements that are in relative motion, or heat exchange of the fluid element with its surroundings on its way to approach thermal equilibrium with the whole fluid. The relativistic Navier–Stokes description given above in Equation (47)—which accounts only for terms that are *linear* in velocity

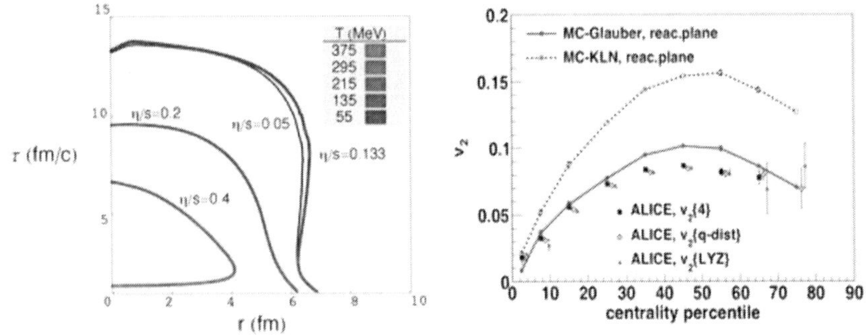

Figure 15. Left: Location of freeze-out surfaces for central Au + Au collisions [108]; **Right**: The centrality dependence of the elliptic flow of charged hadrons from Pb–Pb collisions at = 2.76 TeV [298]. The ALICE data are from Ref. [299].

gradient—leads, unfortunately, to severe problems. Qualitatively, it can be seen from Equation (47) observing what happens if one of the thermodynamic forces or is suddenly switched off/on [191]. In this case, the corresponding thermodynamic flux or Π which is a purely local function of the velocity gradient also instantaneously vanishes/appears. The linear proportionality between dissipative fluxes and forces causes an instantaneous (acausal) influence on the dissipative currents, leading to numerical instabilities [192].

Any numerical implementation of relativistic dissipative fluid dynamics thus requires the inclusion of terms that are second order in gradients [186]. The resulting equations for the dissipative fluxes and Π are relaxation-type equations [187] with microscopic relaxation times and , which encode the time delay between the appearance of thermodynamic gradients that drive the system out of local equilibrium and the associated build-up of dissipative flows in response to these gradients, thereby restoring causality [187,193]. Since the relaxation times must be positive, the Taylor expansion coefficients α and β must all be larger than zero. Accounting for non-zero relaxation times at all stages of the evolution constrains departures from local equilibrium, thereby both stabilizing the theory and improving its quantitative precision.

4.3.2. Blast Wave Parametrization

Interpretation of the results of hydrodynamical calculations or of the experimental data in terms of the collective flow of matter [87,194] is greatly facilitated by the use of the analytical so-called *Blast Wave* (BW) parametrization [195,196,197]. Within the boost-invariant scenario of

Bjorken [129], and for the full azimuthal symmetry which is valid in central collisions of two nuclei, the velocity field of expanding matter is given by

$$u^{\mu}(\rho,\eta) = (\cosh\rho\ \cosh\eta,\ \vec{e}_r \sinh\rho,\ \cosh\rho\ \sinh\eta)\ , \quad (50)$$

where and η are transverse and longitudinal rapidities, respectively, and is the unit vector in the transverse plane. The transverse velocity distribution of the thermalized matter in the region is described by a self-similar profile

$$\beta_T(r) = \beta_S \left(\frac{r}{R}\right)^k \quad , \quad (51)$$

where is the surface velocity, and parameter k is usually given the value $k = 2$ to resemble the solutions of hydrodynamics [196]. The spectrum of locally thermalized matter is constructed as a superposition of the individual thermal components [198]:

$$E\frac{d^3N}{d^3p} = \frac{g}{(2\pi)^3} \int e^{-(u^{\nu}p_{\nu}-\mu)/T_{kin}} p^{\lambda} d\sigma_{\lambda}\ , \quad (52)$$

where σ is the hypersurface defining a borderline between the hydrodynamical behaviour and free-streaming particles (the so-called *freeze-out hypersurface*), and is the *temperature of the kinetic freeze-out*. Boosting each component with the transverse rapidity, one obtains the transverse momentum spectra of particles from the collective radial flow of expanding matter:

$$\frac{dN}{p_T\, dp_T} \propto \int_0^R r\, dr\, m_T I_0\left(\frac{p_T \sinh\rho}{T}\right) K_1\left(\frac{m_T \cosh\rho}{T}\right)\ , \quad (53)$$

where, $m_T = \sqrt{(m^2 + p_T{}^2)}$, $I_0(x)$, and $K_1(x)$ are the Bessel functions.

Formulas for the case of non-central collisions when the transverse shape (51) is controlled not by one but two parameters, and , can be found in Ref. [197]. In full generality, there are eight parameters describing the blast wave parametrization: , and . Here, T is the temperature, and describe the strength of the zero- and second-order oscillation of the transverse rapidity, the parameter corresponds to a surface diffuseness of the emission source, and and are the mean and width of a Gaussian longitudinal proper time freeze-out distribution.

In Figure 9 and Figure 10, the examples of the BW fit analysis are presented. As can be seen from Figure 10, the kinetic freeze-out temperature that determines the shape of the -spectra of particles is strongly anti-correlated with the radial flow velocity : the higher is, the lower is, and vice versa. Nevertheless, the radial flow reveals itself as a shoulder structure at small transverse momenta in the -spectra of Λs, protons, and kaons; see Figure 9. For the pions, there is almost no sensitivity to distinguish between the two cases—a reduction of temperature is almost compensated by the radial flow.

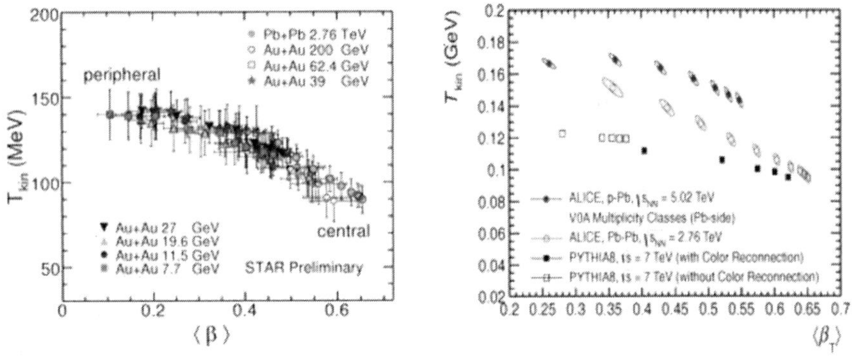

Figure 10. Left: Variation of the blast wave parameters and obtained from the fits to the spectra of pions, kaons, protons, and their anti-particles produced in Au + Au and Pb + Pb collisions at energies = 7.7 GeV – 2.76 TeV and different collision centralities. The centrality increases from left to right for a given energy. Reproduced from Ref. [199]; **Right**: The same, but for Pb + Pb at = 2.76 TeV and *p* + Pb at = 5.02 TeV. MC simulations of *p* + *p* collisions at = 7 TeV using the PYTHIA8 event generator [200] with and without color reconnection are shown as open and filled squares, respectively. Reproduced from Ref. [176].

4.4. Initial State Description of Nuclear and Hadronic Interactions

An indispensable part of the full description of the experimental data from heavy-ion collisions comes not only from the understanding of its dynamics starting from the moment of thermalization, but also at earlier times. In particular, the question of where the observed (local) thermalization of deconfined matter comes from is still quite open [201,202,203,204]. The importance of event-by-event initial state fluctuations on anisotropic collective flow and other final state observables is also worth mentioning [205,206,207,208]. Since these topics currently remain a significant source of uncertainty in predicting the final state observables, in the next two paragraphs we will provide two alternative ways of describing the initial state of the collision.

In high-energy nucleus–nucleus (A + B) interactions, the de Broglie wavelength of the nucleons (N) of the incoming nucleus is much smaller than the inter-nucleon distances inside the partner nucleus. To each incoming nucleon, the positions of the nucleons within the partner nucleus appear to be frozen. After a single elementary NN (elastic or inelastic) collision, both participating nucleons acquire a transverse momentum, which in the majority of cases is very small compared to their longitudinal one, and so the longitudinal momenta before and after the collision are very close to each other . High incident energies and small scattering angles mean that the scattering is dominated by a large orbital momentum ℓ, and so it is convenient to replace the partial-wave expansion of the scattering amplitude by an impact parameter b = (1 +ℓ)/p representation. The A + B collision can thus be described using a semi-classical approach due to Glauber [209,210,211,212], which treats the nuclear collision as multiple NN interactions [213,214]. The nucleons which have suffered at least one NN collision are called the *participants*, and those who have avoided it are called the *spectators* (see Figure 11). The total number of spectators and participants thus adds up to N + = A + B. On the other hand, the total number of collisions suffered by all participants fulfils inequality ≥ /2.

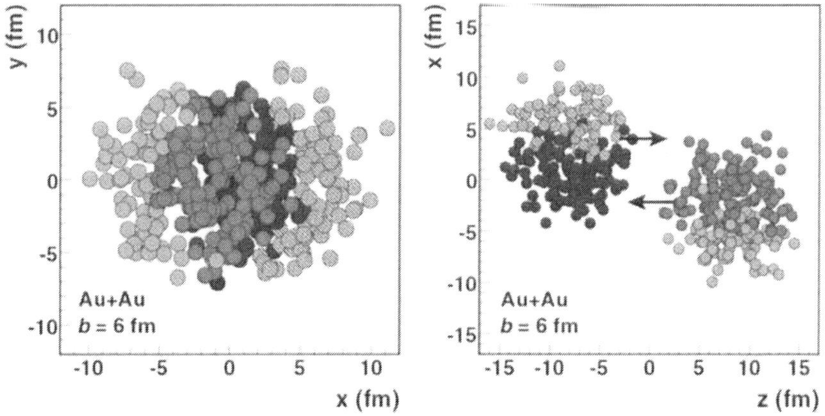

Figure 11. A Glauber Monte-Carlo event (Au + Au at = 200 GeV) viewed in the transverse plane (**left** panel) and along the beam axis (**right** panel). The nucleons are drawn with diameter . Darker disks represent the participants, and lighter disks represent the spectators. Reproduced from Ref. [215].

4.4.1. Glauber Model

Using the nuclear mass number density = A,B and the inelastic NN cross-section , we can express and analytically [9,22,28,215]:

$$N_{\rm part}(b) = \int d^2\mathbf{s}\, T_A(\mathbf{s})\left(1-e^{-\sigma_{NN}^{in}T_B(\mathbf{s})}\right) + \int d^2\mathbf{s}\, T_B(\mathbf{s}-\mathbf{b})\left(1-e^{-\sigma_{NN}^{in}T_A(\mathbf{s})}\right) \quad (54)$$

$$N_{\rm coll}(b) = \int d^2\mathbf{s}\, \sigma_{NN}^{in} T_A(\mathbf{s})T_B(\mathbf{b}-\mathbf{s}) \equiv \sigma_{NN}^{in} T_{AB}(\mathbf{b})\,, \quad (55)$$

where and are the nuclear thickness and the nuclear overlap functions, respectively:

$$T_A(\mathbf{b}) \equiv \int dz \rho_A(z,\mathbf{s})\,, \quad T_{AB}(\mathbf{b}) \equiv \int d^2\mathbf{s}\, T_A(\mathbf{s})T_B(\mathbf{b}-\mathbf{s})\,. \quad (56)$$

The Glauber model calculations are also often carried out via Monte Carlo [215,216,217]. Nucleons inside the colliding nuclei are distributed randomly according to a nuclear density profile. At a given impact parameter, , the impact parameter of all pairs of nucleons is calculated. Interaction occurs when < ; see Figure 11. The calculated (b) and (b) are then used to make a contact with the measured bulk observables, like the multiplicity of charged particles measured by the tracking detectors at midrapidity (Figure 12, left panel), or the energy left by the spectator nucleons in the Zero Degree Calorimeter (Figure 12, right panel). The left panel also explains how different bins in multiplicity of charged particles can be transformed into the bins in collision centrality. The left panel of Figure 13illustrates that even for quite different nuclear systems (Au + Au, Cu + Au, Cu + Cu), the number of participants selects the collisions with almost the same energy density . The right panel of Figure 13 shows that centrality dependence of the yields of EW-interacting particles (like direct photons, , or Z-bosons) is completely determined by the .

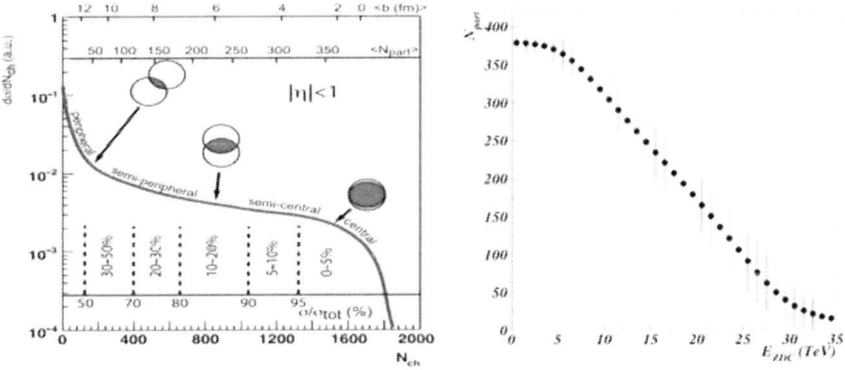

Figure 12. Left: A cartoon showing the centrality definition from the final-state charged particle multiplicity and its correlation with the average impact parameter and the mean number of nucleons participating in the collision < >. Reproduced from Ref. [27]; **Right**: The number of participants as a function in Pb + Pb collision at incident momentum per nucleon 158 GeV/c (= 17.3 GeV) calculated in the Glauber approach. Reproduced from Ref. [218]. The error bars represent the r.m.s. of the distribution at fixed .

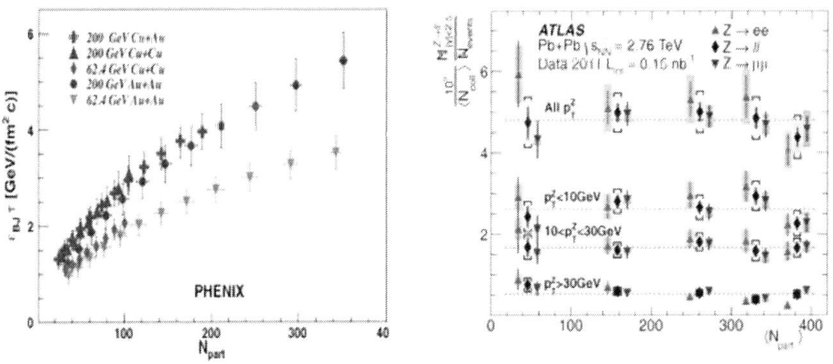

Figure 13. Left: The Bjorken estimate of the initial energy density (Equation (32)) multiplied by the thermalization time τ as a function of for different colliding systems (Au + Au, Cu + Au, Cu + Cu) at two energies (= 62.4 and 200 GeV). Reproduced from Ref. [130]; **Right**: The dependence of Z boson yields divided by average number of collisions < > on the average number of participants < > for Pb + Pb collisions at = 2.76 TeV. Reproduced from Ref. [219].

A generalisation of the Glauber model beyond its original non-relativistic potential description of the scattering process was first formulated by Gribov [220,221], who used the effective field theory to describe multiple interactions proceeding via the Pomeron exchange. The interference terms appearing naturally in this treatment of multiple scattering automatically assure the unitarity of the theory [222]. With the advent of the QCD, the parton-based multiple scattering models became popular [223,224,225,226,227]. Such models allow the treatment of A + B and NN collisions on more equal footing. Their participants could thus be not only nucleons, but also the pQCD partons, the valence quarks [223,228], or any effective sub-nucleon degrees of freedom [229].

4.4.2. QCD Scattering in the Dipole Picture: Initial-State Energy Loss and Shadowing

The color dipole formalism [230,231] provides phenomenological means for the universal treatment of inclusive and diffractive $p + p$, p + A, and A+A collisions at high energies. In reactions involving a hard scale such as the Deep Inelastic Scattering (DIS), Drell–Yan (DY), heavy quark production, etc., this formalism effectively accounts for the higher-order QCD corrections and enables the quantification of the initial state interaction, saturation, gluon shadowing, and nuclear coherence effects [230,232,233,234,235,236,237,238]. At small x, the eigenstates of interaction correspond to color dipoles with a definite transverse separation, the main ingredients of the dipole picture. Provided that the initial projectile state can be considered as a collection of dipoles, any production process in the target rest frame is then viewed as a superposition of universal dipole-target partial amplitudes at different dipole separations , impact parameters , and Bjorken x. Integrating the amplitude over ,

$$\sigma_{\bar{q}q} = \int d^2b \, 2\mathrm{Im} f_{\mathrm{el}}(\mathbf{b}, \mathbf{r}; x) \,, \tag{57}$$

one arrives at the universal dipole cross-section , which is determined phenomenologically. For example, the DIS process in the target rest frame is viewed as a scattering of the "frozen" dipole of size $r\sim$, originating as a fluctuation of the virtual photon off the target nucleon. The DY pair production is considered as a bremsstrahlung of massive (and boson) off a projectile quark before after and the quark scatters off the target [233]. The projectile dipole probes the dense gluonic field in the target at high energies when nonlinear (e.g., saturation) effects due to multiple scatterings can become relevant. The latter, however, cannot be reliably predicted due to unknown soft and non-perturbative QCD dynamics.

In the limit of point-like color-neutral dipole , the corresponding dipole cross-section vanishes quadratically due to color screening, which is known as the color transparency [231,239,240]. A naive saturated shape of the dipole cross-section known as the Golec-Biernat–Wüsthoff ansatz originally fitted to the DIS data [241]

$$\sigma_{q\bar{q}} \propto 1 - e^{-\frac{r^2 Q_s^2(x)}{4}}, \qquad Q_s^2(x) \propto \left(\frac{x_0}{x}\right)^{\lambda}, \tag{58}$$

with the energy-dependent saturation scale denoting a characteristic boundary between the linear and non-linear QCD regimes. Such a simple ansatz has proven to successfully capture the major features of a vast scope of observables in $p + p$ and p + A, while the DGLAP evolution of the gluon density in the target affects the saturation scale at large scales such that a QCD-corrected ansatz is often used in practice [242,243,244,245].

The saturation effects are amplified in nuclear collisions, such that the nuclear saturation scale is expected to be enhanced w.r.t. the nucleon one by a factor of . Then, the dipole–nucleus cross-section can be represented in terms of the forward partial dipole–nucleus amplitude analogically to Equation (57). Evolution of this amplitude in rapidity is predicted, for example, in the Color Glass Condensate formalism (see below) ; while no first-principle calculation capable of predicting the -dependence exists, only phenomenological fits inspired by saturation physics are implemented (aside from what is mentioned above, see Refs. [246,247,248,249,250,251,252]).

In heavy-ion collisions at high energies, due to the absence of final state interactions, the DY pair production process on nuclear targets in the dipole picture is often considered to be an excellent probe accessing the impact parameter dependence of the Initial State Interaction (ISI) effects as well as nuclear shadowing and nuclear broadening—the crucial information that cannot be derived within the parton model (e.g., [253,254,255] and references therein). The ISI effects emerging due to multiple rescattering of projectile partons in a medium prior to a hard scattering are only relevant close to kinematic boundaries due to energy conservation significantly suppressing the nuclear cross-sections [256,257]; e.g., when the Feynman variable and . In order to account for the ISI-induced energy loss in the Glauber approximation, one sums up over initial state interactions in a p + A collision at a given impact parameter b, leading to a nuclear ISI-improved projectile quark PDF [256,257]:

$$q_f(x, Q^2) \Rightarrow q_f^A(x, Q^2, \mathbf{b}) = C_v\, q_f(x, Q^2)\, \frac{e^{-\xi\sigma_{\rm eff} T_A(\mathbf{b})} - e^{-\sigma_{\rm eff} T_A(\mathbf{b})}}{(1-\xi)(1-e^{-\sigma_{\rm eff} T_A(\mathbf{b})})}, \qquad \xi = \sqrt{x_L^2 + x_T^2}, \tag{59}$$

where is the nuclear thickness function defined in Equation (56), is fixed by the Gottfried sum rule, and mb is the effective cross-section controlling the multiple ISI. The ISI-induced energy loss can induce a significant suppression at large , , and forward rapidities. In fact, it has been noticed earlier in production in central collisions [258], and in direct photon production in central collisions [259,260] at midrapidity and large transverse momenta (where no shadowing is expected) by the PHENIX Collaboration.

In the course of propagation, a projectile parton experiences multiple rescatterings in the color medium of a target nucleus. These cause a notable suppression of the gluon radiation and energy loss by the parton—a QCD analog of the Landau–Pomeranchuk–Migdal (LPM) effect in QED [261,262,263] (for more details, see Ref. [264] and references therein). In heavy ion collisions, the LPM effect was studied in Refs. [265,266,267].

In the nuclear DY reaction, for example, the shadowing is controlled by the coherence length corresponding to the mean lifetime of -quark fluctuations

$$l_c = \frac{1}{x_2 m_N} \frac{(M_{l\bar{l}}^2 + p_T^2)(1-\alpha)}{\alpha(1-\alpha)M_{l\bar{l}}^2 + \alpha^2 m_f^2 + p_T^2}, \quad (60)$$

given in terms of the quark mass , nucleon mass , dilepton invariant mass , its transverse momentum , and light-cone momentum fractions of dilepton α and the target gluon . In Ref. [254], it was noticed that the long coherence length (LCL) compared to the nuclear radius (namely,) is practically useful in kinematic domains of RHIC and the LHC experiments.

The quark shadowing in the LCL limit is automatically accounted for by scattering of the leading Fock components in the dipole formula which represents a process-dependent convolution of the light-cone wave function for a given Fock state with a superposition of partial dipole amplitudes. In $p + A$ collisions, the latter are given by the eikonal Glauber–Gribov form [231]:

$$\mathrm{Im} f_{\mathrm{el}}^{A}(\mathbf{b}, \mathbf{r}; x) = 1 - \exp\left(-\frac{1}{2} T_A(\mathbf{b})\, \sigma_{q\bar{q}}(\mathbf{r}, x)\right), \quad (61)$$

which resums high-energy elastic scatterings of the dipole in a nucleus. Such an eikonalisation is justified in the LCL regime where the transverse separation of partons in the projectile multiparton Fock state is "frozen" in the course of propagation of the dipoles through the nuclear matter such that it becomes an eigenvalue of the scattering operator.

At large scales and at small collision energies (GeV), one recovers fm such that the LCL approximation fails and an alternative description is necessary. In Refs. [268,269,270], a universal path-integral approach to multiple dipole interactions with a nuclear target which can be used for any value of has been developed. It is also known as the Green function approach, and consistently incorporates the color transparency and quantum coherence effects responsible for nuclear shadowing. Recently, the Green functions technique accounting for the nuclear attenuation of the colorless dipole in the medium has been employed in Ref. [255] in analysis of the nuclear coherence effects in DY reaction in p + A collisions. In Ref. [271], it was demonstrated that the Green function technique is equivalent to the diagrammatic formulation developed by the Baier–Dokshitzer–Mueller–Peigne–Schiff Collaboration (e.g., Refs. [272,273,274]). In Ref. [275], it was understood that the basic reason for the observed suppression of high- hadron production in heavy ion collisions is the attenuation of early produced colorless dipoles ("pre-hadrons") propagating through a dense absorptive matter, but is not an energy loss because of a much shorter production length than was typically assumed.

In the case of DY, the lowest Fock state scattering is not enough in the LHC kinematics domain, where the higher Fock components containing gluons (e.g., , etc.) also become noticeable, causing an additional suppression known as gluon shadowing (GS). These components are heavier than , and thus have a shorter coherence length. In the high energy limit, the additional suppression factor induced by the GS corrections has been computed in Ref. [276,277,278] using the Green function technique, going beyond the LCL limit as

$$R_G(x, Q^2, \mathbf{b}) \equiv \frac{x g_A(x, Q^2, \mathbf{b})}{A \cdot x g_p(x, Q^2)} \approx 1 - \frac{\Delta\sigma}{\sigma_{tot}^{\gamma^* A}}, \quad (62)$$

in terms of the inelastic correction to the total nuclear DY cross-section , related to the formation of a next-to-leading Fock component. The GS is a leading-twist phenomenon effective at small such that very slowly at .

4.4.3. Color Glass Condensate

In addition to the sQGP matter, another instance where quarks and gluons cannot be treated as independent degrees of freedom is the case of parton coherence. A generalization of pQCD to hard collisions of small-

x ( ) partons (called also a *semi-hard regime*) was first discussed by Gribov, Levin, and Ryskin [279]. The basic failure of the standard DGLAP approach [280,281,282] is that it predicts too-fast increase of small-x parton density with the scale . Consequently, the growth of hadronic cross-sections proceeds at rate which would sooner or later violate unitarity. The proposed solution—parton recombination and saturation—is at variance with the standard assumption that the partons themselves can be considered as independent free particles. The parameter determining the probability of parton–parton recombination is the ratio of the parton–parton cross-section to the square of the average distance between partons. The fact that the cross-section of such a semi-hard process (which now complies with the unitarity) increases rapidly with incident energy gives rise to expectations that (at least asymptotically) bulk particle production in hadron–hadron collisions can be described via pQCD [18].

The modern implementation of the above ideas is the Color Glass Condensate (CGC) formalism [283,284,285,286]—a natural generalization of pQCD to dense partonic systems. When applied to heavy nuclei, it predicts strong color fields in the initial stage of the collision. The strength of the fields is due to the condensation of low-x gluons into single macroscopic (i.e., classical) field state called the CGC. Since the characteristic scale of the parton saturation grows as [18,284], it is enhanced on nuclear targets. According to the CGC-motivated phenomenology, the saturation phenomena are expected to show up (if not already) in + Au collisions at RHIC and , for sure, in nuclear collisions at the LHC. For example, due to the gluon saturation, the growth of the inelastic nucleon–nucleon cross-section with increasing collision energy may result in a broadening of the nucleon density distribution in position space. This in turn leads to a natural smoothing of the initial energy density distribution in the transverse plane of the matter created near midrapidity in heavy-ion collisions [207].

The CGC is described by an effective field theory that separates two kinds of degrees of freedom—fast frozen color sources and slow dynamical color fields. The basic evolution equation of such an effective field theory is an RG equation known as the Jalilian–Marian–Iancu–McLerran–Weigert–Leonidov–Kovner (JIMWLK) equation [287,288,289,290,291,292], which reflects the independence of physical quantities with respect to variations of the cutoff separating these degrees of freedom (for more details, see Ref. [285] and references therein).

A supporting argument for the CGC as a possible state of QCD matter comes from successful analysis of HERA data in terms of geometrical scaling [293]. The geometrical scaling is the statement that the total cross-section depending a priori on two independent variables—the photon virtuality and the Bjorken variable x—is a function of a single variable , where the so-called saturation scale depends nontrivially on x, with dimensions given by a fixed reference scale . However, calculations of Ref. [294] show that the standard linear leading-order DGLAP perturbative evolution is able to explain the geometric scaling. The situation with CGC applicability at current energies is thus unsettled (see also Refs. [253,254]). The experimental data from RHIC and the LHC, as well as exploitation of non-CGC-based models [228] are needed to resolve this problem.

There are two popular representative models of the initial state which are based on the CGC—the KLN model [295] and the IP Glasma model [296]. A Monte Carlo implementation of KLN CGC initial state [174,207,297] is based on the number distribution of gluons produced in the transverse plane given by the -factorisation formula [295]:

$$\begin{aligned}\frac{dN_g}{d^2r_\perp dy} &= \kappa\frac{4N_c}{N_c^2-1}\int\frac{d^2p_\perp}{p_\perp^2}\int\frac{d^2k_\perp}{4}\,\alpha_s(Q^2) \\ &\times\ \phi_A(x_1,(\mathbf{p}_\perp+\mathbf{k}_\perp)^2/4)\,\phi_B(x_2,(\mathbf{p}_\perp-\mathbf{k}_\perp)^2/4)\,.\end{aligned} \tag{63}$$

Here, and y denote the transverse momentum and rapidity of the produced gluons, and / are the light-cone momentum fractions of the colliding gluons. The running coupling is evaluated at the scale = . The gluon distribution function is given by

$$\phi_A(x,k_\perp^2;\mathbf{r}_\perp)\sim\frac{1}{\alpha_s(Q_{s,A}^2)}\frac{Q_{s,A}^2}{\max(Q_{s,A}^2,k_\perp^2)}\,. \tag{64}$$

An overall normalisation factor κ is chosen to fit the multiplicity data in most central Au + Au collisions at RHIC. In the MC-KLN model [295], the saturation momentum is parameterized by assuming that the saturation momentum squared is 2 GeV at in Au + Au collisions at fm at RHIC, where fm ; i.e.,

$$Q_{s,A}^2(x;\mathbf{r}_\perp)=2\,\mathrm{GeV}^2\frac{\rho_A(\mathbf{r}_\perp)}{1.53\ \mathrm{fm}^{-2}}\left(\frac{0.01}{x}\right)^{\lambda}\,. \tag{65}$$

Here, λ is a free parameter which is expected to be in the range of from the global analysis of scattering for [241,293].

The IP-Glasma model [296] solves the classical Yang–Mills equations in which initial charge distributions of two colliding nuclei are sampled from a Gaussian distribution with the impact parameter and Bjorken x-dependent color charge distributions. A parameterization of x and impact parameter dependence of the saturation scale is taken from the IP-Sat (Impact Parameter Saturation) model [242,251]. Fluctuations in the IP-Glasma model have a length scale on the order of the inverse of the saturation scale ~ 0.1–0.2 fm. A comparison of the initial energy density distribution among the IP-Glasma, MC-KLN, and MC-Glauber models is shown in Figure 14.

Figure 14. Examples of the initial energy density distribution from (**left**) the IP-Glasma model at fm, (**middle**) the MC-KLN model, and (**right**) the MC-Glauber model. Reproduced from Ref. [296].

Figure 15 provides two examples of transport model calculations. The left panel shows the location of freeze-out surfaces for central Au + Au collisions at several fixed values of the shear viscosity-to-entropy density ratio obtained from a numerical solution of viscosious hydrodynamics [108]. The shading corresponds to the freeze-out temperature. The freeze-out occurs when the viscous terms become large compared to the ideal terms. Note that hydrodynamics breaks down not only at late, but also at early times (see the curve = 0.4 in Figure 15). The right panel displays the centrality dependence of the elliptic flow coefficient (Equation (66)) for two models for the initial density in the transverse plane—one is motivated by the parton saturation (CGC), and the other exploits nucleons only (Glauber). The calculations [298] were done within a hybrid model, where the expansion of the QGP starting at = 0.6 fm/c is described by ideal hydrodynamics with a state-of-the-art lattice QCD EoS, and the subsequent evolution of hadronic matter below switching temperature = 155 MeV is described using a hadronic cascade model. This nicely illustrates the strength of hydrodynamics—either the viscosity of QGP from RHIC to the LHC increases, or the CGC initial condition is ruled out [298].

4.4.4. Other Initial State Models

For completeness, let us here briefly list some other approaches to initial state description used in the analysis of ultra-relativistic heavy ion collisions. Fluctuating initial conditions given by a multi-phase transport (AMPT) model [165] were applied in an event-by-event partonic transport plus hydrodynamics hybrid approach [166,300,301] to study collective flow. In Refs. [302,303], NLO pQCD together with saturation-like suppression of low-energy partons were used to calculate the initial energy densities and formation times which have been used further on in 3D hydro calculations of space-time evolution of the QCD matter with dissipative fluid dynamics, event by event. A new initial conditions model for high-energy $p + p$, $p +$ A, and A+A collisions that deposit entropy proportional to the generalized mean of nuclear overlap density was introduced in Ref. [304]. The model assumes that N one-on-one nucleon collisions produce the same amount of entropy as a single N-on-N collision.

5. EXPERIMENTAL SIGNATURES OF DECONFINED QCD MATTER

Evolution of the high energy nucleus–nucleus collision is schematically depicted in Figure 16. Two Lorentz-contracted pancakes of nuclear matter collide, thermalize, and form a deconfined QGP medium which expands, cools down, and hadronises to final state hadrons. Experimentally we do not observe each stage separately, but only through the time-integrated final state quantities—the momentum spectra of hadrons, photons, or leptons, particle multiplicities, energy flow, etc. Nevertheless, some time ordering of different processes giving rise to the final state observables exists. At very early collision times when colliding matter thermalizes, the entropy is produced which later—after (almost) isotropic expansion— transforms into particle multiplicities [28,181,205]. Early collision times also favour production of high partons [29,305] or heavy quarks (c, b) [306]. The formation of QGP reveals itself in many ways, including radiation of low momentum direct or virtual photons serving as a thermometers, enhanced production of hadrons containing strange (s) quarks [21,307,308], and melting of or mesons [115,116,306] called quarkonia. The subsequent rapid expansion of deconfined matter having more than ten times the degrees of freedom than the hadronic matter (see Equation (22))—and therefore also much higher internal pressure—produces a strong radial flow which leaves its imprint on the spectra of final state particles and their yields [28,183,205,309,310].

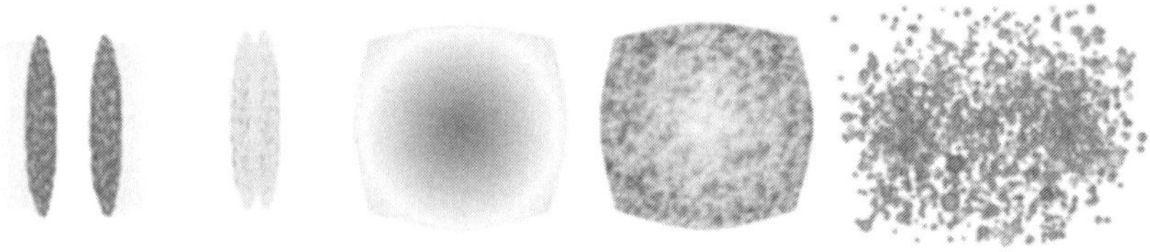

Figure 16. Cartoon of a collision of two ultra-relativistic nuclei. **Left** to **Right**: the two nuclei approach, collide, form a hot and dense equilibrated system, QGP expands and hadronizes, and finally hadrons rescatter and freeze out. The figure is taken from Ref. [311].

In the following, we present several examples of observables related to different stages of dynamics of nucleus–nucleus collisions at high-energies.

5.1. Bulk Observables

Traditionally, the very first measurements of heavy-ion collisions at a new energy regime comprise the charged-particle density at midrapidity , also including its centrality dependence. Its collision-energy dependence for the 5% (6%) most central heavy-ion collisions—normalized per participant pair (i.e., /2)—is presented in Figure 17 (left panel). The right panel of Figure 17 shows that the normalized charged-particle density is rising with centrality, which means that the particle multiplicity at mid-rapidity increases faster than , presumably due to the contribution of hard processes to the particle production [29]. However, this increase is very similar to that observed at the top RHIC energy.

One of the most celebrated predictions of the collective behaviour of matter created in non-central collisions of ultra-relativistic nuclei concerns its evolution in the transverse plane, which results from the pressure gradients due to spatial anisotropy of the initial density profile [183,314] (see Figure 18). The azimuthal anisotropy is usually quantified by the Fourier coefficients [315]:

$$v_n = \langle \cos[n(\phi - \Psi_n)] \rangle , \tag{66}$$

where ϕ is the azimuthal angle of the particle, is the angle of the initial state spatial plane of symmetry, and n is the order of the harmonic. In a non-central heavy ion collision, the beam axis and the impact parameter

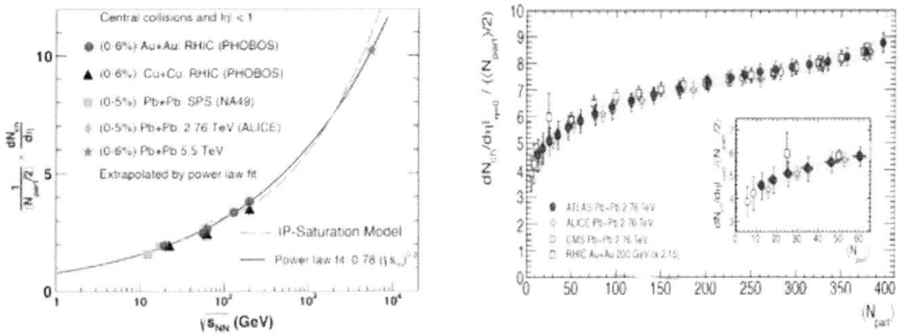

Figure 17. Left: The charged particle rapidity density per participaiting pair, /(0.5), in central Au + Au, Cu + Cu, and Pb + Pb collisions from SPS to LHC energies. The star denotes an extrapolation to Pb + Pb collisions at 5.5 TeV. The IP -saturation model calculation [312] is illustrated by a dashed curve. Adapted from Ref. [30]; **Right**: /(0.5) vs. in Pb + Pb and Au + Au collisions at the LHC and RHIC, respectively. The RHIC data are multiplied by 2.15. The inset shows the < 60 region in more detail. Reproduced from Ref. [313].

define the reaction plane azimuth . For a smooth matter distribution in the colliding nuclei, the plane of symmetry is the reaction plane , and the odd Fourier coefficients are zero by symmetry. However, due to fluctuations in the matter distribution (including contributions from fluctuations in the positions of the participating nucleons in the nuclei—see Figure 11), the plane of symmetry fluctuates event-by-event around the reaction plane. This plane of symmetry is determined by the participating nucleons, and is therefore called the participant plane [316]. Since the planes of symmetry are not known experimentally, the anisotropic flow coefficients are estimated from measured correlations between the observed particles [299,317].

In the following, we shall restrict ourselves to the properties of the Fourier coefficients with and , which provide the dominant contributions to the observed azimuthal *elliptic* and *triangular* asymmetry, respectively. The sensitivity of to initial condition is illustrated on Figure 15 (right panel), where the centrality dependence of the elliptic flow in Pb + Pb collisions at =2.76 TeV is shown. For more details on the corresponding initial state models, see Section 4.4.3.

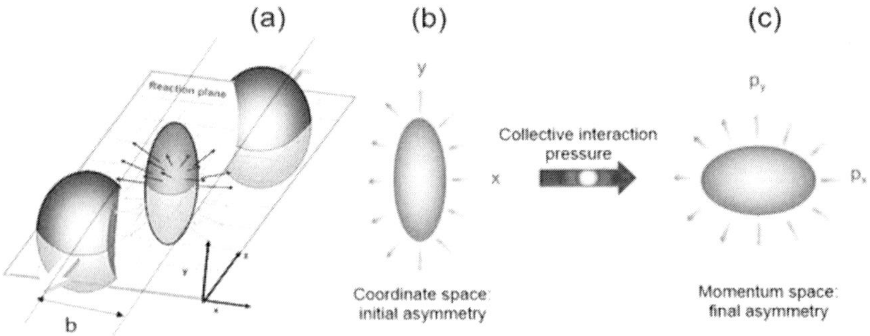

Figure 18. (**a**) A non-central collision of two nuclei leads to an almond-shaped interaction volume; (**b**) This initial spatial anisotropy with respect to the reaction plane translates via pressure gradients into (**c**) a momentum anisotropy of the produced particles. Reproduced from Ref. [30].

The left panel of Figure 19 shows the measured energy dependence of the integrated elliptic flow coefficient in one centrality bin. Starting from ≈ 5 GeV, there is a continuous increase of . Below this energy, two phenomena occur. At very low energies, due to the rotation of the compound system generated in the collision, the emission is in-plane ( ). At the laboratory kinetic energy around 100 MeV/nucleon, the preferred emission turns into out-of-plane, and becomes negative. The slowly moving spectator matter prevents the in-plane emission of participating nucleons or produced pions, which appear to be sqeezed-out of the reaction zone [318]. As the spectators move faster, their shadowing disappears, changing the pattern back to the in-plane emission.

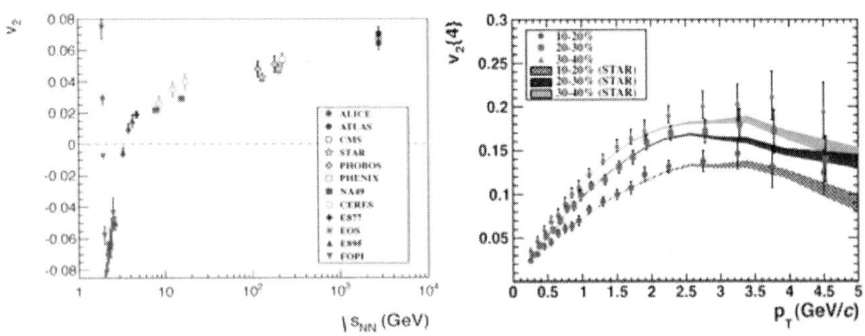

Figure 19. Left: A compilation of data on the dependence of the integrated elliptic flow, v , on the beam energy. The data correspond to (~20%–30%) Au + Au or Pb + Pb most central collisions. Reproduced from Ref. [193]; **Right**: The differential elliptic flow of charged particles. The Pb + Pb collisions at = 2.76 TeV (colored symbols) are compared to Au + Au collisions at = 200 GeV (grey lines). Reproduced from Ref. [299].

Let us note that at RHIC, for the first time, the magnitude of the elliptic flow (Figure 19) was found to be consistent with the EoS expected from the QGP [16,183]. The integrated value of for the produced particles increases by 70% from the top SPS energy to the top RHIC energy (see left panel of Figure 19), and it appears to do so smoothly. In comparison to the elliptic flow measurements in Au + Au collisions at GeV, at the LHC, increases by about 30% at TeV. However, this increase is not seen in the differential elliptic flow of charged particles shown on the right panel of Figure 19. Thus, the bulk medium produced at RHIC and LHC has similar properties, and the 30% increase of between the two energies is due to an enlarged available phase space, resulting in the same increase of the average transverse momentum of particles < > between the RHIC and LHC energies.

As was first noted in Ref. [314], at high energies, only the interactions among the constituents of matter formed in the initially spatially deformed overlap can produce . A transfer of this spatial deformation into momentum space provides a unique signature for re-interactions in the fireball, and proves that the matter has undergone significant nontrivial dynamics between its creation and its freeze-out [183]. The rapid degradation of the initial spatial deformation due to re-scattering causes the "self-quenching" of elliptic flow: if the elliptic flow does not develop early (when the collision fireball was still spatially deformed), it does not develop at all [183]. In particular, the transformation of the rapidly expanding ideal gas of non-interacting quarks and gluons into strongly interacting hadrons is unable to produce a sufficient elliptic flow. The elliptic flow thus reflects the pressure due to re-scattering—the induced expansion and stiffness of the EoS during the earliest collision stages. Its continuous rise with the energy up to its highest value at the LHC indicates that the early pressure also increases.

The energy dependence of the integrated triangular flow coefficient of charged hadrons is shown on the left panel of Figure 20 in four bins of centrality, 0%–5%, 10%–20%, 30%–40%, and 50%–60%. As is sensitive to the fluctuations in the initial matter distribution, it is interesting to observe that at = 7.7 and 11.5 GeV, values of for 50%–60% centrality become consistent with zero. For more central collisions, however, is finite—even at the lowest energies—and changes very little from 7.7 GeV to 19.6 GeV. Above that, it begins to increase more quickly, and roughly linearly with . Generally, one would expect that higher energy collisions producing more particles should be more effective at converting the initial state geometry fluctuations into . Deviations from that expectation could indicate interesting physics, such as a softening of the EoS [113,114] discussed already in Section 3.3. This can be investigated by scaling by the charged particle rapidity density per participating NN pair, = /(0.5), see the left panel of Figure 20. A local minimum of / in the region near 15–20 GeV observed in the centrality range 0%–50% and absent in the more peripheral events could indicate an interesting trend in the pressure developed inside the system.

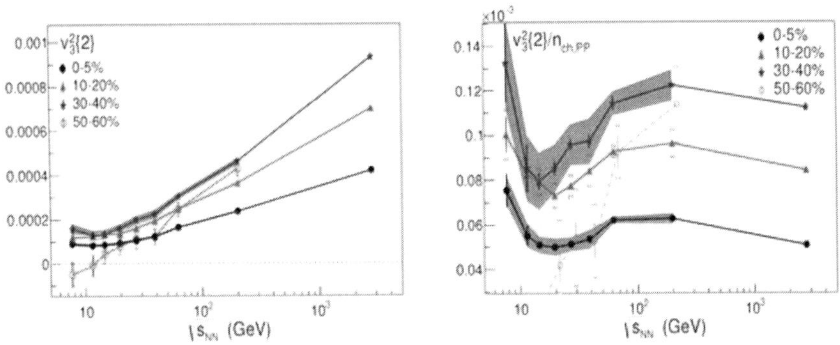

Figure 20. Left: The energy dependence of for four centrality bins. Points at 2.76 TeV corresponding to Pb + Pb [317]; **Right**: divided by pair. Reproduced from Ref. [319].

As was already briefly mentioned in Section 4.4.4, recent years have witnessed a growth of interest in studying collective phenomena on event-by-event basis. In addition to the key publications [165,166,205,206,207,208,300,301,302,303], it is worth mentioning the study [320] where the anisotropic flow coefficients – were computed by combining the IP-Glasma flow with the subsequent relativistic viscous hydrodynamic evolution of matter through the quark–gluon plasma and hadron gas phases. The event-by-event geometric fluctuations in nucleon positions and intrinsic color charge fluctuations at the sub-nucleon scale are expected to result in experimentally measurable event-by-event anisotropic flow coefficients [321]. Let us note that fluctuating initial profiles observed in over-many-events integrated triangular and higher odd flow coefficients are also revealed in difference between various participant plane angles introduced in Equation (66). Correlations between the event plane angles of different harmonic order can not only yield valuable additional insights into the initial conditions [322], but are also experimentally measurable [323]. The same is also true for the correlations between different flow harmonics [324,325,326,327].

Enhanced production of hadrons with the quantum numbers not present in colliding matter is one of the oldest signals of the deconfined QGP medium [307,308]. Measurements of the yields of multistrange baryons were carried out at CERN SPS by WA85, and later on by WA97/NA57 collaborations since the mid-eighties. After 2000, more data came from

RHIC, and starting from 2010 also from the LHC. The current status is summarized in the five panels of Figure 21. In the top left and middle panels (Figure 21a,b), a compilation of the results from SPS, RHIC, and the LHC in terms of *strangeness enhancement* defined as normalized (to *p* + *p* or *p* + Be) yield per participants is presented. On the top right (Figure 21c), the hyperon-to-pion ratios as functions of for Pb + Pb, Au + Au, and *p* + *p* collisions at the LHC and RHIC energies are displayed. The normalized yields are larger than unity for all the particles, and increase with their strangeness content. This behaviour is consistent with the picture of enhanced pair production in a hot and dense QGP medium [307,308].

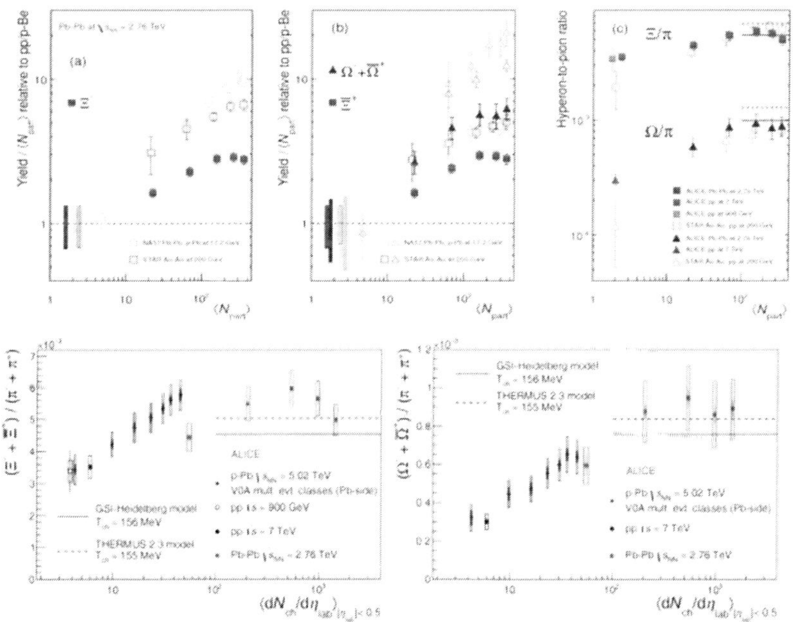

Figure 21. Top: (**a**,**b**) Normalized (to *p* + *p* or *p* + Be) yields of multistrange baryons per participants at midrapidity as a function of for the LHC (full symbols), RHIC, and SPS (open symbols) data. Boxes on the dashed line at unity indicate the statistical and systematic uncertainties on the *p* + *p* or *p* + Be reference; (**c**) Hyperon-to-pion ratios as a function of for Pb + Pb, Au + Au, and *p* + *p* collisions at LHC and RHIC energies. The lines are the thermal model [309] (full line) and [328] (dashed line) predictions. Reproduced from Ref. [329]; **Bottom**: (**left**) (+)/(+) and (**right**) (+)/(+) ratios as functions of for *p* + *p*, *p* + Pb, and Pb + Pb collisions at the LHC. The Pb + Pb data points [329] represent, from left to right, the 60%–80%, 40%–60%, 20%–40%, 10%–20%, and 0%–10% centrality classes. The chemical equilibrium predictions by the GSI-Heidelberg [309] and the THERMUS 2.3 [328] models are represented by the horizontal lines. Reproduced from Ref. [330].

The two bottom plots represent a comparison between the hyperon-to-pion ratios from $p+p$, $p+$ Pb, and Pb + Pb collisions. Interestingly, the ratios in $p+$ Pb collisions increase with multiplicity from the values measured in $p+p$ to those observed in Pb + Pb. The rate of increase is more pronounced for particles with higher strangeness content. Let us note that the Grand canonical statistical description of Pb + Pb data shown as full and dashed lines in Figure 21 may not be appropriate in small multiplicity environments such as those produced in the $p+$ Pb case. It appears that for the latter case, the evolution of hyperon-to-pion ratios with the event multiplicity is qualitatively well described by the Strangeness Canonical model implemented in THERMUS 2.3 [328]. In this case, a local conservation law is applied to the strangeness quantum number within a correlation volume , while treating the baryon and charge quantum numbers grand-canonically within the whole fireball volume V [330].

5.2. Hard Probes

Heavy quarks, quarkonia, and jets—commonly referred to as *hard probes*—are created in the first moments after the collision, and are therefore considered as key probes of the deconfined QCD medium. Production of these high transverse momentum ( ) objects occurs over very short time scales ($\tau \approx \approx 0.1$), and can thus probe the evolution of the medium. Since the production cross-sections of these energetic particles are calculable using pQCD, they have been long recognised as particularly useful "tomographic" probes of the QGP [331,332,333].

Let us start our discussion with the results on the inclusive production of high- hadrons. The latter are interesting on their own because it was there where for the first time the suppression pattern was observed [14,15,16,17]. In an inclusive regime, the comparison between / (the differential yield of high- hadrons or jets per event in A+B collisions) to that in $p+p$ collisions is usually quantified by introducing the nuclear modification factor

$$R_{AB}(p_T,\eta) = \frac{dN^2_{AB}(p_T)/dp_T d\eta}{\langle N_{\mathrm{coll}}\rangle\, dN^2_{pp}(p_T)/dp_T d\eta}\,. \quad (67)$$

For collisions of two nuclei behaving as a simple superposition of nucleon–nucleon collisions, the nuclear modification factor would be = 1. The data of Figure 22 reveal a very different behaviour. The left panel shows a compilation of from Au + Au and Pb + Pb collisions, the right panel the result of from three LHC experiments at the same energy = 5.02

TeV. In the case, the suppression pattern of high- (>2–3 GeV/c) hadrons in the deconfined medium—predicted many years ago [331,332,333] as a *jet quenching effect*—is clearly visible at RHIC and the LHC. However, for proton–nucleus collisions (Figure 22, right panel), no suppression is seen, even at the highest LHC energy. Moreover, in the 5% most central Pb + Pb collisions at the LHC shows a maximal suppression by a factor of 7–8 in the region of 6–9 GeV. This dip is followed by an increase, which continues up to the highest measured at = 5.02 TeV, and approaches unity in the vicinity of = 200 GeV [334].

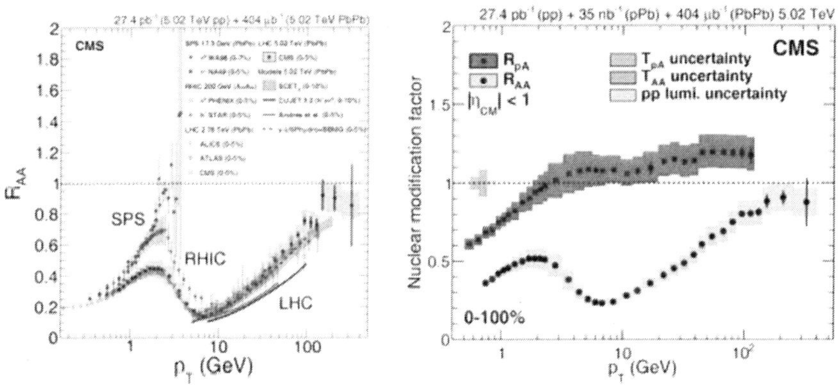

Figure 22. Left: A compilation of the measurements of ( ) for neutral pions ( ), charged hadrons ( ), and charged particles in central heavy-ion collisions at SPS, RHIC, and the LHC. Reproduced from Ref. [334]; **Right**: The nuclear modification factor ( ) in *p* + Pb compared to from Pb + Pb collisions at = 5.02 TeV at the LHC. Reproduced from Ref. [334].

5.2.1. High- Hadrons and Jets

The suppression of high- hadrons in the deconfined medium was thoroughly studied at RHIC using azimuthal correlations between the trigger particle and associated particle; see Figure 23. Near-side peaks in central (0%–5%) Au + Au collisions present in all panels of Figure 23 (left) indicate that the correlation is dominated by jet fragmentation. An away-side peak emerges as is increased. The narrow back-to-back peaks are indicative of the azimuthally back-to-back nature of dijets observed in an elementary parton–parton collision. Contrary to the latter, the transverse-momentum imbalance of particles from the jet fragmentation due to different path lengths of two hard partons in the medium is apparent. The azimuthal angle difference for the highest range (8 < < 15 GeV/c) for mid-central (20%–40%) and central Au+Au collisions—as well as for d+Au collisions—is presented in Figure 23 (right panel). We observe narrow correlation peaks in all three ranges. For each , the nearside peak shows a similar correlation strength above background for the three systems, while the away-side correlation strength decreases from d + Au to central Au + Au. For the d + Au case, the yield of particles on the opposite side = π prevails over the same side. Moreover, for Au + Au collisions, the nearside yields obtained after subtraction of the background contribution due to the elliptic flow show a little centrality dependence, while the away-side yields decrease with increasing centrality [335].

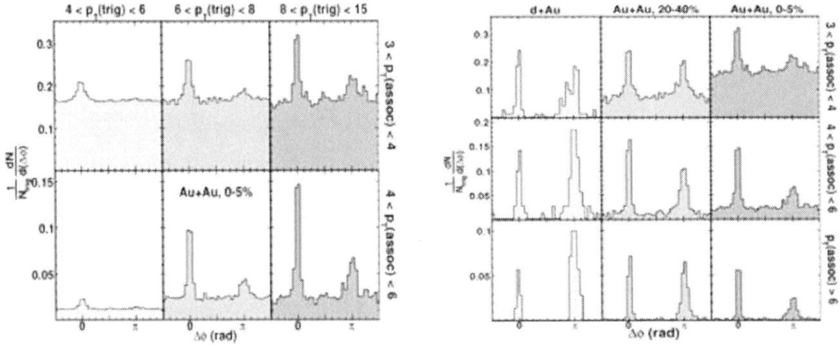

Figure 23. Left: Azimuthal correlation histograms of high- charged hadron pairs normalized per trigger particle for 0%–5% Au + Au events for various and ranges. The yield in the lower left panel is suppressed due to the constraint ; **Right**: The same but for GeV/c, for d + Au, 20%–40% Au + Au and 0%–5% Au + Au events. Reproduced from Ref. [335].

Unfortunately, the advantage of the large yield of dijets is offset by a loss of information about the initial properties of the probes (i.e., prior to their interactions with the medium). Correlating two probes that both undergo an energy loss also induces a selection bias towards scatterings occurring at—and oriented tangential to—the surface of the medium. It is thus interesting to study correlations when one of the particles does not interact strongly with the medium. Triggering on the high- isolated photon (i.e., not from decays) would do the job. While in p + p collisions an emerging quark jet should balance its transverse momentum with the photon, in the heavy-ion collisions, much of its momentum is thermalized while the quark traverses the plasma. This is illustrated in Figure 24 (left panel), where a single hard photon with = 402 GeV emerges unhindered from the de-confined medium produced in Pb+Pb collisions at the LHC. The accompanying quark jet produced via the QCD Compton scattering loses 1/3 of its energy (≈140 GeV!) inside the hot and dense matter.

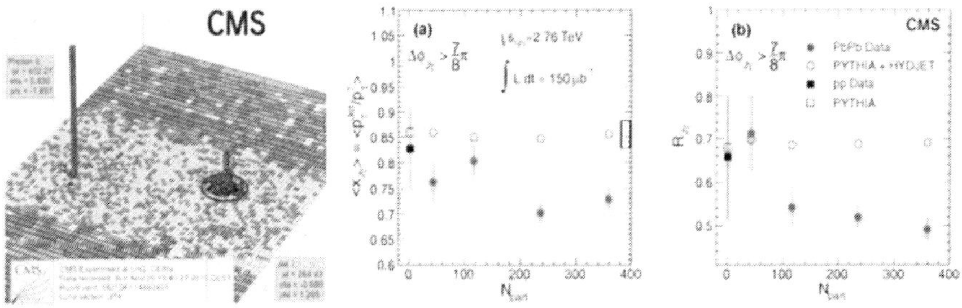

(a) (b) (c)

Figure 24. (**a**): The energy distribution in the plane in a single Pb + Pb event recorder by the CMS detector at LHC (reproduced from Ref. [336]); (**b**) and (**c**): The average ratio of jet transverse momentum to photon transverse momentum and the average fraction R of isolated photons with an associated jet with energy above 30 GeV as functions of number of participants . Photons and jets are emitted almost back-to-back in azimuth . Reproduced from Ref. [337].

The measurement presented on the middle and right panels of Figure 24 shows that for more central Pb + Pb collisions, a significant decrease in the ratio of jet transverse momentum to photon transverse momentum— —relative to the PYTHIA reference [200] is observed. Furthermore, significantly more photons with > 60 GeV/c in Pb + Pb are observed to not have an associated jet with > 30 GeV/c jet, compared to the reference. However, no significant broadening of the photon + jet azimuthal correlation has been observed.

An important progress in the theoretical understanding of the suppression of energetic partons traversing a deconfined matter was the introduction of the diffusion coefficient relevant for the transverse momentum broadening and collisional energy loss of partons (jets) [338]. This quantity, which is commonly referred to as the jet quenching parameter, can be determined either via weak coupling techniques [339,340,341], a combination of lattice simulations and dimensionally-reduced effective theory [342], or from the gauge/gravity duality [343]. Typical estimates for this quantity at RHIC and LHC energies range between 5 and 10 GeV /fm, demonstrating the currently still sizable uncertainties in these calculations.

5.2.2. Quarkonia

Melting of the quarkonia—bound states of heavy quark and anti-quark where q = —due to a colour screening in the deconfined hot and dense medium was proposed thirty years ago as a clear and unambiguous signature of the deconfinement [115]. However, shortly after that, it was noticed that not only diffusion of the heavy quarks from melted quarkonium, but also the drag which charm quarks experience when propagating through the plasma is important [344].

The latter might lead to an enhancement instead of a suppression. This is in variance with the original proposal that the heavy quarks, once screened, simply fly apart. With the advent of the strongly-interacting QGP, the Langevin equation model of quarkonium production was formulated, where the charm quark–antiquark pairs evolve on top of a hydrodynamically expanding fireball [345]. A heavy quark and antiquark interact with each other according to the screened Cornell potential and interact, independently, with the surrounding medium, experiencing both drag and rapidly decorrelating random forces. An extension of this approach to bottomonium production [346] shows that a large fraction of pairs that were located sufficiently close together during the initial hard production will remain correlated in the hot medium for a typical lifetime of the system created in heavy-ion collisions. The distribution of the correlated pair in relative distance is such that it will dominantly form 1S bottomonium. A study of quarkonia production in heavy-ion collisions thus provides an interesting window not only into static, but also into dynamical properties of the hot, dense, and rapidly expanding medium [116,306].

On the left panels of Figure 25, the invariant-mass distributions of pairs (di-muons) produced in the $p + p$ (a) and Pb + Pb (b) collisions at the LHC are presented. A prominent peak due to the production of the heavy quarkonium state, the bottomonium , can be clearly seen in both $p + p$ and Pb + Pb data. Peaks from the higher excited states of Υ, , and , although discernible in the $p + p$ case, are barely visible in the Pb + Pb data. More quantitative information on this effect can be found in the right panels of Figure 25, where the centrality dependence of the double ratio (top) and of the nuclear modification factors of and (bottom) are displayed. Let us note that the observed suppression of the relative yield is in agreement with the expectations that different quarkonium states will dissociate at different temperatures with a suppression pattern ordered sequentially with the binding energy; i.e., the difference between the mass of a given quarkonium and twice the mass of the lightest meson containing the corresponding heavy quark [347]. Moreover, the observed pattern is now also confirmed in Pb + Pb collisions at = 5.02 TeV [348]. The double ratio is significantly below unity at all centralities, and no variation with kinematics is observed, confirming a strong Υ suppression in heavy-ion collisions at the LHC.

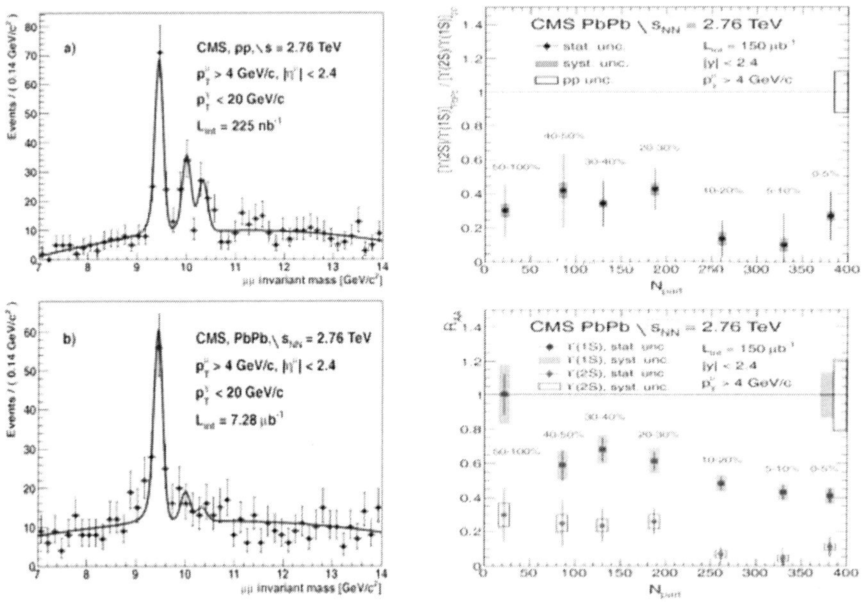

Figure 25. Left: The di-muon invariant-mass distributions from the (**a**) $p + p$ and (**b**) Pb + Pb data at = 2.76 TeV; **Right**: The centrality dependence (**top**) of the double ratio and (**bottom**) of the nuclear modification factors for the and states. The event centrality bins are indicated by percentage intervals. Reproduced from Ref. [349].

5.3. Penetrating Probes

Electromagnetic probes like photons [350,351] and di-leptons [352] (for recent developments, see Refs. [253,254,255]) have long been expected to provide crucial information on the properties of QGP. The absence of strong final-state interactions makes them an ideal *penetrating probe* of strongly-interacting matter [353]. In collisions of ultra-relativistic nuclei, the photons and leptons can be produced either in the initial hard collisions between partons of the incident nuclei (e.g., , , or), or radiated from the thermally equilibrated partons and hadrons or via hadronic decays. The direct photons are defined to be all produced photons, except those from hadron decays in the last stage of the collision. The high- isolated photon can be used to estimate the momentum of the associated parton, allowing a characterisation of the in-medium parton energy loss (see Figure 24). The prompt photons also carry information about the initial state and its possible modifications in nuclei, and should thus be one of the best probes of the gluon saturation. The thermal photons emitted from the produced matter in nuclear collisions carry information on the temperature of QGP.

The first observation of direct photons in ultra-relativistic heavy-ion collisions has been made by the CERN SPS experiment WA98 [354]. In 10% most central Pb + Pb collisions at = 17.2 GeV, they observed a clear excess of direct photons in the range of > 1.5 GeV/c which was not present in more peripheral collisions. The extraction of the direct photon signal (which is extremely difficult) was described in-depth in Ref. [355]. The data from RHIC and the LHC are presented in Figure 26 on the left and right panels, respectively.

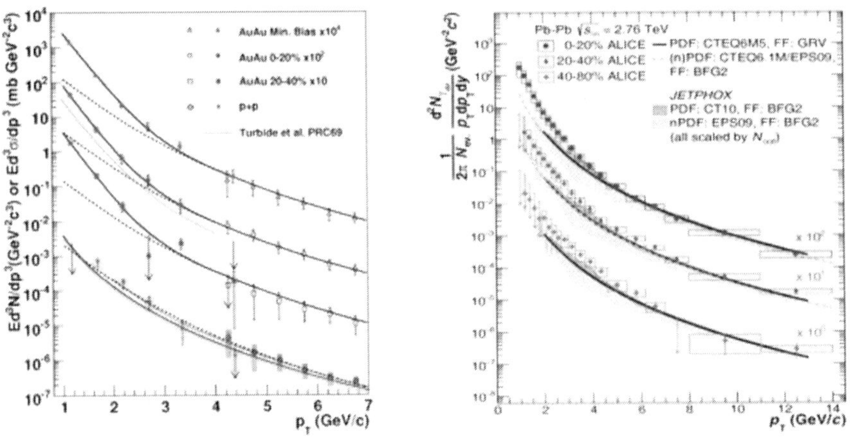

Figure 26. Left: The invariant cross-section ($p + p$) and invariant yield (Au + Au) of direct photons as functions of . The three curves on the $p + p$ data represent the NLO pQCD calculations, and the dashed curves show a modified power-law fit to the $p + p$ data, scaled by . The dashed (**black**) curves are the same, but with the exponential plus scaled $p + p$ data. The dotted (red) curve near the 0%–20% centrality data is a theory calculation. Reproduced from Ref. [356]; **Right**: The direct photon spectra in Pb + Pb collisions at = 2.76 TeV for the 0%–20% (scaled by a factor of 100), the 20%–40% (scaled by a factor of 10), and the 40%–80% centrality classes compared to the NLO pQCD predictions for the direct photon yield in collisions at the same energy, scaled by a number of binary nucleon collisions for each centrality class. Reproduced from Ref. [357].

6. NEW DEVELOPMENTS

6.1. Search for the Critical Point of QCD Phase Diagram

The search for the (tri)critical point (CEP) in the phase diagram—where the phase transition between the QGP and hadron matter changes from the first- to the second-order one—represents one of the most active fields of contemporary ultra-relativistic heavy-ion physics, both experimentally [59,358,359,360] and theoretically [58,63,361,362]. In order to gain more insight into the CEP location, quite advanced techniques from condensed matter physics, such as Finite-Size Scaling (FSS) analysis of data [363] or thermal fluctuations characterized by the appropriate cumulants of the partition function [361,362,364] are being exploited.

The search for the CEP exploiting the potential of the RHIC accelerator complex was mounted by the STAR and PHENIX collaborations in 2010 within the *Beam Energy Scan* (BES) program [59]. Going down from the RHIC maximum energy = 200 GeV, they have scanned the available phase space down to = 7.7 GeV. Some of the results from that scan were already mentioned in previous sections, and can be found in Figure 8, Figure 10, Figure 17, Figure 19 and Figure 20. In the following, we therefore restrict ourselves to the measurements which provide a direct link to the lattice results discussed in Section 3.4. As already mentioned there, the study of ratios of the Taylor expansion coefficients given by Equation (31)—which are also known also as susceptibilities—seems to be very attractive, since both the temperature and volume dependences drop out. In particular, the ratio calculated from the moments of the net-baryon multiplicity has different values for the hadronic and partonic phases [364]. For HRG, it equals unity, but is expected to deviate from unity near the CEP. Other interesting ratios of the net-baryon charge moments which can be expressed using mean ( ), variance ( ), skewness ( ), and kurtosis ( ) of the net baryon number distributions read

$$R_{ij}^{B} \equiv \frac{\chi_i^B}{\chi_j^B} \;,\; R_{12}^{B} = \frac{M_B}{\sigma_B^2} \;,\; R_{31}^{B} = \frac{S_B \sigma_B^3}{M_B} \;,\; R_{42}^{B} = k_B \sigma_B^2 \,. \tag{68}$$

Experimentally, the net-baryon number fluctuations and their cumulants are not accessible, and so one has to resort to measurements of the cumulants of the net-proton number fluctuations [365,366]. On the other hand, the electric charge fluctuations are experimentally accessible [367]. This is illustrated in Figure 27, where the measurements from the

STAR experiment at RHIC are shown. Generally, study of the and rapidity acceptance dependence for the moments of the net-proton distributions shows that the larger the acceptance is, the larger are the deviations from unity. Preliminary results from the STAR collaboration [368] reveal that increasing the transverse momentum acceptance of protons and anti-protons from 0.4 < < 0.8 GeV/c to 0.4 < < 2.0 GeV/c leads to a pronounced non-monotonic structure in the energy dependence of of net-proton distributions from 5% most central Au + Au collisions. At energies above 39 GeV, the values of are close to unity, while for energies below 39 GeV, it shows a significant deviation below unity around 19.6 and 27 GeV and a large increase above unity observed at 7.7 GeV.

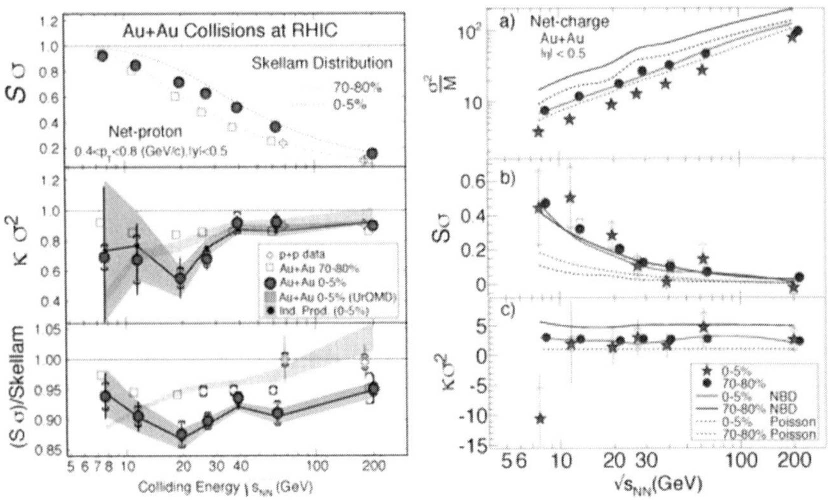

Figure 27. The energy dependence of appropriate combinations of the moments—mean (M), variance ( ), skewness (S), and kurtosis (κ)—of the multiplicity of conserved charges at mid-rapidity in Au + Au collisions for seven energies ranging from = 7.7 to 200 GeV, and for two centralities, the most central (0%–5%) and peripheral (70%–80%) bins. **Left**: The collision energy and centrality dependence of the net-proton and from Au + Au and $p + p$ collisions at RHIC (adapted from Ref. [365]); **Right**: The beam-energy dependence of the net-charge multiplicity distributions moments **(a)**, **(b)**, and **(c)**, after all corrections. Reproduced from Ref. [367]. The results from the Poisson and the Negative Binomial Distribution (NBD) baselines are superimposed. The values of for the Poisson baseline are always unity.

6.2. Collectivity in Small Systems

Recent years have witnessed a surprising development in multiparticle dynamics of high multiplicity $p + p$ [369,370,371] and $p +$ A [372,373,374,375,376,377] collisions. It all started in 2010 with the observation of *ridge-like* structures in $p + p$ collisions by the CMS experiment at the LHC [369]. The surprise was due to the fact that a very similar effect was found just a few years before in heavy-ion collisions: first in Au + Au collisions at = 200 GeV at RHIC [378,379], and later on also in Pb + Pb collisions at the LHC [380] and in Cu + Cu collisions at = 62.4 GeV and 200 GeV at RHIC [381].

In heavy-ion collisions, it was found that pairs of particles are preferentially emitted with small relative azimuthal angles ( ). Surprisingly, this preference persists even when the particles are separated by large pseudo-rapidity (η) gaps ( ). These long-range correlations—known as the ridge—have been traced to the conversion of density anisotropies in the initial overlap of the two nuclei into momentum space correlations through subsequent interactions in the expansion [382].

In $p + p$ minimum bias collisions at the LHC, the peak in the correlation function of particles with > 0.1 GeV/c observed at small angular differences (, ≈ 0, see Figure 28a) is due to several effects: resonance decays, Bose–Einstein correlations, and near-side jet fragmentation. The fragmentation due to back-to-back jets is visible as a broad elongated ridge around ≈ π. The pattern does not change much even when selecting the events with very high multiplicity (see Figure 28c). The cut on the multiplicity enhances the relative contribution of high jets, which fragment into a large number of particles, and therefore, has a qualitatively similar effect on the shape as the particle cut 1 < < 3 GeV/c on minimum bias events (see Figure 28b). However, using now the particle cut 1 < < 3 GeV/c in conjunction with a high multiplicity cut changes the picture dramatically (see Figure 28d). A novel feature never seen before in $p + p$collisions at lower energies shows up—a clear and significant ridge-like structure at ≈ 0 extending to of at least four units [369].

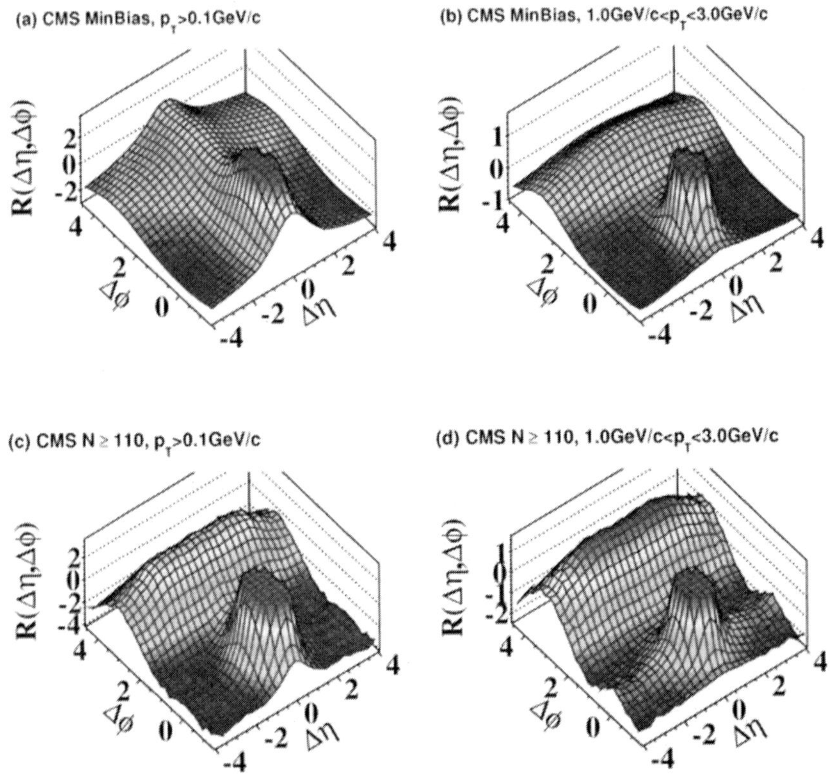

Figure 28. The 2-D two-particle correlation functions for $p + p$ at = 7 TeV: (**a**) minimum bias events with > 0.1 GeV/c; (**b**) minimum bias events with 1 < < 3 GeV/c; (**c**) high multiplicity ( ) events with > 0.1 GeV/c; and (**d**) high multiplicity ( ) events with 1 < < 3 GeV/c. The sharp near-side peak from jet correlations is cut off in order to better illustrate the structure outside that region. Reproduced from Ref. [369].

Let us note that for two particles with approximately the same energy ≈ E, the correlations at ≈ are by the uncertainty relation connected to the correlations in coordinate space. While for pions with ≈ 1 and ≈ 0.1 GeV/c, we have ≈ 1 fm, for ≈ 4 and ≈ 1 GeV/c, one gets ≈ 0.02 fm. It is obvious that at such small inter-parton distances, one enters the realm of the initial state description of nuclear collisions when the density of matter even inside the proton fluctuates—see Section 4.4.3. The CGC scenario was recently exploited in Ref. [383] to predict long-range photon-jet correlations in $p + p$ and p + A collisions at near-side for low transverse momenta of the produced photon and jet in high-multiplicity events.

The relevance of the saturation approach is further supported by the observation of the same ridge-like structure in $p + p$ collisions at = 13 TeV [370,371], see Figure 29. For the associated yield of long-range near-side correlations for high-multiplicity events (N >110) peaks in the region GeV/c, see Figure 29a —the yield reaches a maximum around ≈ 1 GeV/c and decreases with increasing . No center-of-mass energy dependence is visible. The multiplicity dependence of the associated yield for GeV/c particle pairs is shown in Figure 29b. For low-multiplicity events, the associated yield is consistent with zero. At higher multiplicity, the ridge-like correlation emerges, with an approximately linear rise of the associated yield with multiplicity for . Let us note that within the CGC models, the observation that the integrated near-side yield as a function of multiplicity is independent of collision energy is a natural consequence of the fact that multiparticle production is driven by a single semi-hard saturation scale [384].

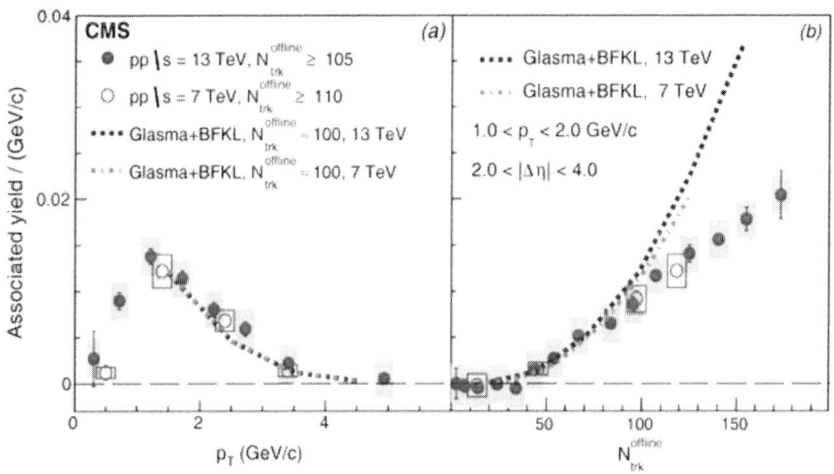

Figure 29. The associated yield for the near-side of the correlation function averaged over for $p + p$ data at = 13 TeV (filled circles) and 7 TeV (open circles) [370]. Panel (**a**) shows the associated yield as a function of for events with . In panel (**b**), the associated yield for 1 < < 2 GeV/c is shown as a function of multiplicity— . The selection applies to both particles in each pair. Curves represent the predictions of the gluon saturation model [384].

Another interesting phenomenon observed during recent years is a flow-like pattern in super-central $p + p$ and $p +$ Pb collisions at the LHC [373,374,385,386,387,388] and in d + Au, He + Au, and + Au collisions at RHIC [376,377]. These collisions not only reveal a similar elliptic flow , but in some cases, also anisotropy previously observed only in collisions of large nuclei. The measurement of higher-order cumulants of the azimuthal distributions has strengthened the collective nature interpretation of the anisotropy seen in p + Pb collisions. Moreover, the collective radial flow analysis of $p + p$ data enabled authors of Ref. [389] to claim that in this case the fireball explosions start with a very small initial size, well below 1 fm. This raises questions about whether the perfect liquid sQGP is also formed in these much smaller systems [390,391,392,393,394,395]. Are these flow-like structures only similar in appearance to what one observes in heavy-ion collisions, or do they have the same physical origin [396,397,398]? Obviously, answering these questions may help us to understand the emergence of collective phenomena in strongly-interacting systems in general. To make further progress on these fundamental questions, more analyses—both experimental and theoretical—are really needed [399].

7. CONCLUSIONS

In this review, we have tried to present an up-to-date phenomenological summary of a relatively new and rapidly developing field of contemporary physics—the physics of Quark Gluon Plasma. We have also explored its broader ramifications when discussing matter under extreme conditions, strongly interacting plasmas, physics of strong electromagnetic fields, history of ultra-relativistic heavy-ion collisions, relativistic hydrodynamics, the role of the QCD ground state, and QCD saturation phenomena. In the experimental part, we were overwhelmed by a huge amount of results, and so, in order to keep this review of tolerable length, we had to skip several quite important topics. To name just a few—flow and suppression of identified particles [30,400], femtoscopy [401,402], or identified particle yields and chemical freeze-out conditions [29,30,309]. We hope that the interested reader will find in the given references enough information to pursue a deeper study of these important subjects. However, in skipping these topics, we hope that we have not given up our main goal—to give the reader a possibility to see the QGP landscape at large. At last, we would like to say that it is never enough to stress how important this field is also for other branches of physics, and so we finish with yet another argument. The study of ultra-relativistic heavy-ion collisions appears so far to be our only way of studying the phase transitions in non-Abelian gauge theories (most likely taken place in the early universe) under laboratory conditions.

ACKNOWLEDGMENTS

This work was supported in part by the Swedish Research Council, contract number 621-2013-428, the grant LG 13031 of the Ministry of Education of the Czech Republic and by the grant 13-20841S of the Czech Science Foundation (GACR).

CONFLICTS OF INTEREST

The authors declare no conflict of interest.

REFERENCES

1. Collins, J.C.; Perry, M.J. Superdense Matter: Neutrons or Asymptotically Free Quarks? *Phys. Rev. Lett.* **1975**, *34*, 1353. [] []
2. Cabibbo, N.; Parisi, G. Exponential hadronic spectrum and quark liberation. *Phys. Lett. B* **1975**, *59*, 67–69. [] []
3. Gell-Mann, M. A Schematic Model of Baryons and Mesons. *Phys. Lett.* **1964**, *8*, 214–215. [] []
4. Zweig, G. An SU(3) Model for Strong Interaction Symmetry and Its Breaking. Version 1. CERN-TH-401. Available online: http://inspirehep.net/record/11881/files/CM-P00042883.pdf (accessed on 17 January 2017).
5. Fritzsch, H.; Gell-Mann, M.; Leutwyler, H. Advantages of the Color Octet Gluon Picture. *Phys. Lett. B* **1973**, *47*, 365–368. [] []
6. Gross, D.J.; Wilczek, F. Ultraviolet Behavior of Nonabelian Gauge Theories. *Phys. Rev. Lett.* **1973**, *30*, 1343–1346. [] []
7. Politzer, H.D. Reliable Perturbative Results for Strong Interactions? *Phys. Rev. Lett.* **1973**, *30*, 1346–1349. [] []
8. Nambu, Y. The Confinement of Quarks. *Sci. Am.* **1976**, *235*, 48–70. [] []
9. Yagi, K.; Hatsuda, T.; Miake, Y. *Quark-Gluon Plasma: From Big Bang to Little Bang (Cambridge Monographs on Particle Physics, Nuclear Physics and Cosmology)*; Cambridge University Press: Cambridge, UK, 2005. []
10. Schmidt, H.R.; Schukraft, J. The Physics of ultrarelativistic heavy ion collisions. *J. Phys. G* **1993**, *19*, 1705–1795. [] []
11. Stock, R. Nucleus-Nucleus Collisions and the QCD Matter Phase Diagram. In *Landolt-Börnstein—Group I Elementary Particles, Nuclei and Atoms, Vol. 21A, Chapter 7*; Springer: Berlin/Heidelberg, Germany, 2009. []
12. Schukraft, J.; Stock, R. *Toward the Limits of Matter: Ultra-Relativistic Nuclear Collisions at CERN*; World Scientific: Singapore, 2015. []
13. Heinz, U.W.; Jacob, M. *Evidence for a New State of Matter: An Assessment of the Results from the CERN Lead Beam Program*; Theoretical Physics Division: Geneva, Switzerland, 2000. []
14. Arsene, I.; Bearden, I.G.; Beavis, D.; Besliu, C.; Budick, B.; Bøggild, H.; Chasman, C.; Christensen, C.H.; Christiansen, P.; Cibor, J.; et al. Quark gluon plasma and color glass condensate at RHIC? The Perspective from the BRAHMS experiment. *Nucl. Phys. A* **2005**, *757*, 1–27. [] []

15. Back, B.B.; Baker, M.D.; Ballintijn, M.; Barton, D.S.; Becker, B.; Betts, R.R.; Bickley, A.A.; Bindel, R.; Budzanowski, A.; Busza, W.; et al. The PHOBOS perspective on discoveries at RHIC. *Nucl. Phys. A* **2005**, *757*, 28–101. [] []
16. Adams, J.; Aggarwal, M.M.; Ahammed, Z.; Amonett, J.; Anderson, B.D.; Arkhipkin, D.; Averichev, G.S.; Badyal, S.K.; Bai, Y.; Balewski, J.; et al. Experimental and theoretical challenges in the search for the quark gluon plasma: The STAR Collaboration's critical assessment of the evidence from RHIC collisions. *Nucl. Phys. A* **2005**, *757*, 102–183. [] []
17. Adcox, K.; Adler, S.S.; Afanasiev, S.; Aidala, C.; Ajitanand, N.N.; Akiba, Y.; Al-Jamel, A.; Alexander, J.; Amirikas, R.; Aoki, K.; et al. Formation of dense partonic matter in relativistic nucleus-nucleus collisions at RHIC: Experimental evaluation by the PHENIX collaboration. *Nucl. Phys. A* **2005**, *757*, 184–283. [] []
18. Gyulassy, M.; McLerran, L. New forms of QCD matter discovered at RHIC. *Nucl. Phys. A* **2005**, *750*, 30–63. [] []
19. Kapusta, J.; Müller, B.; Rafelski, J. *Quark-Gluon Plasma: Theoretical Foundations. An Annotated Reprint Collection*; Elsevier: Amsterdam, The Netherlands, 2003. []
20. Csernai, L.P. *Introduction to Relativistic Heavy Ion Collisions*; Wiley: Chichester, UK, 1994. []
21. Letessier, J.; Rafelski, J. *Hadrons and Quark-Gluon Plasma*; Cambridge Monographs on Particle Physics, Nuclear Physics and Cosmology; Cambridge University Press: Cambridge, UK, 2002. []
22. Vogt, R. *Ultrarelativistic Heavy-Ion Collisions*; Elsevier Science: Amsterdam, The Netherlands, 2007. []
23. Florkowski, W. *Phenomenology of Ultra-Relativistic Heavy-Ion Collisions*; World Scientific: Singapore, 2010. []
24. Rak, J.; Tannenbaum, M.J. *High pT Physics in the Heavy Ion Era*; Cambridge Monographs on Particle Physics, Nuclear Physics and Cosmology; Cambridge University Press: Cambridge, UK, 2013. []
25. Braun-Munzinger, P.; Stachel, J. The quest for the quark-gluon plasma. *Nature* **2007**, *448*, 302–309.
26. Shuryak, E. Physics of Strongly coupled Quark-Gluon Plasma. *Prog. Part. Nucl. Phys.* **2009**, *62*, 48–101. [] []
27. Sarkar, S.; Satz, H.; Sinha, B. The physics of the quark-gluon plasma. *Lect. Notes Phys.* **2010**, *785*, 1–369. []
28. Florkowski, W. Basic phenomenology for relativistic heavy-ion collisions. *Acta Phys. Pol. B* **2014**, *45*, 2329–2354. [] []
29. Roland, G.; Safarik, K.; Steinberg, P. Heavy-ion collisions at the LHC. *Prog. Part. Nucl. Phys.* **2014**, *77*, 70–127. [] []
30. Nouicer, R. New State of Nuclear Matter: Nearly Perfect Fluid of Quarks and Gluons in Heavy Ion Collisions at RHIC Energies. *Eur. Phys. J. Plus* **2016**, *131*, 70. [] []
31. Zeldovich, Y.B.; Raizer, Y.P. *Physics of Shock Waves and High-Temperature Hydrodynamics Phenomena*; Academic Press: New York, NY, USA; London, UK, 1966. []
32. Fortov, V.I. *Extreme States of Matter on Earth and in the Cosmos*; Springer: Berlin/Heidelberg, Germany, 2011. []
33. Oppenheimer, J.R.; Volkoff, G.M. On Massive Neutron Cores. *Phys. Rev.* **1939**, *55*, 374–381. [] []
34. Landau, L.D. On the theory of stars. *Physik. Zeits. Sowjetunion* **1932**, *1*, 285–288. []
35. Zeldovich, Y.B. The equation of state at ultrahigh densities and its relativistic limitations. *J. Exp. Theor. Phys.* **1962**, *14*, 1143–1147. []
36. Baym, G.; Chin, S.A. Can a neutron star be a giant MIT bag? *Phys. Lett. B* **1976**, *62*, 241–244. [] []
37. Gamow, G. Expanding universe and the origin of elements. *Phys. Rev.* **1946**, *70*, 572–573. [] []

38. Penzias, A.A.; Wilson, R.W. A Measurement of excess antenna temperature at 4080-Mc/s. *Astrophys. J.* **1965**, *142*, 419–421. [] []
39. Sakharov, A.D. Maximum temperature of thermal radiation. *JETP Lett.* **1966**, *3*, 288–289. []
40. Huang, K.; Weinberg, S. Ultimate temperature and the early universe. *Phys. Rev. Lett.* **1970**, *25*, 895–897. [] []
41. Hagedorn, R. Statistical thermodynamics of strong interactions at high-energies. *Nuovo Cim. Suppl.* **1965**, *3*, 147–186. []
42. Rafelski, J. *Melting Hadrons, Boiling Quarks—From Hagedorn Temperature to Ultra-Relativistic Heavy-Ion Collisions at CERN: With a Tribute to Rolf Hagedorn*; Springer: Cham, Switzerland, 2016. []
43. Majumder, A.; Muller, B. Hadron mass spectrum from lattice QCD. *Phys. Rev. Lett.* **2010**, *105*, 252002. []
44. Zeldovich, Y.B. The hot model of the universe and the elementary particles. *Acta Phys. Hung.* **1967**, *22*, 51–58. [] []
45. Zeldovich, Y.B. Hot model of Universe. *Sov. Phys. Uspechi* **1967**, *9*, 602–617. [] []
46. Linde, A.D. Phase Transitions in Gauge Theories and Cosmology. *Rep. Prog. Phys.* **1979**, *42*, 389–437. [] []
47. Bailin, D.; Love, A. *Cosmology in Gauge Field Theory and String Theory*; IOP: Bristol, UK, 2004. []
48. Boyanovsky, D.; de Vega, H.J.; Schwarz, D.J. Phase transitions in the early and the present universe. *Ann. Rev. Nucl. Part. Sci.* **2006**, *56*, 441–500. [] []
49. Trodden, M. Electroweak baryogenesis. *Rev. Mod. Phys.* **1999**, *71*, 1463–1500. [] []
50. Kajantie, K.; Laine, M.; Rummukainen, K.; Shaposhnikov, M.E. Is there a hot electroweak phase transition at m(H) larger or equal to m(W)? *Phys. Rev. Lett.* **1996**, *77*, 2887–2890. []
51. Susskind, L. Lattice Models of Quark Confinement at High Temperature. *Phys. Rev. D* **1979**, *20*, 2610–2618. [] []
52. Petreczky, P. Lattice QCD at non-zero temperature. *J. Phys. G* **2012**, *39*, 093002. [] []
53. Aoki, Y.; Endrodi, G.; Fodor, Z.; Katz, S.D.; Szabo, K.K. The order of the quantum chromodynamics transition predicted by the standard model of particle physics. *Nature* **2006**, *443*, 675–678. [] []
54. Bhattacharya, T.; Buchoff, M.I.; Christ, N.H.; Ding, H.-T.; Gupta, R.; Jung, C.; Karsch, F.; Lin, Z.; Mawhinney, R.; McGlynn, G.; et al. QCD Phase Transition with Chiral Quarks and Physical Quark Masses. *Phys. Rev. Lett.* **2014**, *113*, 082001. []
55. Cheng, M.; Christ, N.H.; Datta, S.; van der Heide, J.; Jung, C.; Karsch, F.; Kaczmarek, O.; Laermann, E.; Mawhinney, R.D.; Miao, C.; et al. QCD equation of state with almost physical quark masses. *Phys. Rev. D* **2008**, *77*, 014511. [] []
56. Rischke, D. The Quark-Gluon Plasma in Equilibrium. *Prog. Part. Nucl. Phys.* **2004**, *52*, 197–296. [] []
57. McLerran, L.; Pisarski, R.D. Phases of Dense Quarks at Large N_c. *Nucl. Phys. A* **2007**, *796*, 83–100. [] []
58. Buballa, M. NJL-model analysis of dense quark matter. *Phys. Rept.* **2005**, *407*, 205–376. [] []
59. Aggarwal, M.M.; Ahammed, Z.; Alakhverdyants, A.V.; Alekseev, I.; Anderson, B.D.; Arkhipkin, D.; Averichev, G.S.; Balewski, J.; Barnby, L.S.; Baumgart, S.; et al. An Experimental Exploration of the QCD Phase Diagram: The Search for the Critical Point and the Onset of De-confinement. *arXiv*, 2010; arXiv:1007.2613. []
60. Itoh, N. Concept of quark matter was mentioned as early as 1970 by Itoh in the context of neutron stars. *Prog. Theor. Phys.* **1970**, *44*, 291–292. [] []
61. Shuryak, E.V. Theory of Hadronic Plasma. *Sov. Phys. JETP* **1978**, *47*, 212–219. []
62. Ichimaru, S. Strongly coupled plasmas: High-density classical plasmas and degenerate electron liquids. *Rev. Mod. Phys.* **1982**, *54*, 1017–1059. [] []

63. Braun-Munzinger, P.; Koch, V.; Schäfer, T.; Stachel, J. Properties of hot and dense matter from relativistic heavy ion collisions. *Phys. Rep.* **2016**, *621*, 76–126. [] []
64. Braun-Munzinger, P.; Wambach, J. The Phase Diagram of Strongly-Interacting Matter. *Rev. Mod. Phys.* **2009**, *81*, 1031–1050. [] []
65. Fukushima, K.; Hatsuda, T. The phase diagram of dense QCD. *Rept. Prog. Phys.* **2011**, *74*, 014001. [] []
66. Bailin, D.; Love, A. Superfluidity and Superconductivity in Relativistic Fermion Systems. *Phys. Rep.* **1984**, *107*, 325–385. [] []
67. Buballa, M.; Hosek, J.; Oertel, M. Anisotropic admixture in color superconducting quark matter. *Phys. Rev. Lett.* **2003**, *90*, 182002. [
68. Alford, M.G.; Rajagopal, K.; Wilczek, F. Color flavor locking and chiral symmetry breaking in high density QCD. *Nucl. Phys. B***1999**, *537*, 443–458. [] []
69. Becker, W. *Neutron Stars and Pulsars*; Springer: Berlin, Germany, 2009. []
70. Migdal, A.B. Pion Fields in Nuclear Matter. *Rev. Mod. Phys.* **1978**, *50*, 107–172. [] []
71. D'Enterria, D.; Ballintijn, M.; Bedjidian, M.; Hofman, D.; Kodolova, O.; Loizides, C.; Lokthin, I.P.; Lourenço, C.; Mironov, C.; et al. CMS physics technical design report: Addendum on high density QCD with heavy ions. *J. Phys. G* **2007**, *34*, 2307. []
72. Water Structure and Science. Available online: http://www1.lsbu.ac.uk/water/water_phase_diagram.html (accessed on 17 January 2017).
73. Shifman, M.A.; Vainshtein, A.I.; Zakharov, V.I. QCD and Resonance Physics. Theoretical Foundations. *Nucl. Phys. B* **1979**, *147*, 385–447. [] []
74. Shifman, M.A.; Vainshtein, A.I.; Zakharov, V.I. QCD and Resonance Physics: Applications. *Nucl. Phys. B* **1979**, *147*, 448–518. [] []
75. Schäfer, T.; Shuryak, E.V. Instantons in QCD. *Rev. Mod. Phys.* **1998**, *70*, 323–426. [] []
76. Diakonov, D. Instantons at work. *Prog. Part. Nucl. Phys.* **2003**, *51*, 173–222. [] []
77. Diakonov, D. Topology and confinement. *Nucl. Phys. Proc. Suppl.* **2009**, *195*, 5–45. [] []
78. Belavin, A.A.; Polyakov, A.M.; Schwartz, A.S.; Tyupkin, Y.S. Pseudoparticle Solutions of the Yang-Mills Equations. *Phys. Lett. B* **1975**, *59*, 85–87. [] []
79. Vainshtein, A.I.; Zakharov, V.I.; Novikov, V.A.; Shifman, M.A. ABC's of Instantons. *Sov. Phys. Usp.* **1982**, *25*, 195–215. [] []
80. Matinyan, S.G.; Savvidy, G.K. Vacuum Polarization Induced by the Intense Gauge Field. *Nucl. Phys. B* **1978**, *134*, 539–545. [] []
81. Pasechnik, R. Quantum Yang–Mills Dark Energy. *Universe* **2016**, *2*, 4. [] []
82. Pasechnik, R.; Beylin, V.; Vereshkov, G. Dark Energy from graviton-mediated interactions in the QCD vacuum. *J. Cosmol. Astropart. Phys.* **2013**, *1306*, 11. [] []
83. Pasechnik, R.; Beylin, V.; Vereshkov, G. Possible compensation of the QCD vacuum contribution to the dark energy. *Phys. Rev. D* **2013**, *88*, 023509. [] []
84. Pasechnik, R.; Prokhorov, G.; Teryaev, O. Mirror QCD and Cosmological Constant. *arXiv*, 2016; arXiv:1609.09249. []
85. Pagels, H.; Tomboulis, E. Vacuum of the Quantum Yang-Mills Theory and Magnetostatics. *Nucl. Phys. B* **1978**, *143*, 485–502. [] []
86. Prokhorov, G.; Pasechnik, R.; Vereshkov, G. Dynamics of wave fluctuations in the homogeneous Yang-Mills condensate. *J. High Energy Phys.* **2014**, *2014*, 3. []
87. Abelev, B.; Adam, J.; Adamová, D.; Adare, A.M.; Aggarwal, M.M.; Rinella, G.A.; Agnello, M.; Agocs, A.G.; Agostinelli, A.; Ahammed, Z.; et al. Centrality dependence of π, K, p production in Pb-Pb collisions at = 2.76 TeV. *Phys. Rev. C* **2013**, *88*, 044910. [] []
88. Abelev, B.; Adam, J.; Adamová, D.; Adare, A.M.; Aggarwal, M.M.; Rinella, G.A.; Agnello, M.; Agocs, A.G.; Agostinelli, A.; Ahammed, Z.; et al. Two- and three-pion quantum statistics correlations in Pb-Pb collisions at = 2.76 TeV at the CERN Large Hadron Collider. *Phys. Rev. C* **2014**, *89*, 024911. [] []
89. Adam, J.; Adamová, D.; Aggarwal, M.M.; Rinella, G.A.; Agnello, M.; Agrawal, N.; Ahammed, Z.; Ahmad, S.; Ahn, S.U.; Aiola, S.; et al. Multipion Bose-Einstein

correlations in pp, p-Pb, and Pb-Pb collisions at energies available at the CERN Large Hadron Collider. *Phys. Rev. C* **2016**, *93*, 054908. [] []
90. Petrán, M.; Letessier, J.; Petráček, V.; Rafelski, J. Hadron production and quark-gluon plasma hadronization in Pb-Pb collisions at = 2.76 TeV. *Phys. Rev. C* **2013**, *88*, 034907. [] []
91. Csorgo, T.; Csernai, L.P. Quark-gluon plasma freezeout from a supercooled state? *Phys. Lett. B* **1994**, *333*, 494–499. [] []
92. Shuryak, E. Heavy Ion Collisions: Achievements and Challenges. *arXiv*, 2014; arXiv:1412.8393. []
93. Blaizot, J.P.; Gelis, F.; Liao, J.F.; McLerran, L.; Venugopalan, R. Bose-Einstein Condensation and Thermalization of the Quark Gluon Plasma. *Nucl. Phys. A* **2012**, *873*, 68–80. [] []
94. Begun, V.; Florkowski, W.; Rybczynski, M. Explanation of hadron transverse-momentum spectra in heavy-ion collisions at = 2.76 TeV within chemical non-equilibrium statistical hadronization model. *Phys. Rev. C* **2014**, *90*, 014906. [] []
95. Begun, V.; Florkowski, W.; Rybczynski, M. Transverse-momentum spectra of strange particles produced in Pb+Pb collisions at = 2.76 TeV in the chemical non-equilibrium model. *Phys. Rev. C* **2014**, *90*, 054912. [] []
96. Begun, V.; Florkowski, W. Bose-Einstein condensation of pions in heavy-ion collisions at the CERN Large Hadron Collider (LHC) energies. *Phys. Rev. C* **2015**, *91*, 054909. [] []
97. Begun, V. Fluctuations as a test of chemical non-equilibrium at the LHC. *Phys. Rev. C* **2016**, *94*, 054904. [] []
98. Plumer, M.; Raha, S.; Weiner, R.M. How Free Is The Quark-Gluon-Plasma? *Nucl. Phys. A* **1984**, *418*, 549C–557C. [] []
99. Rischke, D.H.; Gorenstein, M.I.; Schafer, A.; Stoecker, H.; Greiner, W. Nonperturbative effects in the SU(3) gluon plasma. *Phys. Lett. B* **1992**, *278*, 19–23. [] []
100. Bannur, V.M. Equation of state for a nonideal quark gluon plasma. *Phys. Lett. B* **1995**, *362*, 7–10. [] []
101. Thoma, M.H. The quark–gluon plasma liquid. *J. Phys. G* **2005**, *31*, L7–L11. [] []
102. Bannur, V.M. Strongly coupled quark gluon plasma (SCQGP). *J. Phys. G* **2006**, *32*, 993–1002. [] []
103. Bonitz, M.; Henning, C.; Block, D. Complex plasmas: A laboratory for strong correlations. *Rep. Prog. Phys.* **2010**, *73*, 066501. [] []
104. Giorgini, S.; Pitaevskii, L.P.; Stringari, S. Theory of ultracold atomic Fermi gases. *Rev. Mod. Phys.* **2008**, *80*, 1215–1274. [] []
105. Cao, C.; Elliott, E.; Joseph, J.; Wu, H.; Petricka, J.; Schäfer, T.; Thomas, J.E. Universal Quantum Viscosity in a Unitary Fermi Gas. *Science* **2011**, *331*, 58–61. []
106. Johnson, C.V.; Steinberg, P. What black holes teach about strongly coupled particles. *Phys. Today* **2010**, *63*, 29–33. [] []
107. Kovtun, P.; Son, D.T.; Starinets, A.O. Viscosity in strongly interacting quantum field theories from black hole physics. *Phys. Rev. Lett.* **2005**, *94*, 111601. [] []
108. Schafer, T.; Teaney, D. Nearly Perfect Fluidity: From Cold Atomic Gases to Hot Quark Gluon Plasmas. *Rep. Prog. Phys.* **2009**, *72*, 126001. [] []
109. Thomas, J.E. The nearly perfect Fermi gas. *Phys. Today* **2010**, *63*, 34–37. [] []
110. Gupta, S. Analyticity and the phase diagram of QCD. *Phys. Lett. B* **2004**, *588*, 136–144. [] []
111. Bazavov, A.; Bhattacharya, T.; DeTar, C.; Ding, H.-T.; Gottlieb, S.; Gupta, R.; Hegde, P.; Heller, U.M.; Karsch, F.; Laermann, E.; et al. Equation of state in (2+1)-flavor QCD. *Phys. Rev. D* **2014**, *90*, 094503. [] []
112. Landau, L.D.; Lifshitz, E.M. *Statistical Physics, Part 1*; Pergamon Press: London, UK, 1980. []
113. Shuryak, E.V.; Zhirov, O.V. Is the explosion of a quark-gluon plasma found? *Phys. Lett. B* **1986**, *171*, 99–102. [] []

114. Hung, C.M.; Shuryak, E.V. Hydrodynamics near the QCD phase transition: Looking for the longest-lived fireball. *Phys. Rev. Lett.* **1995**, *75*, 40]

115. Matsui, T.; Satz, H. J/ψ Suppression by Quark-Gluon Plasma Formation. *Phys. Lett. B* **1986**, *178*, 416–422. [] []

116. Brambilla, N.; Eidelman, S.; Heltsley, B.K.; Vogt, R.; Bodwin, G.T.; Eichten, E.; Frawley, A.D.; Meyer, A.B.; Mitchell, R.E.; Papadimitriou, V.; et al. Heavy quarkonium: Progress, puzzles, and opportunities. *Eur. Phys. J. C* **2011**, *71*, 1534. [] []

117. Borsányi, S.; Fodor, Z.; Katz, S.D.; Pásztor, A.; Szabó, K.K.; Török, C. Static Q pair free energy and screening masses from correlators of Polyakov loops: continuum extrapolated lattice results at the QCD physical point. *J. High Energy Phys.* **2015**, *1504*, 138. [] []

118. Bazavov, A.; Petreczky, P.; Velytsky, A. Quarkonium at Finite Temperature. In *Quark-gluon Plasma 4*; World Scientific Publishing Company: Singapore, 2009. []

119. Bazavov, A.; Bhattacharya, T.; Cheng, M.; DeTar, C.; Ding, H.-T.; Gottlieb, S.; Gupta, R.; Hegde, P.; Heller, U.M.; Karsch, F.; et al. The chiral and deconfinement aspects of the QCD transition. *Phys. Rev. D* **2012**, *85*, 054503. [] []

120. Ding, H.T.; Karsch, F.; Mukherjee, S. Thermodynamics of strong-interaction matter from Lattice QCD. *Int. J. Mod. Phys. E***2015**, *24*, 1530007. [] []

121. Scheid, W.; Muller, H.; Greiner, W. Nuclear Shock Waves in Heavy-Ion Collisions. *Phys. Rev. Lett.* **1974**, *32*, 741. [] []

122. Sobel, M.I.; Bethe, H.A.; Siemens, P.J.; Bondorf, J.P. Shock waves in colliding nuclei. *Nucl. Phys. A* **1975**, *251*, 502–529. [] []

123. Chapline, G.F.; Johnson, M.H.; Teller, E.; Weiss, M.S. Highly excited nuclear matter. *Phys. Rev. D* **1973**, *8*, 4302. [] []

124. Nagamiya, S.; Gyulassy, M. High-energy Nuclear Collisions. *Adv. Nucl. Phys.* **1984**, *13*, 201–315. [] []

125. Stock, R. Particle Production In High-Energy Nucleus Nucleus Collisions. *Phys. Rep.* **1986**, *135*, 259–315. [] []

126. Lee, T.D.; Wick, G.C. Vacuum Stability and Vacuum Excitation in a Spin 0 Field Theory. *Phys. Rev. D* **1974**, *9*, 2291–2316. [] []

127. Harrison, M.; Peggs, S.G.; Roser, T. The RHIC accelerator. *Ann. Rev. Nucl. Part. Sci.* **2002**, *52*, 425–469. [] []

128. Evans, L.; Bryant, P. LHC Machine. *J. Instrum.* **2008**, *3*, S08001. [] []

129. Bjorken, J.D. Highly Relativistic Nucleus-Nucleus Collisions: The Central Rapidity Region. *Phys. Rev. D* **1983**, *27*, 140–151. [] []

130. Adare, A.; Afanasiev, S.; Aidala, C.; Ajitanand, N.N.; Akiba, Y.; Akimoto, R.; Al-Bataineh, H.; Alexander, J.; Alfred, M.; Al-Jamel, A.; et al. Transverse energy production and charged-particle multiplicity at midrapidity in various systems from = 7.7 to 200 GeV. *Phys. Rev. C* **2016**, *93*, 024901. [] []

131. Chatrchyan, S.; Khachatryan, V.; Sirunyan, A.M.; Tumasyan, A.; Adam, W.; Bergauer, T.; Dragicevic, M.; Erö, J.; Fabjan, C.; et al. Measurement of the pseudorapidity and centrality dependence of the transverse energy density in PbPb collisions at = 2.76 TeV. *Phys. Rev. Lett.* **2012**, *109*, 152303.

132. Mitchelll, J.T. Transverse Energy Measurements from the Beam Energy Scan in PHENIX. *Nucl. Phys. A* **2016**, *956*, 842–845. [] []

133. Fischer, W.; Baltz, A.J.; Blaskiewicz, M.; Gassner, D.; Drees, K.A.; Luo, Y.; Minty, M.; Thieberger, P.; Wilinski, M.; Pshenichnov, I.A. Measurement of the total cross section of uranium-uranium collisions at = 192.8 GeV. *Phys. Rev. C* **2014**, *89*, 014906. [] []

134. Fermi, E. On the Theory of the impact between atoms and electrically charged particles. *Z. Phys.* **1924**, *29*, 315–327. [] []

135. Von Weizsacker, C.F. Radiation emitted in collisions of very fast electrons. *Z. Phys.* **1934**, *88*, 612–625. []

136. Williams, E.J. Nature of the high-energy particles of penetrating radiation and status of ionization and radiation formulae. *Phys. Rev.* **1934**, *45*, 729–730. [] []
137. Jackson, J.D. *Classical Electrodynamics*; Wiley: Hoboken, NJ, USA, 1998. []
138. Baur, G.; Hencken, K.; Trautmann, D.; Sadovsky, S.; Kharlov, Y. Coherent gamma gamma and gamma-A interactions in very peripheral collisions at relativistic ion colliders. *Phys. Rep.* **2002**, *364*, 359–450. [] []
139. Bertulani, C.A.; Klein, S.R.; Nystrand, J. Physics of ultra-peripheral nuclear collisions. *Ann. Rev. Nucl. Part. Sci.* **2005**, *55*, 271–310. [] []
140. Baltz, A.J.; Baur, G.; d'Enterria, D.; Frankfurt, L.; Gelis, F.; Guzey, V.; Hencken, K.; Kharlov, Y.; Klasen, M.; Klein, S.R. The Physics of Ultraperipheral Collisions at the LHC. *Phys. Rep.* **2008**, *458*, 1–171. [] []
141. Klein, S.R. Ultra-peripheral Collisions at RHIC: An Experimental Overview. In Proceedings of the Initial States 2014 Conference, Napa, CA, USA, 3–7 December 2014.
142. Tuchin, K. Particle production in strong electromagnetic fields in relativistic heavy-ion collisions. *Adv. High Energy Phys.* **2013**, *2013*, 490495. [] []
143. Kouveliotou, C.; Duncan, R.C.; Thompson, C. Magnetars. *Sci. Am.* **2003**, *288N2*, 24. [] []
144. Kharzeev, D.E.; McLerran, L.D.; Warringa, H.J. The effects of topological charge change in heavy ion collisions: 'Event by event P and CP violation. *Nucl. Phys. A* **2008**, *803*, 227–253. [] []
145. Ding, H.-T.; Francis, A.; Kaczmarek, O.; Karsch, F.; Laermann, E.; Soeldner, W. Thermal dilepton rate and electrical conductivity: An analysis of vector current correlation functions in quenched lattice QCD. *Phys. Rev. D* **2011**, *83*, 034504. [] []
146. Aarts, G.; Allton, C.; Amato, A.; Giudice, P.; Hands, S.; Skullerud, J.I. Electrical conductivity of the quark-gluon plasma across the deconfinement transition. *Phys. Rev. Lett.* **2013**, *111*, 172001. []
147. Aarts, G.; Allton, C.; Amato, A.; Giudice, P.; Hands, S.; Skullerud, J.I. Electrical conductivity and charge diffusion in thermal QCD from the lattice. *J. High Energy Phys.* **2015**, *1502*, 186. [] []
148. Filip, P. Decay of Resonaces in Strong Magnetic Field. *J. Phys. Conf. Ser.* **2015**, *636*, 012013. [] []
149. Kharzeev, D.E.; Liao, J.; Voloshin, S.A.; Wang, G. Chiral magnetic and vortical effects in high-energy nuclear collisions—A status report. *Prog. Part. Nucl. Phys.* **2016**, *88*, 1–28. [] []
150. Gursoy, U.; Kharzeev, D.; Rajagopal, K. Magnetohydrodynamics, charged currents and directed flow in heavy ion collisions. *Phys. Rev. C* **2014**, *89*, 054905. [] []
151. Kharzeev, D.; Pisarski, R.D.; Tytgat, M.H.G. Possibility of spontaneous parity violation in hot QCD. *Phys. Rev. Lett.* **1998**, *81*, 512–515. [] []
152. Abelev, B.I.; Aggarwal, M.M.; Ahammed, Z.; Alakhverdyants, A.V.; Anderson, B.D.; Arkhipkin, D.; Averichev, G.S.; Balewski, J.; Barannikova, O.; Barnby, L.S.; et al. Azimuthal Charged-Particle Correlations and Possible Local Strong Parity Violation. *Phys. Rev. Lett.* **2009**, *103*, 251601. []
153. Adamczyk, L.; Adkins, J.K.; Agakishiev, G.; Aggarwal, M.M.; Ahammed, Z.; Alekseev, I.; Alford, J.; Aparin, A.; Arkhipkin, D.; Aschenauer, E.C.; et al. Observation of charge asymmetry dependence of pion elliptic flow and the possible chiral magnetic wave in heavy-ion collisions. *Phys. Rev. Lett.* **2015**, *114*, 252302.
154. Adamczyk, L.; Adkins, J.K.; Agakishiev, G.; Aggarwal, M.M.; Ahammed, Z.; Alekseev, I.; Alford, J.; Anson, C.D.; Aparin, A.; Arkhipkin, D.; et al. Beam-energy dependence of charge separation along the magnetic field in Au+Au collisions at RHIC. *Phys. Rev. Lett.* **2014**, *113*, 052302.
155. Abelev, B.; Adam, J.; Adamova, D.; Adare, A.M.; Aggarwal, M.M.; Rinella, G.A.; Agocs, A.G.; Agostinelli, A.; Salazar, S.A.; Ahammed, Z.; et al. Charge separation

relative to the reaction plane in Pb-Pb collisions at = 2.76 TeV. *Phys. Rev. Lett.* **2013**, *110*, 012301. []
156. Khachatryan, V.; Sirunyan, A.M.; Tumasyan, A.; Adam, W.; Asilar, E.; Bergauer, T.; Brandstetter, J.; Brondolin, E.; Dragicevic, M.; Ero, J.; et al. Observation of charge-dependent azimuthal correlations in pPb collisions and its implication for the search for the chiral magnetic effect. *arXiv*, 2016; arXiv:1610.00263. []
157. Schlichting, S.; Pratt, S. Charge conservation at energies available at the BNL Relativistic Heavy Ion Collider and contributions to local parity violation observables. *Phys. Rev. C* **2011**, *83*, 014913. [] []
158. Skokov, V.; Sorensen, P.; Koch, V.; Schlichting, S.; Thomas, J.; Voloshin, S.; Wang, G.; Yee, H.U. Chiral Magnetic Effect Task Force Report. *arXiv*, 2016; arXiv:1608.00982. []
159. Becattini, F.; Karpenko, I.; Lisa, M.; Upsal, I.; Voloshin, S. Global Hyperon Polarization at Local Thermodynamic Equilibrium with Vorticity, Magnetic Field and Feed-dow. *arXiv*, 2016; arXiv:1608.00982. []
160. Aichelin, J. 'Quantum' molecular dynamics: A Dynamical microscopic n body approach to investigate fragment formation and the nuclear equation of state in heavy ion collisions. *Phys. Rep.* **1991**, *202*, 233–360. [] []
161. Werner, K. Strings, pomerons, and the venus model of hadronic interactions at ultrarelativistic energies. *Phys. Rep.* **1993**, *232*, 87–299. [] []
162. Gyulassy, M.; Wang, X.N. HIJING 1.0: A Monte Carlo program for parton and particle production in high-energy hadronic and nuclear collisions. *Comput. Phys. Commun.* **1994**, *83*, 307–331. [] []
163. Bass, S.A.; Belkacem, M.; Bleicher, M.; Brandstetter, M.; Bravina, L.; Ernst, C.; Gerland, L.; Hofmann, M.; Hofmann, S.; Konopka, J.; et al. Microscopic models for ultrarelativistic heavy ion collisions. *Prog. Part. Nucl. Phys.* **1998**, *41*, 255–369. [] []
164. Humanic, T.J. Constraining a simple hadronization model of relativistic heavy ion collisions using hadronic observables. *Phys. Rev. C* **1998**, *57*, 866–876. [] []
165. Lin, Z.W.; Ko, C.M.; Li, B.A.; Zhang, B.; Pal, S. A Multi-phase transport model for relativistic heavy ion collisions. *Phys. Rev. C* **2005**, *72*, 064901. [] []
166. Xu, H.J.; Li, Z.; Song, H. High-order flow harmonics of identified hadrons in 2.76A TeV Pb + Pb collisions. *Phys. Rev. C* **2016**, *93*, 064905. [] []
167. Roesler, S.; Engel, R.; Ranft, J. The Monte Carlo Event Generator DPMJET-III. In Proceedings of the Monte Carlo 2000, Lisbon, Portugal, 23–26 October 2000.
168. Pierog, T.; Karpenko, I.; Katzy, J.M.; Yatsenko, E.; Werner, K. EPOS LHC: Test of collective hadronization with data measured at the CERN Large Hadron Collider. *Phys. Rev. C* **2015**, *92*, 034906. [] []
169. Akamatsu, Y.; Inutsuka, S.I.; Nonaka, C.; Takamoto, M. A new scheme of causal viscous hydrodynamics for relativistic heavy-ion collisions: A Riemann solver for quark-gluon plasma. *J. Comput. Phys.* **2014**, *256*, 34–54. [] []
170. Bozek, P.; Wyskiel-Piekarska, I. Particle spectra in Pb-Pb collisions at = 2.76 TeV. *Phys. Rev. C* **2012**, *85*, 064915. [] []
171. De Souza, R.D.; Koide, T.; Kodama, T. Hydrodynamic Approaches in Relativistic Heavy Ion Reactions. *Prog. Part. Nucl. Phys.* **2016**, *86*, 35–85. [] []
172. Kolb, P.F.; Sollfrank, J.; Heinz, U.W. Anisotropic transverse flow and the quark hadron phase transition. *Phys. Rev. C* **2000**, *62*, 054909. [] []
173. Lokhtin, I.P.; Malinina, L.V.; Petrushanko, S.V.; Snigirev, A.M.; Arsene, I.; Tywoniuk, K. Heavy ion event generator HYDJET++ (HYDrodynamics plus JETs). *Comput. Phys. Commun.* **2009**, *180*, 779–799. [] []
174. Hirano, T.; Huovinen, P.; Murase, K.; Nara, Y. Integrated Dynamical Approach to Relativistic Heavy Ion Collisions. *Prog. Part. Nucl. Phys.* **2013**, *70*, 108–158. [] []
175. Shen, C.; Qiu, Z.; Song, H.; Bernhard, J.; Bass, S.; Heinz, U. The iEBE-VISHNU code package for relativistic heavy-ion collisions. *Comput. Phys. Commun.* **2016**, *199*, 61–85. [] []

176. Abelev, B.B.; Adam, J.; Adamova, D.; Adare, A.M.; Aggarwal, M.M.; Rinella, G.A.; Agnello, M.; Agocs, A.G.; Agostinelli, A.; Ahammed, Z.; et al. Multiplicity Dependence of Pion, Kaon, Proton and Lambda Production in p-Pb Collisions at = 5.02 TeV. *Phys. Lett. B* **2014**, *728*, 25–38. [] []
177. Adler, S.S.; Afanasiev, S.; Aidala, C.; Ajitanand, N.N.; Akiba, Y.; Alexander, J.; Amirikas, R.; Aphecetche, L.; Aronson, S.H.; Averbeck, R.; et al. Identified charged particle spectra and yields in Au+Au collisions at S(NN)**1/2 = 200-GeV. *Phys. Rev. C***2004**, *69*, 034909. [] []
178. Nonaka, C.; Bass, S.A. Space-time evolution of bulk QCD matter. *Phys. Rev. C* **2007**, *75*, 014902. [] []
179. Muller, B. Quark matter '99—Theoretical summary: What next? *Nucl. Phys. A* **1999**, *661*, 272–281. [] []
180. Kittel, W.; De Wolf, E.A. *Soft Multihadron Dynamics*; World Scientific: Singapore, 2005. []
181. Landau, L.D. On the multiparticle production in high-energy collisions. *Izv. Akad. Nauk Ser. Fiz.* **1953**, *17*, 51. []
182. Weiner, R.M. Experiment and Theory in Computations of the He Atom Ground State. *Int. J. Mod. Phys. E* **2006**, *15*, 37–56. [] []
183. Heinz, U.W. Early collective expansion: Relativistic hydrodynamics and the transport properties of QCD matter. In *Landolt-Bornstein*; Springer: Berlin, Germany, 2010; Volume 23, pp. 240–292. []
184. Stöcker, H.; Greiner, W. High-Energy Heavy Ion Collisions: Probing The Equation Of State Of Highly Excited Hadronic Matter. *Phys. Rep.* **1986**, *137*, 277–392. [] []
185. Hirano, T.; van der Kolk, N.; Bilandzic, A. Hydrodynamics and Flow. *Lect. Notes Phys.* **2010**, *785*, 139–178. []
186. Gale, C.; Jeon, S.; Schenke, B. Hydrodynamic Modeling of Heavy-Ion Collisions. *Int. J. Mod. Phys. A* **2013**, *28*, 1340011. [] []
187. Jaiswal, A.; Roy, V. Relativistic hydrodynamics in heavy-ion collisions: General aspects and recent developments. *Adv. High Energy Phys.* **2016**, *2016*, 9623034. [] []
188. Eckart, C. The Thermodynamics of irreversible processes. 3. Relativistic theory of the simple fluid. *Phys. Rev.* **1940**, *58*, 919–924. [] []
189. Nakamura, A.; Sakai, S. Transport coefficients of gluon plasma. *Phys. Rev. Lett.* **2005**, *94*, 072305. []
190. Karsch, F.; Kharzeev, D.; Tuchin, K. Universal properties of bulk viscosity near the QCD phase transition. *Phys. Lett. B* **2008**, *663*, 217–221. [] []
191. Muronga, A. Causal theories of dissipative relativistic fluid dynamics for nuclear collisions. *Phys. Rev. C* **2004**, *69*, 034903. [] []
192. Hiscock, W.A.; Lindblom, L. Stability and causality in dissipative relativistic fluids. *Ann. Phys.* **1983**, *151*, 466–496. [] []
193. Heinz, U.; Snellings, R. Collective flow and viscosity in relativistic heavy-ion collisions. *Ann. Rev. Nucl. Part. Sci.* **2013**, *63*, 123–151. [] []
194. Abelev, B.I.; Aggarwal, M.M.; Ahammed, Z.; Anderson, B.D.; Arkhipkin, D.; Averichev, G.S.; Bai, Y.; Balewski, J.; Barannikova, O.; Barnby, L.S.; et al. Systematic Measurements of Identified Particle Spectra in *pp*, *d*+ Au and Au+Au Collisions from STAR. *Phys. Rev. C* **2009**, *79*, 034909. [] []
195. Siemens, P.J.; Rasmussen, J.O. Evidence for a blast wave from compress nuclear matter. *Phys. Rev. Lett.* **1979**, *42*, 880–887. [] []
196. Schnedermann, E.; Sollfrank, J.; Heinz, U.W. Thermal phenomenology of hadrons from 200-A/GeV S+S collisions. *Phys. Rev. C* **1993**, *48*, 2462–2475. [] []
197. Retiere, F.; Lisa, M.A. Observable implications of geometrical and dynamical aspects of freeze out in heavy ion collisions. *Phys. Rev. C* **2004**, *70*, 044907. [] []
198. Cooper, F.; Frye, G. Comment on the Single Particle Distribution in the Hydrodynamic and Statistical Thermodynamic Models of Multiparticle Production. *Phys. Rev. D* **1974**, *10*, 186–189. [] []

199. Kumar, L. Systematics of Kinetic Freeze-out Properties in High Energy Collisions from STAR. *Nucl. Phys. A* **2014**, *931*, 1114–1119. [] []
200. Sjostrand, T.; Mrenna, S.; Skands, P.Z. A Brief Introduction to PYTHIA 8.1. *Comput. Phys. Commun.* **2008**, *178*, 852–867. [] []
201. Baier, R.; Mueller, A.H.; Schiff, D.; Son, D.T. 'Bottom up' thermalization in heavy ion collisions. *Phys. Lett. B* **2001**, *502*, 51–58. [] []
202. Xu, Z.; Greiner, C. Thermalization of gluons in ultrarelativistic heavy ion collisions by including three-body interactions in a parton cascade. *Phys. Rev. C* **2005**, *71*, 064901. [] []
203. Kurkela, A.; Moore, G.D. Thermalization in Weakly Coupled Nonabelian Plasmas. *J. High Energy Phys.* **2011**, *1112*, 44. [] []
204. Romatschke, P. Do Nuclear Collisions Create a Locally Equilibrated Quark-Gluon Plasma? *Eur. Phys. J. C* **2016**, *77*, 21. [] []
205. Broniowski, W.; Chojnacki, M.; Florkowski, W.; Kisiel, A. Uniform Description of Soft Observables in Heavy-Ion Collisions at = 200 GeV. *Phys. Rev. Lett.* **2008**, *101*, 022301. [] []
206. Petersen, H.; Coleman-Smith, C.; Bass, S.A.; Wolpert, R. Constraining the initial state granularity with bulk observables in Au+Au collisions at = 200 GeV. *J. Phys. G* **2011**, *38*, 045102. [] []
207. Heinz, U.; Moreland, J.S. Energy dependent growth of the nucleon and hydrodynamic initial conditions. *Phys. Rev. C* **2011**, *84*, 054905. []
208. Noronha-Hostler, J.; Betz, B.; Noronha, J.; Gyulassy, M. Event-by-event hydrodynamics + jet energy loss: A solution to the $R_{AA} \otimes v_2$ puzzle. *Phys. Rev. Lett.* **2016**, *116*, 252301. []
209. Glauber, R.J. Cross-sections in deuterium at high-energies. *Phys. Rev.* **1955**, *100*, 242–248. [] []
210. Czyz, W.; Maximon, L.C. High-energy, small angle elastic scattering of strongly interacting composite particles. *Ann. Phys.* **1969**, *52*, 59–121. [] []
211. Glauber, R.J.; Matthiae, G. High-energy scattering of protons by nuclei. *Nucl. Phys. B* **1970**, *21*, 135–157. [] []
212. Glauber, R.J. Quantum Optics and Heavy Ion Physics. *Nucl. Phys. A* **2006**, *774*, 3–13. [] []
213. Joachain, C.J.; Quigg, C. Multiple scattering expansions in several particle dynamics. *Rev. Mod. Phys.* **1974**, *46*, 279–324. [] []
214. Franco, V.; Varma, G.K. Collisions Between Composite Particles at Medium and High-Energies. *Phys. Rev. C* **1978**, *18*, 349–370. [] []
215. Miller, M.L.; Reygers, K.; Sanders, S.J.; Steinberg, P. Glauber modeling in high energy nuclear collisions. *Ann. Rev. Nucl. Part. Sci.* **2007**, *57*, 205–243. [] []
216. Broniowski, W.; Rybczynski, M.; Bozek, P. GLISSANDO: Glauber initial-state simulation and more. *Comput. Phys. Commun.* **2009**, *180*, 69–83. [] []
217. Loizides, C.; Nagle, J.; Steinberg, P. Improved version of the PHOBOS Glauber Monte Carlo. *SoftwareX* **2015**, *1*, 13–18. [] []
218. Abreu, M.C.; Alessandro, B.; Alexa, C.; Arnaldi, R.; Atayan, M.; Baglin, C.; Baldit, A.; Bedjidian, M.; Beole, S.; Boldea, V. The Dependence of the anomalous J/psi suppression on the number of participant nucleons. *Phys. Lett. B* **2001**, *521*, 195–203. [] []
219. Aad, G.; Abajyan, T.; Abbott, B.; Abdallah, J.; Khalek, S.A.; Abdelalim, A.A.; Abdinov, O.; Aben, R.; Abi, B.; Abolins, M.; et al. Measurement of *Z* boson Production in Pb+Pb Collisions at = 2.76 TeV with the ATLAS Detector. *Phys. Rev. Lett.* **2013**, *110*, 022301.
220. Gribov, V.N. A Reggeon Diagram Technique. *Sov. Phys. JETP* **1968**, *26*, 414–422. []
221. Gribov, V.N. Glauber corrections and the interaction between high-energy hadrons and nuclei. *Sov. Phys. JETP* **1969**, *29*, 483–487. []
222. Abramovsky, V.A.; Gribov, V.N.; Kancheli, O.V. Character of Inclusive Spectra and Fluctuations Produced in Inelastic Processes by Multi-Pomeron Exchange. *Yad. Fiz.* **1973**, *18*, 595–616. []

223. Braun, V.M.; Shabelski, Y.M. Multiple Scattering Theory for Inelastic Processes. *Int. J. Mod. Phys. A* **1988**, *3*, 2417–2501. [] []
224. Drescher, H.J.; Hladik, M.; Ostapchenko, S.; Pierog, T.; Werner, K. Parton based Gribov-Regge theory. *Phys. Rep.* **2001**, *350*, 93–289. [] []
225. Ostapchenko, S. Monte Carlo treatment of hadronic interactions in enhanced Pomeron scheme: I. QGSJET-II model. *Phys. Rev. D* **2011**, *83*, 014018. [] []
226. Pasechnik, R.; Enberg, R.; Ingelman, G. QCD rescattering mechanism for diffractive deep inelastic scattering. *Phys. Rev. D***2010**, *82*, 054036. [] []
227. Pasechnik, R.; Enberg, R.; Ingelman, G. Diffractive deep inelastic scattering from multiple soft gluon exchange in QCD. *Phys. Lett. B* **2011**, *695*, 189–193. [] []
228. Nemchik, J.; Petracek, V.; Potashnikova, I.K.; Sumbera, M. Nuclear suppression at large forward rapidities in d-Au collisions at relativistic and ultrarelativistic energies. *Phys. Rev. C* **2008**, *78*, 025213. [] []
229. Loizides, C. Glauber modeling of high-energy nuclear collisions at the subnucleon level. *Phys. Rev. C* **2016**, *94*, 024914. [] []
230. Nikolaev, N.N.; Zakharov, B.G. The triple-pomeron regime and structure function of the pomeron in diffractive deep inelastic scattering at very smallx. *Phys. C* **1994**, *64*, 631–652. [] []
231. Kopeliovich, B.Z.; Lapidus, L.I.; Zamolodchikov, A.B. Dynamics Of Color In Hadron Diffraction On Nuclei. *JETP Lett.* **1981**, *33*, 595–597. []
232. Nikolaev, N.N.; Zakharov, B.G. The Pomeron in diffractive deep inelastic scattering. *J. Exp. Theor. Phys.* **1994**, *78*, 598–618. []
233. Kopeliovich, B.Z. *Proceedings of the International Workshop XXIII on Gross Properties of Nuclei and Nuclear Excitations, Hirschegg, Austria, 1995*; Feldmeyer, H., Norenberg, W., Eds.; Gesellschaft Schwerionenforschung: Darmstadt, Germany, 1995. []
234. Brodsky, S.J.; Hebecker, A.; Quack, E. The Drell-Yan process and factorization in impact parameter space. *Phys. Rev. D* **1997**, *55*, 2584–2590. [] []
235. Kopeliovich, B.Z.; Schafer, A.; Tarasov, A.V. Bremsstrahlung of a quark propagating through a nucleus. *Phys. Rev. C* **1999**, *59*, 1609–1619. [] []
236. Kopeliovich, B.Z.; Schafer, A.; Tarasov, A.V. The color dipole picture of the Drell-Yan process. *Phys. Lett. B* **2001**, *503*, 91–98. [] []
237. Nikolaev, N.N.; Piller, G.; Zakharov, B.G. Quantum coherence in heavy flavor production on nuclei. *Z. Phys. A* **1996**, *354*, 99–105. [] []
238. Kopeliovich, B.Z.; Tarasov, A.V. Gluon shadowing and heavy flavor production off nuclei. *Nucl. Phys. A* **2002**, *710*, 180–217. [] []
239. Bertsch, G.; Brodsky, S.J.; Goldhaber, A.S.; Gunion, J.F. Diffractive Excitation in Quantum Chromodynamics. *Phys. Rev. Lett.***1981**, *47*, 297. [] []
240. Brodsky, S.J.; Mueller, A. Using nuclei to probe hadronization in QCD. *Phys. Lett. B* **1988**, *206*, 685–690. [] []
241. Golec-Biernat, K.J.; Wusthoff, M. Saturation effects in deep inelastic scattering at low Q^2 and its ... and its implications on diffraction. *Phys. Rev. D* **1998**, *59*, 014017. [] []
242. Bartels, J.; Golec-Biernat, K.; Kowalski, H. Modification of the saturation model: Dokshitzer-Gribov-Lipatov- Altarelli-Parisi evolution. *Phys. Rev. D* **2002**, *66*, 014001. [] []
243. Kowalski, H.; Motyka, L.; Watt, G. Exclusive diffractive processes at HERA within the dipole picture. *Phys. Rev. D* **2006**, *74*, 074016. [] []
244. Watt, G.; Kowalski, H. Impact parameter dependent colour glass condensate dipole model. *Phys. Rev. D* **2008**, *78*, 014016. [] []
245. Rezaeian, A.H.; Siddikov, M.; van de Klundert, M.; Venugopalan, R. NΣc and NΣb resonances in the quark-delocalization color-screening model. *Phys. Rev. D* **2013**, *87*, 034002. [] []
246. Iancu, E.; Itakura, K.; Munier, S. Saturation and BFKL dynamics in the HERA data at small-x. *Phys. Lett. B* **2004**, *590*, 199–208. [] []

247. Kharzeev, D.; Kovchegov, Y.V.; Tuchin, K. Nuclear modification factor in image collisions: Onset of suppression in the color glass condensate. *Phys. Lett. B* **2004**, *599*, 23–31. [] []
248. Kharzeev, D.; Kovchegov, Y.V.; Tuchin, K. The color glass condensate and hadron production in the forward region. *Nucl. Phys. A* **2006**, *765*, 464–482. []
249. Boer, D.; Utermann, A.; Wessels, E. Geometric Scaling at RHIC and LHC. *Phys. Rev. D* **2008**, *77*, 054014. [] []
250. Soyez, G. Saturation QCD predictions with heavy quarks at HERA. *Phys. Lett. B* **2007**, *655*, 32–38. [] []
251. Kowalski, H.; Teaney, D. Impact parameter dipole saturation model. *Phys. Rev. D* **2003**, *68*, 114005. [] []
252. Rezaeian, A.; Schmidt, I. Impact-parameter dependent color glass condensate dipole model and new combined HERA data. *Phys. Rev. D* **2013**, *88*, 074016. [] []
253. Basso, E.; Goncalves, V.P.; Nemchik, J.; Pasechnik, R.; Sumbera, M. Drell-Yan phenomenology in the color dipole picture revisited. *Phys. Rev. D* **2016**, *93*, 034023. [] []
254. Basso, E.; Goncalves, V.P.; Krelina, M.; Nemchik, J.; Pasechnik, R. Nuclear effects in Drell-Yan pair production in high-energy *pA* collisions. *Phys. Rev. D* **2016**, *93*, 094027. [] []
255. Basso, E.; Goncalves, V.P.; Krelina, M.; Nemchik, J.; Pasechnik, R. Drell-Yan process in *pA* collisions: The exact treatment of coherence effects. *Phys. Rev. D* **2016**, *94*, 114009. []
256. Kopeliovich, B.Z.; Nemchik, J.; Potashnikova, I.K.; Johnson, M.B.; Schmidt, I. Breakdown of QCD factorization at large Feynman x. *Phys. Rev. C* **2005**, *72*, 054606. [] []
257. Kopeliovich, B.Z.; Nemchik, J.; Potashnikova, I.K.; Schmidt, I. Energy conservation in high-pT nuclear reactions. *Int. J. Mod. Phys. E* **2014**, *23*, 1430006. [] []
258. Adler, S.S.; Afanasiev, S.; Aidala, C.; Ajitanand, N.N.; Akiba, Y.; Al-Jamel, A.; Alexander, J.; Aoki, K.; Aphecetche, L.; Armendariz, R.; et al. Centrality Dependence of π 0 and η Production at Large Transverse Momentum in s N N = 200 GeV d+ Au Collisions. *Phys. Rev. Lett.* **2007**, *98*, 172302. [] []
259. Afanasiev, S.; Aidala, C.; Ajitanand, N.N.; Akiba, Y.; Al-Jamel, A.; Alexander, J.; Aoki, K.; Aphecetche, L.; Armendariz, R.; Aronson, S.H.; et al. Measurement of Direct Photons in Au+Au Collisions at √sNN = 200 GeV. *Phys. Rev. Lett.* **2012**, *109*, 152302. [] []
260. Sakaguchi, T. Measurement of electro-magnetic radiation at RHIC-PHENIX. *Nucl. Phys. A* **2008**, *805*, 355–360. [] []
261. Landau, L.D.; Pomeranchuk, I. Limits of applicability of the theory of bremsstrahlung electrons and pair production at high-energies. *Dokl. Akad. Nauk Ser. Fiz.* **1953**, *92*, 535–536. []
262. Landau, L.D.; Pomeranchuk, I. Electron cascade process at very high-energies. *Dokl. Akad. Nauk Ser. Fiz.* **1953**, *92*, 735–738. []
263. Migdal, A.B. Bremsstrahlung and pair production in condensed media at high-energies. *Phys. Rev.* **1956**, *103*, 1811–1820. [] []
264. Baier, R.; Shiff, D.; Zakharov, B.G. Energy loss in perturbative QCD. *Ann. Rev. Nucl. Part. Sci.* **2000**, *50*, 37–69. [] []
265. Gyulassy, M.; Wang, X.-N. Multiple collisions and induced gluon Bremsstrahlung in QCD. *Nucl. Phys. B* **1994**, *420*, 583–614. [] []
266. Wang, X.-N.; Gyulassy, M.; Plumer, M. Landau-Pomeranchuk-Migdal effect in QCD and radiative energy loss in a quark-gluon plasma. *Phys. Rev. D* **1995**, *51*, 3436–3446. [] []
267. Wiedemann, U.A.; Gyulassy, M. Transverse momentum dependence of the Landau-Pomeranchuk-Migdal effect. *Nucl. Phys. B***1999**, *560*, 345–382. [] []
268. Zakharov, B.G. Applicability of the eikonal approximation for calculation of the probability of passage of ultrarelativistic positronium through matter. *Sov. J. Nucl. Phys.* **1987**, *46*, 92–95. []

269. Zakharov, B.G. Radiative energy loss of high-energy quarks in finite-size nuclear matter and quark-gluon plasma. *J. Exp. Theor. Phys. Lett.* **1997**, *65*, 615–620. [] []
270. Zakharov, B.G. Transverse Spectra of Radiation Processes in Medium. *J. Exp. Theor. Phys. Lett.* **1999**, *70*, 176–182. [] []
271. Baier, R.; Dokshitzer, Y.L.; Mueller, A.H.; Schiff, D. Angular dependence of the radiative gluon spectrum and the energy loss of hard jets in QCD media. *Phys. Rev. C* **1999**, *60*, 064902. [] []
272. Baier, R.; Dokshitzer, Y.L.; Mueller, A.H.; Schiff, D. Induced Gluon Radiation in a QCD Medium. *Phys. Lett. B* **1995**, *345*, 277–286. [] []
273. Baier, R.; Dokshitzer, Y.L.; Mueller, A.H.; Peigne, S.; Schiff, D. The Landau-Pomeranchuk-Migdal effect in QED. *Nucl. Phys. B* **1996**, *478*, 577–597. [] []
274. Baier, R.; Dokshitzer, Y.L.; Mueller, A.H.; Schiff, D. Medium-induced radiative energy loss; equivalence between the BDMPS and Zakharov formalisms. *Nucl. Phys. B* **1998**, *531*, 403–425. [] []
275. Nemchik, J.; Pasechnik, R.; Potashnikova, I. A heuristic description of high-p_T hadron production in heavy ion collisions. *Eur. Phys. J. C* **2015**, *75*, 95. [] []
276. Kopeliovich, B.Z.; Schaefer, A.; Tarasov, A.V. Nonperturbative Effects in Gluon Radiation and Photoproduction of Quark Pairs. *Phys. Rev. D* **2000**, *62*, 054022. [] []
277. Kopeliovich, B.Z.; Nemchik, J.; Schaefer, A. Color transparency versus quantum coherence in electroproduction of vector mesons off nuclei. *Phys. Rev. C* **2002**, *65*, 035201. [] []
278. Kopeliovich, B.Z.; Nemchik, J.; Potashnikova, I.K.; Schmidt, I. Gluon Shadowing in DIS off Nuclei. *J. Phys. G* **2008**, *35*, 115010. [] []
279. Gribov, L.V.; Levin, E.M.; Ryskin, M.G. Semihard Processes In QCD. *Phys. Rep.* **1983**, *100*, 1–150. [] []
280. Dokshitzer, Y.L. Calculation of the Structure Functions for Deep Inelastic Scattering and e+ e- Annihilation by Perturbation Theory in Quantum Chromodynamics. *Sov. Phys. JETP* **1977**, *46*, 641–653. []
281. Gribov, V.N.; Lipatov, L.N. Deep inelastic e p scattering in perturbation theory. *Sov. J Nucl. Phys.* **1972**, *15*, 438–450. []
282. Altarelli, G.; Parisi, G. Asymptotic Freedom in Parton Language. *Nucl. Phys. B* **1977**, *126*, 298–318. [] []
283. McLerran, L.D.; Venugopalan, R. Gluon distribution functions for very large nuclei at small transverse momentum. *Phys. Rev. D* **1994**, *49*, 3352–3355. [] []
284. Jalilian-Marian, J.; Kovchegov, Y.V. Saturation physics and deuteron gold collisions at RHIC. *Prog. Part. Nucl. Phys.* **2006**, *56*, 104–231. [] []
285. Gelis, F.; Iancu, E.; Jalilian-Marian, J.; Venugopalan, R. The Color Glass Condensate. *Ann. Rev. Nucl. Part. Sci.* **2010**, *60*, 463–489. [] []
286. Kharzeev, D.; Levin, E.; Nardi, M. Color glass condensate at the LHC: Hadron multiplicities in pp, pA and AA collisions. *Nucl. Phys. A* **2005**, *747*, 609–629. [] []
287. Jalilian-Marian, J.; Kovner, A.; Leonidov, A.; Weigert, H. The BFKL equation from the Wilson renormalization group. *Nucl. Phys. B* **1997**, *504*, 415–431. [] []
288. Jalilian-Marian, J.; Kovner, A.; Leonidov, A.; Weigert, H. The Wilson renormalization group for low x physics: Towards the high density regime. *Phys. Rev. D* **1998**, *59*, 014014. [] []
289. Kovner, A.; Milhano, J.G.; Weigert, H. Relating different approaches to nonlinear QCD evolution at finite gluon density. *Phys. Rev. D* **2000**, *62*, 114005. [] []
290. Weigert, H. Unitarity at small Bjorken x. *Nucl. Phys. A* **2002**, *703*, 823–860. [] []
291. Iancu, E.; Leonidov, A.; McLerran, L.D. Nonlinear gluon evolution in the color glass condensate. 1. *Nucl. Phys. A* **2001**, *692*, 583–645. [] []
292. Ferreiro, E.; Iancu, E.; Leonidov, A.; McLerran, L. Nonlinear gluon evolution in the color glass condensate. 2. *Nucl. Phys. A***2002**, *703*, 489–538. [] []
293. Golec-Biernat, K.J.; Wusthoff, M. Saturation in diffractive deep inelastic scattering. *Phys. Rev. D* **1999**, *60*, 114023. [] []

294. Caola, F.; Forte, S. Geometric Scaling from GLAP evolution. *Phys. Rev. Lett.* **2008**, *101*, 022001. []
295. Kharzeev, D.; Levin, E.; Nardi, M. The Onset of classical QCD dynamics in relativistic heavy ion collisions. *Phys. Rev. C* **2005**, *71*, 054903. [] []
296. Schenke, B.; Tribedy, P.; Venugopalan, R. Fluctuating Glasma initial conditions and flow in heavy ion collisions. *Phys. Rev. Lett.* **2012**, *108*, 252301. [] []
297. Drescher, H.-J.; Nara, Y. Effects of fluctuations on the initial eccentricity from the Color Glass Condensate in heavy ion collisions. *Phys. Rev. C* **2007**, *75*, 034905. [] []
298. Hirano, T.; Huovinen, P.; Nara, Y. Elliptic flow in Pb+Pb collisions at = 2.76 TeV: Hybrid model assessment of the first data. *Phys. Rev. C* **2011**, *84*, 011901. [] []
299. Aamodt, K.; Abelev, B.; Quintana, A.A.; Adamova, D.; Adare, A.M.; Aggarwal, M.M.; Rinella, G.A.; Agocs, A.G.; Salazar, S.A.; Ahammed, Z. Elliptic flow of charged particles in Pb-Pb collisions at 2.76 TeV. *Phys. Rev. Lett.* **2010**, *105*, 252302. [] []
300. Pang, L.; Wang, Q.; Wang, X.N. Effects of initial flow velocity fluctuation in event-by-event (3+1)D hydrodynamics. *Phys. Rev. C* **2012**, *86*, 024911. [] []
301. Bhalerao, R.S.; Jaiswal, A.; Pal, S. Collective flow in event-by-event partonic transport plus hydrodynamics hybrid approach. *Phys. Rev. C* **2015**, *92*, 014903. [] []
302. Paatelainen, R.; Eskola, K.J.; Niemi, H.; Tuominen, K. Fluid dynamics with saturated minijet initial conditions in ultrarelativistic heavy-ion collisions. *Phys. Lett. B* **2014**, *731*, 126–130. [] []
303. Niemi, H.; Eskola, K.J.; Paatelainen, R. Event-by-event fluctuations in a perturbative QCD + saturation + hydrodynamics model: Determining QCD matter shear viscosity in ultrarelativistic heavy-ion collisions. *Phys. Rev. C* **2016**, *93*, 024907. [] []
304. Moreland, J.S.; Bernhard, J.E.; Bass, S.A. Alternative ansatz to wounded nucleon and binary collision scaling in high-energy nuclear collisions. *Phys. Rev. C* **2015**, *92*, 011901. [] []
305. Bielcikova, J. Jets and correlations in heavy-ion collisions. In Proceedings of the 2015 European Physical Society Conference on High Energy Physics, Vienna, Austria, 22–29 July 2015.
306. Andronic, A.; Arleo, F.; Arnaldi, R.; Beraudo, A.; Bruna, E.; Caffarri, D.; del Valle, Z.C.; Contreras, J.G.; Dahms, T.; Dainese, A.; et al. Heavy-flavour and quarkonium production in the LHC era: From proton-proton to heavy-ion collisions. *Eur. Phys. J. C***2016**, *76*, 107. [] []
307. Rafelski, J.; Muller, B. Strangeness Production in the Quark-Gluon Plasma. *Phys. Rev. Lett.* **1982**, *48*, 1066. [] []
308. Koch, P.; Muller, B.; Rafelski, J. Strangeness in Relativistic Heavy Ion Collisions. *Phys. Rep.* **1986**, *142*, 167–262. [] []
309. Andronic, A.; Braun-Munzinger, P.; Stachel, J. Thermal hadron production in relativistic nuclear collisions: The Hadron mass spectrum, the horn, and the QCD phase transition. *Phys. Lett. B* **2009**, *673*, 142–145. [] []
310. Teaney, D.; Lauret, J.; Shuryak, E.V. Flow at the SPS and RHIC as a quark gluon plasma signature. *Phys. Rev. Lett.* **2001**, *86*, 4783–4786. []
311. Modeling Relativistic Heavy-Ion Collisions. Available online: https://www.phy.duke.edu/modeling-relativistic-heavy-ion-collisions (accessed on 17 January 2017).
312. Tribedy, P.; Venugopalan, R. QCD saturation at the LHC: Comparisons of models to p + p and A + A data and predictions for p + Pb collisions. *Phys. Lett. B* **2012**, *710*, 125–133. [] []
313. Aad, G.; Abbott, B.; Abdallah, J.; Khalek, S.A.; Abdelalim, A.A.; Abdesselam, A.; Abdinov, O.; Abi, B.; Abolins, M.; AbouZeid, O.S.; et al. Measurement of the centrality dependence of the charged particle pseudorapidity distribution in lead-lead collisions at = 2.76 TeV with the ATLAS detector. *Phys. Lett. B* **2012**, *710*, 363–382. [] []

314. Ollitrault, J.Y. Anisotropy as a signature of transverse collective flow. *Phys. Rev. D* **1992**, *46*, 229–245. [] []
315. Poskanzer, A.M.; Voloshin, S.A. Methods for analyzing anisotropic flow in relativistic nuclear collisions. *Phys. Rev. C* **1998**, *58*, 1671–1678. [] []
316. Manly, S. System size, energy and pseudorapidity dependence of directed and elliptic flow at RHIC. *Nucl. Phys. A* **2006**, *774*, 523–526. [] []
317. Aamodt, K.; Abelev, B.; Quintana, A.A.; Adamova, D.; Adare, A.M.; Aggarwal, M.M.; Rinella, G.A.; Agocs, A.G.; Agostinelli, A.; Salazar, S.A.; et al. Higher harmonic anisotropic flow measurements of charged particles in Pb-Pb collisions at = 2.76 TeV. *Phys. Rev. Lett.* **2011**, *107*, 032301. []
318. Venema, L.B.; Braak, H.; Löhner, H.; Raschke, A.E.; Siemssen, R.H.; Šumbera, M.; Wilschut, H.W.; Berg, F.D.; Kühn, W.; Metag, V.; et al. Azimuthal asymmetry of neutral pion emission in Au+Au reactions at 1 GeV/nucleon. *Phys. Rev. Lett.* **1993**, *71*, 835–838. []
319. Adamczyk, L.; Adkins, J.K.; Agakishiev, G.; Aggarwal, M.M.; Ahammed, Z.; Alekseev, I.; Aparin, A.; Arkhipkin, D.; Aschenauer, E.C.; Attri, A.; et al. Beam Energy Dependence of the Third Harmonic of Azimuthal Correlations in Au+Au Collisions at RHIC. *Phys. Rev. Lett.* **2016**, *116*, 112302. []
320. Gale, C.; Jeon, S.; Schenke, B.; Tribedy, P.; Venugopalan, R. Event-by-event anisotropic flow in heavy-ion collisions from combined Yang-Mills and viscous fluid dynamics. *Phys. Rev. Lett.* **2013**, *110*, 012302. []
321. Aad, G.; Abajyan, T.; Abbott, B.; Abdallah, J.; Abdel Khalek, S.; Abdelalim, A.A.; Abdinov, O.; Aben, R.; Abi, B.; Abolins, M. Measurement of the distributions of event-by-event flow harmonics in lead-lead collisions at = 2.76 TeV with the ATLAS detector at the LHC. *J. High Energy Phys.* **2013**, *1311*, 183. [] []
322. Qiu, Z.; Heinz, U. Hydrodynamic event-plane correlations in Pb+Pb collisions at = 2.76A TeV. *Phys. Lett. B* **2012**, *717*, 261–265. [] []
323. Aad, G.; Abbott, B.; Abdallah, J.; Khalek, S.A.; Abdinov, O.; Aben, R.; Abi, B.; Abolins, M.; Abouzeld, O.; Abramowicz, H.; et al. Measurement of event plane correlations in = 2.76 TeV lead-lead collisions with the ATLAS detector. *Phys. Rev. C* **2014**, *90*, 024905. [] []
324. Qian, J.; Heinz, U. Hydrodynamic flow amplitude correlations in event-by-event fluctuating heavy-ion collisions. *Phys. Rev. C***2016**, *94*, 024910. [] []
325. Giacalone, G.; Yan, L.; Noronha-Hostler, J.; Ollitrault, J.Y. Symmetric cumulants and event-plane correlations in Pb + Pb collisions. *Phys. Rev. C* **2016**, *94*, 014906. [] []
326. Aad, G.; Abbott, B.; Abdallah, J.; Khalek, S.A.; Abdelalim, A.A.; Abdesselam, A.; Abdinov, O.; Abi, B.; Abolins, M.; AbouZeid, O.S.; et al. Measurement of the correlation between flow harmonics of different order in lead-lead collisions at = 2.76 TeV with the ATLAS detector. *Phys. Rev. C* **2015**, *92*, 034903. [] []
327. Aamodt, K.; Abelev, B.; Quintana, A.A.; Adamova, D.; Adare, A.M.; Aggarwal, M.M.; Rinella, G.A.; Agocs, A.G.; Agostinelli, A.; Salazar, S.A.; et al. Correlated event-by-event fluctuations of flow harmonics in Pb-Pb collisions at = 2.76 TeV. *Phys. Rev. Lett.* **2016**, *117*, 182301. []
328. Wheaton, S.; Cleymans, J. THERMUS: A Thermal model package for ROOT. *Comput. Phys. Commun.* **2009**, *180*, 84–106. [] []
329. Abelev, B.; Adam, J.; Adamová, D.; Adare, A.M.; Aggarwal, M.M.; Rinella, G.A.; Agnello, M.; Agocs, A.G.; Agostinelli, A.; Ahammed, Z.; et al. Multi-strange baryon production at mid-rapidity in Pb-Pb collisions at = 2.76 TeV. *Phys. Lett. B* **2014**, *728*, 216–227. [] []
330. Adam, J.; Adamová, D.; Aggarwal, M.M.; Rinella, G.A.; Agnello, M.; Agrawal, N.; Ahammed, Z.; Ahmad, S.; Ahn, S.U.; Aiola, S.; et al. Multi-strange baryon production in p-Pb collisions at = 5.02 TeV. *Phys. Lett. B* **2016**, *758*, 389–401. []
331. Bjorken, J.D. Energy Loss of Energetic Partons in Quark-Gluon Plasma: Possible Extinction of High p(t) Jets in Hadron-Hadron Collisions. Available

online: http://lss.fnal.gov/archive/1982/pub/Pub-82-059-T.pdf (accessed on 17 January 2017).
332. Appel, D.A. Jets as a Probe of Quark-Gluon Plasmas. *Phys. Rev. D* **1986**, *33*, 717–722. [] []
333. Gyulassy, M.; Plumer, M. Jet Quenching in Dense Matter, The quenching of hard jets in ultrarelativistic nuclear collisions is estimated emphasizing its sensitivity to possible changes in the energy loss mechanism in a quark-gluon plasma. *Phys. Lett. B***1990**, *243*, 432–438. [] []
334. Khachatryan, V.; Sirunyan, A.M.; Tumasyan, A.; Adam, W.; Asilar, E.; Bergauer, T.; Brandstetter, J.; Brondolin, E.; Dragicevic, M.; Erö, J.; et al. Charged-Particle Nuclear Modification Factors in PbPb and pPb Collisions at = 5.02 TeV. *arXiv*, 2016; arXiv:1611.01664. []
335. Adams, J.; Aggarwal, M.M.; Ahammed, Z.; Amonett, J.; Anderson, B.D.; Anderson, M.; Arkhipkin, D.; Averichev, G.S.; Bai, Y.; Balewski, J. Direct Observation of Dijets in Central Au+Au Collisions at = 200 GeV. *Phys. Rev. Lett.* **2006**, *97*, 162301. [] []
336. D'Enterria, D. Perturbative probes of QCD matter at the Large Hadron Collider. In Proceedings of the 6th International Conference on Quarks and Nuclear Physics, Palaiseau, Paris, France, 16–20 April 2012.
337. Chatrchyan, S.; Khachatryan, V.; Sirunyan, A.M.; Tumasyan, A.; Adam, W.; Bergauer, T.; Dragicevic, M.; Erö, J.; Fabjan, C.; Friedl, M. Studies of jet quenching using isolated-photon+jet correlations in PbPb and *pp* collisions at = 2.76 TeV. *Phys. Lett. B***2013**, *718*, 773–794. [] []
338. Baier, R.; Dokshitzer, Y.L.; Mueller, A.H.; Peigne, S.; Schiff, D. Radiative energy loss and p(T) broadening of high-energy partons in nuclei. *Nucl. Phys. B* **1997**, *484*, 265–282. [] []
339. Caron-Huot, S. O(g) plasma effects in jet quenching. *Phys. Rev. D* **2009**, *79*, 065039. [] []
340. Laine, M. A non-perturbative contribution to jet quenching. *Eur. Phys. J. C* **2012**, *72*, 2233. [] []
341. Blaizot, J.P.; Mehtar-Tani, Y. Renormalization of the jet-quenching parameter. *Nucl. Phys. A* **2014**, *929*, 202–229. [] []
342. Panero, M.; Rummukainen, K.; Schfer, A. Lattice Study of the Jet Quenching Parameter. *Phys. Rev. Lett.* **2014**, *112*, 162001. [] []
343. Liu, H.; Rajagopal, K.; Wiedemann, U.A. Calculating the jet quenching parameter from AdS/CFT. *Phys. Rev. Lett.* **2006**, *97*, 182301. []
344. Svetitsky, B. Diffusion of charmed quarks in the quark-gluon plasma. *Phys. Rev. D* **1988**, *37*, 2484. [] []
345. Young, C.; Shuryak, E. Charmonium in strongly coupled quark-gluon plasma. *Phys. Rev. C* **2009**, *79*, 034907. [] []
346. Petreczky, P.; Young, C. Sequential Bottomonium Production at High Temperatures. *arXiv*, 2016; arXiv:1606.08421. []
347. Digal, S.; Petreczky, P.; Satz, H. Quarkonium feed down and sequential suppression. *Phys. Rev. D* **2001**, *64*, 094015. [] []
348. CMS Collaboration. *Strong Suppression of Υ Excited States in PbPb Collisions at* $\sqrt{sNN}$ *= 5.02 TeV*; CMS-PAS-HIN-16-008; CMS Collaboration: Cessy, France, 2016. []
349. Chatrchyan, S.; Khachatryan, V.; Sirunyan, A.M.; Tumasyan, A.; Adam, W.; Aguilo, E.; Bergauer, T.; Dragicevic, M.; Erö, J.; Fabjan, C.; et al. Observation of sequential Upsilon suppression in PbPb collisions. *Phys. Rev. Lett.* **2012**, *109*, 222301. [] []
350. Shuryak, E.V. Quark-Gluon Plasma and Hadronic Production of Leptons, Photons and Psions. *Phys. Lett. B* **1978**, *78*, 150–153. [] []
351. Feinberg, E.L. Direct Production of Photons and Dileptons in Thermodynamical Models of Multiple Hadron Production. *Nuovo Cim. A* **1976**, *34*, 391–412. []
352. Domokos, G.; Goldman, J.I. Quark-matter diagnostics. *Phys. Rev. D* **1981**, *23*, 203. [] []

353. Stankus, P. Direct photon production in relativistic heavy-ion collisions. *Ann. Rev. Nucl. Part. Sci.* **2005**, *55*, 517–554. [] []
354. Aggarwal, M.M.; Agnihotri, A.; Ahammed, Z.; Angelis, A.L.S.; Antonenko, V.; Arefiev, V.; Astakhov, V.; Avdeitchikov, V.; Awes, T.C.; Baba, P.V.K.S.; et al. Observation of direct photons in central 158-A-GeV Pb-208 + Pb-208 collisions. *Phys. Rev. Lett.* **2000**, *85*, 3595. []
355. Aggarwal, M.M.; Aggarwal, M.M.; Agnihotri, A.; Ahammed, Z.; Angelis, A.L.S.; Antonenko, V.; Areev, V.; Astakhov, V.; Avdeitchikov, V.; Awes, T.C.; Baba, P.V.K.S.; et al. Direct Photon Production in 158-A-GeV Pb-208 + Pb-208 Collisions. *Phys. Rev. Lett.* **2000**, *85*, 3595–3599.
356. Adare, A.; Afanasiev, S.; Aidala, C.; Ajitanand, N.N.; Akiba, Y.; Al-Bataineh, H.; Alexander, J.; Al-Jamel, A.; Aoki, K.; Aphecetche, L.; et al. Enhanced Production of Direct Photons in Au+Au Collisions at = 200 GeV and Implications for the Initial Temperature. *Phys. Rev. Lett.* **2010**, *104*, 132301. []
357. Adam, J.; Adamova, D.; Aggarwal, M.M.; Aglieri Rinella, G.; Agnello, M.; Agrawal, N.; Ahammed, Z.; Ahn, S.U.; Aiola, S.; Akindinov, A. Direct photon production in Pb-Pb collisions at = 2.76 TeV. *Phys. Lett. B* **2016**, *754*, 235–248. [] []
358. Sumbera, M. Soft Physics at RHIC. *EPJ Web. Conf.* **2012**, *28*, 03006. [] []
359. Sumbera, M. Results from STAR Beam Energy Scan Program. *Acta Phys. Pol. Supp.* **2013**, *6*, 429–436. [] []
360. Adare, A.; Afanasiev, S.; Aidala, C.; Ajitanand, N.N.; Akiba, Y.; Akimoto, R.; Al-Bataineh, H.; Alexander, J.; Al-Ta'ani, H.; Angerami, A.; et al. Measurement of higher cumulants of net-charge multiplicity distributions in Au+Au collisions at = 7.7 – 200 GeV. *Phys. Rev. C* **2016**, *93*, 011901. [] []
361. Stephanov, M.A.; Rajagopal, K.; Shuryak, E.V. Signatures of the tricritical point in QCD. *Phys. Rev. Lett.* **1998**, *81*, 4816–4819. [] []
362. Asakawa, M.; Kitazawa, M. Fluctuations of conserved charges in relativistic heavy ion collisions: An introduction. *Prog. Part. Nucl. Phys.* **2016**, *90*, 299–342. [] []
363. Lacey, R.A. Indications for a Critical End Point in the Phase Diagram for Hot and Dense Nuclear Matter. *Phys. Rev. Lett.* **2015**, *114*, 142301.
364. Gavai, R.V.; Gupta, S. Lattice QCD predictions for shapes of event distributions along the freezeout curve in heavy-ion collisions. *Phys. Lett. B* **2011**, *696*, 459–463. [] []
365. Adamczyk, L.; Adkins, J.K.; Agakishiev, G.; Aggarwal, M.M.; Ahammed, Z.; Alekseev, I.; Alford, J.; Anson, C.D.; Aparin, A.; Arkhipkin, D.; et al. Energy Dependence of Moments of Net-proton Multiplicity Distributions at RHIC. *Phys. Rev. Lett.* **2014**, *112*, 032302. [] []
366. Aggarwal, M.M.; Ahammed, Z.; Alakhverdyants, A.V.; Alekseev, I.; Alford, J.; Anderson, B.D.; Arkhipkin, D.; Averichev, G.S.; Balewski, J.; Barnby, L.S.; et al. Higher Moments of Net-proton Multiplicity Distributions at RHIC. *Phys. Rev. Lett.* **2010**, *105*, 022302. []
367. Adamczyk, L.; Adkins, J.K.; Agakishiev, G.; Aggarwal, M.M.; Ahammed, Z.; Alekseev, I.; Alford, J.; Anson, C.D.; Aparin, A.; Arkhipkin, D.; et al. Beam energy dependence of moments of the net-charge multiplicity distributions in Au+Au collisions at RHIC. *Phys. Rev. Lett.* **2014**, *113*, 092301. [] [
368. Luo, X. Energy Dependence of Moments of Net-Proton and Net-Charge Multiplicity Distributions at STAR. In Proceedings of the 9th International Workshop on Critical Point and Onset of Deconfinement, Bielefeld, Germany, 17–21 November 2014.
369. Khachatryan, V.; Sirunyan, A.M.; Tumasyan, A.; Adam, W.; Bergauer, T.; Dragicevic, M.; Erö, J.; Fabjan, C.; Friedl, M.; Frühwirth, R.; et al. Observation of Long-Range Near-Side Angular Correlations in Proton-Proton Collisions at the LHC. *J. High Energy Phys.* **2010**, *1009*, 91. [] []
370. Khachatryan, V.; Sirunyan, A.M.; Tumasyan, A.; Adam, W.; Asilar, E.; Bergauer, T.; Brandstetter, J.; Brondolin, E.; Dragicevic, M.; Erö, J.; et al. Measurement of long-

range near-side two-particle angular correlations in pp collisions at = 13 TeV. *Phys. Rev. Lett.* **2016**, *116*, 172302. []

371. Aad, G.; Abbott, B.; Abdallah, J.; Abdinov, O.; Aben, R.; Abolins, M.; AbouZeid, O.S.; Abramowicz, H.; Abreu, H.; Abreu, R.; et al. Observation of Long-Range Elliptic Azimuthal Anisotropies in = 13 and 2.76 TeV *pp* Collisions with the ATLAS Detector. *Phys. Rev. Lett.* **2016**, *116*, 172301. []

372. Aad, G.; Abajyan, T.; Abbott, B.; Abdallah, J.; Abdel Khalek, S.; Abdelalim, A.A.; Abdinov, O.; Aben, R.; Abi, B.; Abolins, M.; et al. Observation of Associated Near-Side and Away-Side Long-Range Correlations in = 5.02 TeV Proton-Lead Collisions with the ATLAS Detector. *Phys. Rev. Lett.* **2013**, *110*, 182302. [] []

373. Chatrchyan, S.; Khachatryan, V.; Sirunyan, A.M.; Tumasyan, A.; Adam, W.; Aguilo, E.; Bergauer, T.; Dragicevic, M.; Erö, J.; Fabjan, C.; et al. Observation of long-range near-side angular correlations in proton-lead collisions at the LHC. *Phys. Lett. B***2013**, *718*, 795–814. [] []

374. Khachatryan, V.; Sirunyan, A.M.; Tumasyan, A.; Adam, W.; Bergauer, T.; Dragicevic, M.; Erö, J.; Friedl, M.; Frühwirth, R.; Ghete, V.M.; et al. Evidence for Collective Multiparticle Correlations in p-Pb Collisions. *Phys. Rev. Lett.* **2015**, *115*, 012301. [] []

375. Abelev, B.; Adam, J.; Adamova, D.; Adare, A.M.; Aggarwal, M.; Rinella, G.A.; Agnello, M.; Agocs, A.G.; Agostinelli, A.; Ahammed, Z.; et al. Long-range angular correlations on the near and away side in *p*-Pb collisions at = 5.02 TeV. *Phys. Lett. B***2013**, *719*, 29–41. [] []

376. Adare, A.; Afanasiev, S.; Aidala, C.; Ajitanand, N.N.; Akiba, Y.; Akimoto, R.; Al-Bataineh, H.; Alexander, J.; Alfred, M.; Al-Ta'ani, H.; Andrews, K.R. Measurements of elliptic and triangular flow in high-multiplicity ^{3}He+Au collisions at = 200 GeV. *Phys. Rev. Lett.* **2015**, *115*, 142301. []

377. Orjuela Koop, J.D.; Adare, A.; McGlinchey, D.; Nagle, J.L. Azimuthal anisotropy relative to the participant plane from a multiphase transport model in central p + Au , d + Au , and ^{3}He + Au collisions at = 200 GeV. *Phys. Rev. C* **2015**, *92*, 054903. [] []

378. Abelev, B.I.; Aggarwal, M.M.; Ahammed, Z.; Alakhverdyants, A.V.; Anderson, B.D.; Arkhipkin, D.; Averichev, G.S.; Balewski, J.; Barannikova, O.; Barnby, L.S.; Baudot, J.; et al. Long range rapidity correlations and jet production in high energy nuclear collisions. *Phys. Rev. C* **2009**, *80*, 064912. [] []

379. Alver, B.; Back, B.B.; Baker, M.D.; Ballintijn, M.; Barton, D.S.; Betts, R.R.; Bickley, A.A.; Bindel, R.; Busza, W.; Carroll, A.; et al. High transverse momentum triggered correlations over a large pseudorapidity acceptance in Au+Au collisions at s(NN)**1/2 = 200 GeV. *Phys. Rev. Lett.* **2010**, *104*, 062301.

380. Aamodt, K.; Abelev, B.; Quintana, A.A.; Adamova, D.; Adare, A.M.; Aggarwal, M.M.; Rinella, G.A.; Agocs, A.G.; Agostinelli, A.; Salazar, S.A.; et al. Harmonic decomposition of two-particle angular correlations in Pb-Pb collisions at = 2.76 TeV. *Phys. Lett. B* **2012**, *708*, 249–264. [] []

381. Agakishiev, G.; Aggarwal, M.M.; Ahammed, Z.; Alakhverdyants, A.V.; Alekseev, I.; Alford, J.; Anderson, B.D.; Anson, C.D.; Arkhipkin, D.; Averichev, G.S.; et al. System size and energy dependence of near-side di-hadron correlations. *Phys. Rev. C***2012**, *85*, 014903. [] []

382. Voloshin, S.A. Two particle rapidity, transverse momentum, and azimuthal correlations in relativistic nuclear collisions and transverse radial expansion. *Nucl. Phys. A* **2005**, *749*, 287–290. [] []

383. Rezaeian, A.H. Photon-jet ridge at RHIC and the LHC. *Phys. Rev. D* **2016**, *93*, 094030. [] []

384. Dusling, K.; Tribedy, P.; Venugopalan, R. Energy dependence of the ridge in high multiplicity proton-proton collisions. *Phys. Rev. D* **2016**, *93*, 014034. [] []

385. Aad, G.; Abajyan, T.; Abbott, B.; Abdallah, J.; Khalek, S.A.; Abdelalim, A.A.; Abdinov, O.; Aben, R.; Abi, B.; Abolins, M.; et al. Measurement with the ATLAS

detector of multi-particle azimuthal correlations in p+Pb collisions at = 5.02 TeV. *Phys. Lett. B* **2013**, *725*, 60–78. []

386. Abelev, B.; Adam, J.; Adamová, D.; Adaredu, A.M.; Aggarwalcc, M.M.; Rinellaag, G.A.; Ci, M.A.; Agocsdt, A.G.; Agostinelliy, A.; Ahammed, Z. Long-range angular correlations of ß, K and p in p-Pb collisions at = 5.02 TeV. *Phys. Lett. B* **2013**, *726*, 164–177. [] []

387. Abelev, B.; Adam, J.; Adamová, D.; Aggarwal, M.M.; Rinella, G.A.; Agnello, M.; Agostinelli, A.; Agrawal, N.; Ahammed, Z.; Ahmad, N.; et al. Multiparticle azimuthal correlations in p -Pb and Pb-Pb collisions at the CERN Large Hadron Collider. *Phys. Rev. C* **2014**, *90*, 054901. [] []

388. Khachatryan, V.; Sirunyan, A.M.; Tumasyan, A.; Adam, W.; Bergauer, T.; Dragicevic, M.; Erö, J.; Fabjan, C.; Friedl, M.; Frühwirth, R.; et al. Long-range two-particle correlations of strange hadrons with charged particles in pPb and PbPb collisions at LHC energies. *Phys. Lett. B* **2015**, *742*, 200–224. [] []

389. Hirono, Y.; Shuryak, E. Femtoscopic signature of strong radial flow in high-multiplicity *pp* collisions. *Phys. Rev. C* **2015**, *91*, 054915. [] []

390. Bozek, P.; Broniowski, W. Correlations from hydrodynamic flow in p-Pb collisions. *Phys. Lett. B* **2013**, *718*, 1557. [] []

391. Bozek, P.; Broniowski, W.; Torrieri, G. Mass hierarchy in identified particle distributions in proton-lead collisions. *Phys. Rev. Lett.* **2013**, *111*, 172303. [] []

392. Werner, K.; Bleicher, M.; Guiot, B.; Karpenko, I.; Pierog, T. Evidence for Flow from Hydrodynamic Simulations of *p* Pb Collisions at 5.02 TeV from v_2 Mass Splitting. *Phys. Rev. Lett.* **2014**, *112*, 232301.

393. Romatschke, P. Light-Heavy Ion Collisions: A window into pre-equilibrium QCD dynamics? *Eur. Phys. J. C* **2015**, *75*, 305. [] []

394. Niemi, H. Recent developments in hydrodynamics and collectivity in small systems. In Proceedings of the European Physical Society Conference on High Energy Physics (EPS-HEP2015), Vienna, Austria, 22–29 July 2015.

395. Koblesky, T. Collectivity in Small QCD Systems. In Proceedings of the CIPANP2015 Conference, Vail, CO, USA, 19–24 May 2015.

396. Schenke, B.; Venugopalan, R. Eccentric protons? Sensitivity of flow to system size and shape in p+p, p+Pb and Pb+Pb collisions. *Phys. Rev. Lett.* **2014**, *113*, 102301.

397. Bzdak, A.; Ma, G.L. Elliptic and triangular flow in *p*+Pb and peripheral Pb+Pb collisions from parton scatterings. *Phys. Rev. Lett.* **2014**, *113*, 252301. [] []

398. Bozek, P.; Bzdak, A.; Ma, G.L. Rapidity dependence of elliptic and triangular flow in proton–nucleus collisions from collective dynamics. *Phys. Lett. B* **2015**, *748*, 301–305. [] []

399. Antinori, F.; Becattini, F.; Braun-Munzinger, P.; Chujo, T.; Hamagaki, H.; Harris, J.; Heinz, U.; Hippolyte, B.; Hirano, T.; Jacak, B.; et al. Thoughts on Heavy-Ion Physics in the High Luminosity Era: The Soft Sector. *arXiv*, 2016; arXiv:1604.03310. []

400. Bala, R.; Bautista, I.; Bielcikova, J.; Ortiz, A. Heavy-ion physics at the LHC: Review of Run I results. *Int. J. Mod. Phys. E* **2016**, *25*, 1642006. [] []

401. Sinyukov, Y.M.; Akkelin, S.V.; Karpenko, I.A.; Shapoval, V.M. Femtoscopic and Nonfemtoscopic Two-Particle Correlations in $A + A$ and $p + p$ Collisions at RHIC and LHC Energies. *Adv. High Energy Phys.* **2013**, *2013*, 198928. [] []

402. Lisa, M. Timescales in heavy ion collisions. *Acta Phys. Pol. B* **2016**, *47*, 1847–1855. [] []

NOTES

1. The string picture of hadrons also makes it possible to explain the transition from a hadronic to a QGP state of matter at finite temperature T as a phase transition from the ordered state to the disordered state (see Ref. [9], p. 44). With T approaching the critical temperature of deconfinement , the effective string tension decreases, and q and attached to its end points lose their correlation.

2. For a representative collection of papers tracing the development of theoretical ideas on the QCD deconfining phase transition before nineties , see Ref. [19].

CHAPTER 4

Laser-Driven Ion Acceleration From Relativistically Transparent Nanotargets

B M Hegelich[1,2,5], I Pomerantz[1], L Yin[2], H C Wu[2], D Jung[2,3,4], B J Albright[2], D C Gautier[2], S Letzring[2], S Palaniyappan[2], R Shah[2], K Allinger[3], R Hörlein[4], J Schreiber[3,4], D Habs[3,4], J Blakeney[1], G Dyer[1], L Fuller[1],E Gaul[1], E Mccary[1], A R Meadows[1], C Wang[1], T Ditmire[1] and J C Fernandez[2]

[1] *The University of Texas, Austin, TX 78712, USA*

[2] *Los Alamos National Laboratory, Los Alamos, NM 87545, USA*

[3] *Department fur Physik, Ludwig-Maximilians-Universit ¨ at M¨ unchen, ¨ D-85748 Garching, Germany 4 Max-Planck-Institut fur Quantenoptik, D-85748 Garching, Germany*

[4] *The University of Texas, Austin, TX 78712, USA*

ABSTRACT

Here we present experimental results on laser-driven ion acceleration from relativistically transparent, overdense plasmas in the break-out afterburner (BOA) regime. Experiments were preformed at the Trident ultra-high contrast laser facility at Los Alamos National Laboratory, and at the Texas Petawatt laser facility, located in the University of Texas at Austin. It is shown that when the target becomes relativistically transparent to the laser, an epoch of dramatic acceleration of ions occurs that lasts until the electron density in the expanding target reduces to the critical density in the non-relativistic limit. For given laser parameters, the optimal target thickness yielding the highest maximum ion energy is one in which this time window for ion acceleration overlaps with the intensity peak of the laser pulse. A simple analytic model of relativistically induced transparency is presented for plasma expansion at the time-evolving sound speed, from which these times may be estimated. The maximum ion energy attainable is controlled by the finite acceleration volume and time over which the BOA acts.

1. INTRODUCTION

An ultrahigh-intensity laser interacting with matter accelerates electrons to nearly the speed of light. These, in turn, can be used to accelerate protons and heavier ions. Since electromagnetic fields in laser plasmas can be more than a million times stronger than in conventional accelerators, lasers have the potential to shrink accelerators from kilometer- to room size, becoming suitable for universities, industry and hospitals. How-ever, the maximum ion energy observed seems to have saturated and remains too low for many applications. We present a detailed experimental demonstration of a new laser acceleration regime, called break-out afterburner (BOA) acceleration, with much higher performance than other acceleration mechanisms. With the ultrahigh pulse contrast at ultrahigh power at the Los Alamos National Laboratory (LANL) Trident laser and nm-thick diamond targets, the energy of the laser-accelerated carbon ions increases by an order of magnitude, exceeding the required energy for ion fast ignition in inertial fusion energy production and enabling for the first time biological experiments with laser-driven carbon ions relevant to hadron tumor therapy. Laser-ion accelerators have made rapid progress over the last decade. Since the demonstration of directed multi-MeV proton beams accelerated by ultrahigh intensity lasers [1, 2], these beams have been studied intensely, motivated by the potential of shrinking kilometers of conventional accelerator hardware to a room-sized device that can be used for applications in nuclear physics, energy, material science and medicine. Characterization and control of many beam parameters (e.g. divergence [3, 4], emittance [5, 6], species [7]) has been demonstrated, as has a technique for laser-based, mono-energetic ion acceleration [8–10]. While much new knowledge has been gained, peak particle energy has stagnated. The maximum published proton energy of 60 MeV was reached in the first experiments at the Nova PW laser in 1999 utilizing the 'target normal sheath acceleration' (TNSA) process [11]. While this is sufficient for proton fast-ignition, it is not enough for proton therapy (250 MeV protons are required). For heavier ions like carbon and oxygen the maximum published ion energies are in the range of 3–5 MeV amu^{-1} [12]. The required energies for ion-based ($Z > 1$) fast ignition (IFI) is ~40 MeV amu^{-1} [13–15], and for ion-based tumor therapy [16] energies of 400 MeV amu^{-1} are required. A more effective acceleration process is needed. Radiation pressure acceleration (RPA) [17–20] is one proposed mechanism, and while we were able to show a first onset of RPA experimentally [21], the most promising 'light sail' regime [22] of RPA requires very high laser intensities (> 10^{22} W cm^{-2}) and short pulses (< 100 fs) to prevent large scale target deformation during the acceleration phase. The BOA [23–26] is another candidate process with the advantage that lower intensities (~10^{20} W cm^{-2}) and longer pulses (~500 fs) may be used, which for current laser systems contain one–two orders of magnitude more energy. While the underlying physics in these regimes differ, they share features distinguishing them from TNSA: in TNSA, ion acceleration is decoupled from the laser by the thick (> μm) target [10]. In contrast, in the RPA and BOA, the laser interacts directly with electrons co-propagating with the beam ions. The accelerating fields are larger than in TNSA and persist for longer, thus providing a larger field gradient and a longer effective acceleration distance. Also, in RPA and BOA the high-z species see the same or even higher accelerating fields than the accompanying protons, whereas in TNSA the protons shield the fields from the heavier species. Furthermore, as explained below and in [27], the different dynamics of BOA leads to a fundamentally different angular distribution of the ions compared with TNSA. The main technical challenge is to field a target that remains overdense to the laser when the peak of the pulse arrives, yet is thin enough for the dynamics of the mechanism to proceed. This requires ultrahigh contrast laser pulses and stable, free-standing, nanometer targets with high aspect ratio (~10^5) and ~nanometer thickness. We report on the first experimental realization and comprehensive understanding of the BOA, which proceeds as follows [23, 25]: an ultrahigh intensity pulse interacts with an ultrathin (~ few skindepth) solid density (n_e ~10^{23} cm^{-3}) target. The laser heats the electrons to high γ and drives a hydrodynamic expansion of the target. At time t_1, the target becomes relativistically transparent ($n_e/n_{cr}\gamma \approx 1$) and the BOA phase begins [26]. The laser propagates through the target, pushing electrons forward and creating electrostatic fields that efficiently transfer energy from the electrons to the ions; this energy is immediately replenished by the laser. The BOA persists until the density of the expanding target becomes too low for efficient coupling, which occurs at time t_2, when the target becomes classically underdense ($n_e/n_{cr} < 1$). In our experiments, the BOA boosts the peak carbon ion energy to ~44 MeV nucleon^{-1}, i.e. >0.5 GeV, ten times higher than the highest published energy to date for laser-accelerated carbons with TNSA. This is achieved with a laser power of ~150 TW at an intensity of ~2×10^{20} W cm^{-2}, well below the peak of current technology at 1.1 PW [28] and 2×10^{22} W cm^{-2} [29], respectively. The peak particle energy achieved is already suitable for many applications, including

IFI [30–32]. This mechanism scales to multi-GeV ion energy with currently available PW-class systems, bringing laser-driven accelerators substantially closer to deployment in advanced accelerators [33], new diagnostics [34] and medical physics applications [35].

2. EXPERIMENTAL SETUP AND RESULTS

In our experiments, the laser parameters were held nominally constant, while the target thickness was systematically varied to span the full range of the BOA regime. The technical difficulties in realizing RPA and BOA in particular are twofold: the target needs to be thin enough that the electrons within the laser spot can be heated to relativistic transparency; and it must remain classically overdense to enable efficient coupling and energy transfer from laser-to-electrons (and thus, to ions). To realize these conditions experimentally, two advances were necessary: ultra-thin, nm-scale, free-standing robust targets and sufficiently high laser contrast to prevent premature ionization and expansion of the target before the peak of the pulse. The first was met by utilizing nm-thin diamond like carbon (DLC) foils, produced at LMU Munich [36] using a specialized cathodic arc discharge [37], which yields targets from 3–60 nm thickness. DLC is a form of amorphous carbon with a high content of sp^3-bonds (50–75% for the targets used), so the foils possess exceptional mechanical robustness. Furthermore the DLC foils are optically transparent, which increases, relative to metallic targets, resilience against target heating and vaporization making them less susceptible to laser prepulses. A second set of targets was produced by chemical vapor deposition (CVD)6, yielding ~1 μm thickness targets that were then ion etched as far down as ~75 nm, giving a total thickness range in the experiments from 3 to 2000 nm with a nominal uncertainty of 3 nm. The thickness variations are largest at the limits of the respective methods, i.e. at 10 and 50 nm for DLC targets, and 100 nm for CVD targets, leading to larger fluctuations. To realize the second condition, extraordinarily high contrast laser pulses are required. The laser contrast C_{time} is the ratio of the pedestal/prepulse intensity at the time given by the subscript to the peak intensity of the main pulse. For a typical ultra-high intensity system that ratio is $\sim 10^{-6}$ at ~1 ns before the 500 fs peak pulse. For a focused peak pulse intensity of $I \sim 10^{20}$ W cm^{-2}, the prepulse intensity will therefore be $\sim 10^{14}$ W cm^{-2}, creating a preplasma and destroying the target before the peak pulse hits. At $I > 10^{20}$ W cm^{-2} contrast ratios of $C_{1\text{ns}} < 10^{-11}$ and $C_{5\text{ps}} < 10^{-6}$ are required to observe BOA. At the Trident laser facility [38], we realized these ultra-high contrast conditions by using a new cleaning system based on an optical parametric pre-pulse eliminator [39, 40].

Based on our understanding of BOA we expect to observe the following behavior:

i. An increased maximum ion energy compared to the TNSA regime.

ii. A broken angular symmetry of the beam, with the fastest ions emitted off-axis in a plane orthogonal to the laser polarization.

iii. Higher maximum velocities for ions of larger mass (here: carbon ions and protons) at optimal BOA conditions.

iv. The existence of an optimal target thickness for a given set of laser parameters and target density.

Experiments with ultra-thin targets have been performed before but in a completely different regime both in terms of laser and plasma parameters [41]. Here we describe two experimental campaigns: the first was performed at the Trident ultra-high contrast laser facility at LANL, and the second at the Texas Petawatt (TPW) laser facility [42], located in the University of Texas at Austin. On the TPW experiments we used a double plasma mirror (PM) setup [43] to improve the contrast on-target, a method that we also used on our earlier experiments on Trident [44].

2.1. Texas Petawatt experiments

In the TPW experiments, an $f/3$ off-axis parabolic mirror is used to focus the 1053 nm wavelength laser pulses with pulse durations of 160 fs (full-length at half-maximum (FWHM)) and typical energies of 90 J7 to a spot containing ~30% of the energy within a 5 μm radius. The laser pulse duration and beam parameters were carefully recorded during the run. The experimental setup is depicted in figure 1(a). The focusing beam is reflected twice from a pair of PMs, positioned at 45° to the beam normal and parallel to each other. The PMs were positioned so that the spot size on the first PM is 12 mm. Both anti-reflective coated and non-coated substrates were used in the experiment. The energy throughput of the PM pair was measured by picking off a part of the expanding beam into an energy meter. The throughput of the PMs was found to have typical values of 69%, in agreement with previously published data [43, 45].

Plastic (CH_2) and DLC foils in a thickness range of 40–2000 nm were used as targets. Nuclear activation of a copper foil stack [46] and a wide angle magnetic spectrometer (iWasp) [47] were used to record both the spatial and energy distribution of the emitted ions. Following a shot, the copper foils were placed on top of an electron-sensitive Fuji image plate (IP) for a period of 40 min ($\sim\tau_{1/2}$). Figure 2(b) shows typical IP scans of a six-foil stack. The emitted positrons deposit a signal proportional to the activity level onto the IP. Tabulated energy loss values for protons in copper [48]

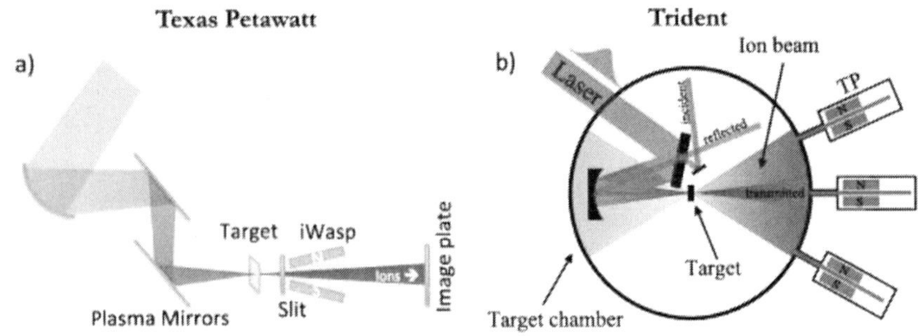

FIGURE 1. Experimental setup (a) at the TPW experiment: the focusing beam is reflected twice by a pair of PMs, before impinging on the target foil. Ions emitted from the back of the target enter the iWasp through a thin slit, and are deflected upward in the magnetic field. The ion signal is recorded using an image plate positioned behind the spectrometer. (b) at Trident: TP spectrometers on-axis and off-axis at 8.5° and a variety of secondary diagnostics, such as electron spectrometers and laser pulse diagnostics.

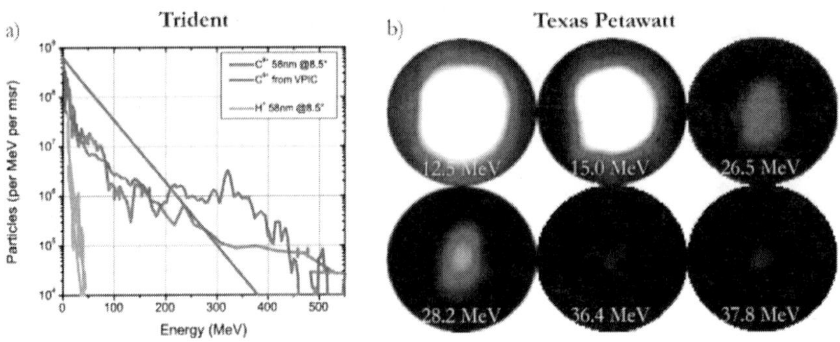

FIGURE 2. (a) Ion spectra from BOA acceleration using a 58 nm DLC target, optimized for carbon acceleration under Trident conditions (this shot: 90 J, 540 fs, 2×10^{20} W cm^{-2}): the highest ion energies where observed at an 8.5° angle in the plane orthogonal to the laser polarization (red line) and agree well with a two-dimensional (2D) VPIC simulation with the same shot parameters (green line). The accelerated protons (grey) obtain the same energy per nucleon, i.e. the same velocity, as predicted for the BOA regime. (b) IP scans of a stack of six Cu foils which were positioned 10 mm behind a 185 nm CH_2 target at the TPW. The most probable energy for the activation of each foil is indicated. The signal corresponds to an angular proton beam divergence range of 15° (for 12.5 MeV protons) to 5° (for 37.8 MeV protons).

were used to calculate the most probable proton energy required to activate each foil in the stack (indicated in the figure).

An ion-sensitive Fuji IP [49] positioned downstream to the iWasp was used to record the signal of the deflected ions. The signal of x-ray passing through the iWasp slit recorded on the IP provided an absolute reference to determine the amount of deflection. Calibration shots in which a CR-39 [50] plastic detector was placed in front of the IP provided an absolute calibration for the deflected ion energy.

2.2. Trident experiments

The Trident laser delivers pulses of nominally 150 TW at a repetition rate of ~1 shot per hour, i.e. a very limited number of shots in an experiment. The data discussed here were gathered in three independent beam times in April 2009, March 2010 and April 2011, resulting in a total of ~200. Pulse energy and pulse duration are measured for every shot and shot-to-shot fluctuations between 60 and 100 J and 500 and 900 fs are observed. The pulse is focused with an $f/3$ optic to a spot containing ~45% of the energy within a 3.5 μm radius, and ~10% of the energy within a 1 μm radius, resulting in an average intensity over the 7 μm spot of I_{ave} ~0.8–2.3 × 10^{20} W cm^{-2} and a peak intensity $I_{peak} \approx 2.1 - 6.4 \times 10^{20}$ W cm^{-2}. The focus longitudinal position varied by ±30 μm initially and < ± 10 μm in later campaigns and could not be measured on the shot; this roughly doubled the uncertainty in the actual on-target intensity. In the standard configuration for this experiment, as depicted in figure 1(b), the pulse hits the target at normal incidence (0°) and is linearly polarized in the vertical plane.

The main diagnostic consists of two Thomson parabola (TP) ion spectrometers [51], specifically designed for resolution and maximum trace separation at high particle energies [52]. They are fielded at 0° and 8.5° in the horizontal plane downstream behind the target with a stacked arrangement of CR-39 [50] and Fuji IPs [49] as detectors. The CR-39 was then analyzed by a specialized automated analysis system [53]. Secondary diagnostics include film stacks to record the ion beam profile, electron-, visible light and UV spectrometers and a proton/electron spectrometer [54] at 6°. A full aperture backscatter diagnostic [55] recorded pictures of the target at the time of the interaction, confirming that 3 nm targets were still overdense at peak pulse interaction. In the last campaign we also fielded a single shot frequency-resolved optical gating device (FROG) [56] to measure the shape, duration and phase of the pulse as it is reflected from the target. This technique, described in [26], gives a direct measurement of relativistic transparency by characterizing the effect on the driving laser pulse. In a separate experiment we also fielded a FROG in transmission to get an even more complete characterization [26]. This was not possible here because the required solid angle was taken up by the ion diagnostics.

As we show here, all four signatures have been observed in our recent experiments, being consistent with full BOA acceleration.

(1) *Increased maximum energies.* Ion energies in excess of 0.5 GeV were observed for target thicknesses of ~50–250 nm. Figure 2 shows carbon C^{6+} and proton spectra from a 58 nm DLC target. The carbon ions exceed an energy of 0.5 GeV, an order of magnitude higher than reported previously for a Trident-class laser working in the TNSA regime. This represents a huge step toward particle energies required for hadron therapy (2–3 GeV). At an energy of 0.5 GeV, ~10^8 carbon ion are within a 10 MeV bin. Scaling this number for the requirements of hadron therapy will require a few-Hz repetition rate system. These ion energies also exceed the requirements for IFI [31, 32], although scaling to the required ~10^{14} carbon ions per shot forms a substantial challenge for current laser technology.

(2) *Angular distribution.* The highest ion energies of 550 MeV are observed at the 8.5° TP (red curve on figure 3(b)), while the cutoff energy on axis (black curve) is significantly lower at 230 MeV. These results are in agreement with 2D and three-dimensional (3D) particle-in-cell (PIC) simulations, using the VPIC code [57]. To resolve the dynamics accurately, the simulations must resolve the initial plasma skin depth $\delta = c/\omega_{pe}$ ($\omega^2_{pe} = 4\pi n_e e^2/m_e$) and initial electron Debye length $\lambda_d = v_{the}/\omega_{pe}$, where v_{the} is electron thermal speed, which, given the large simulation volume over ps simulation time duration, requires the largest supercomputers and most efficient codes. To compare to the experimental results, we simulated a laser pulse with a wavelength $\lambda = 1\ \mu$m, a temporal profile $I(t) = I_0 \sin^2(t\pi/2\tau)$, $\tau = 540$ fs (FWHM) and a transverse profile $E_y \approx \exp(-z^2/w^2)$ with $w = 5.12\ \mu$m, incident on a solid density C^{6+} target with $n_e/n_{cr} = 821$ (2.8 g cc^{-1}) mixed with 20% protons in number density. A typical 2D simulation domain is 50 μm × 50 μm in (x,z). To resolve the Debye length, the electron temperature is set to $T_e = 18$ keV or 165 keV initially, which at $n_e = 821 n_{cr}$ corresponds to a Debye length of λ_d ~1 or ~3.3 nm, respectively. In 2D, the cell size is 0.3 or 0.44 × 1.76λ_d along x and z, with 1000 particles $cell^{-1}$. VPIC simulations are performed over a range of target thickness from 4 to 2000 nm to compare with the experimental scaling. In figure 2, the simulated ion energy spectrum for a 58 nm target at t_2 (green line) is overlaid with the corresponding experimental spectrum (red line), showing good quantitative agreement. In the experiments and even in the 2D simulations we find that the highest ion energies are found oblique to the laser propagation. To understand why, we turn to a 3D VPIC simulation of the BOA. This simulation was performed well in advance of the experiment with parameters slightly different from the actual shot

parameters8. Figure 3(a) shows iso-surfaces of C^{6+}kinetic energy between 500 and 700 MeV together with iso-surfaces of E^2y. Contours of carbon kinetic energy in the x–z plane are displayed in the inset in a cut plane located along the laser axis. A contour of accelerating electric field is also shown in the orthogonal cut plane in the inset; the peak field co-moves with the carbon ion density peak. The complete ion acceleration dynamics are shown in a supplementary simulation movie accompanying figure 3 (available from stacks.iop.org/NJP/15/085015/mmedia). From the 3D image, the highest-energy ions are off-axis in lobes oriented along the direction perpendicular to the laser polarization. Ion spectra corresponding to the observation angles of the experiment from this simulation are shown in the figure 3(b). The simulated spectra (right) qualitatively reproduce the observed experimental results (left). The ion beam lobes result from the ponderomotive force on the electrons during the BOA, which expels electrons radially from the center of the laser pulse and creates a slowly varying perpendicular momentum profile $\vec{p}^{\,s}_{e\perp}$ that is nearly azimuthally symmetric. (Here '⊥' indicates the plane perpendicular to the laser propagation). The time-averaged electron fluid density evolves according to the continuity equation $\partial \langle n_e \rangle / \partial t = -\nabla \langle n_e \vec{p}_e m_e^{-1} \gamma^{-1} \rangle$, where brackets indicate an average over a laser cycle, n_e is the electron density in the laboratory frame, and $\vec{p}_e$ is the electron fluid momentum. The quantity $\gamma^{-1} = \{1 + [(\vec{p}_{e\perp} + e\vec{A}/c)^2 + p^2_{e\parallel}]/(m_e c)^2\}^{-1/2}$ contains a dot product of slow and fast components of the perpendicular electron momentum, which in turn produces a smaller time-averaged γ^{-1} along the laser polarization than perpendicular to it. While single-particle relativistic orbits of electrons in the laser field would yield no breaking of angular symmetry, in dense plasma, where collective effects manifest, the relativistic fluid kinematics change: an impulse applied to a population of moving electrons in the direction of their motion produces a smaller change in fluid radial speed than an impulse of the same magnitude applied perpendicular to their motion. In the time-averaged fluid dynamics, this difference leads to a larger flux of electrons perpendicular to the laser polarization and, thus, to the production of lobes in the electron density. Space-charge electric fields from these electron density lobes generate corresponding lobes in the most energetic portion of the ion beams. More details on the relativistic kinematics may be found in [27].

(3) *Mass dependence of the acceleration process.* A further strong indicator that we are in the BOA regime is that the carbon velocities by far exceed 1/2 the proton velocity, the theoretical limit of TNSA; in some cases carbon ions even outran the protons. Figure 2 shows an example where the fastest carbon ions gained ~44 MeV amu^{-1} while the protons (gray curve) gained ~42 MeV amu^{-1} corresponding to ratio of carbon-to-proton velocity

of $v_{ion}/v_{proton} = 1 \pm 0.1$. In TNSA this ratio can never exceed 0.5, whereas $v_{ion}/v_{proton} > 0.5$ is a characteristic feature of BOA, as shown in [25]. Since the target is transparent and all bulk ions take part in the interaction, the maximum fields are set up by the electrons and the ion species that contributes most of the charge, and move with those distributions. Less abundant ion species, especially if they are lighter, show a different dynamics can see lower fields for part of the time. This situation is impossible to obtain in pure TNSA, where only the few atomic ion layers on the rear surface take part in the interaction. In TNSA, all accelerated ions see the same *quasi-static* field. The higher charge-to-mass ratio of protons ensures that they obtain at least twice the velocity of the next heavier fully stripped ion species. In practice, the difference is even higher, with the largest reported fraction of $v_{ion}/v_{proton} = 0.16$ in [7].

(4) *Scaling and analytic model.* We expect a characteristic optimal target thickness for the BOA that depends on the times t_1 and t_2, when the target becomes relativistically transparent and classically underdense, respectively. This expectation is based on an extensive parameter scan using VPIC simulations and on a simple analytical model for the maximum ion energies. It is confirmed by the aggregate data gathered over three independent beam times. In the simulations, >80% of the ion acceleration in the BOA occurs between times t_1, when the target becomes relativistically transparent ($n_e = \gamma n_{cr}$) and t_2, when the electron density in the expanding target reduces to $n_e \approx n_{cr}$. Figure 4(a) shows maximum carbon ion energy as a function of time overlaid with a temporal profile of the laser intensity. The carbon energy gain is small before t_1 and after t_2 (not shown). For time $t_1 < t < t_2$, the target density decreases and γ increases, as in figure 4(b). Peak laser intensity falls between t_1 and t_2, so the average laser intensity, the energy source for the BOA, is maximized.

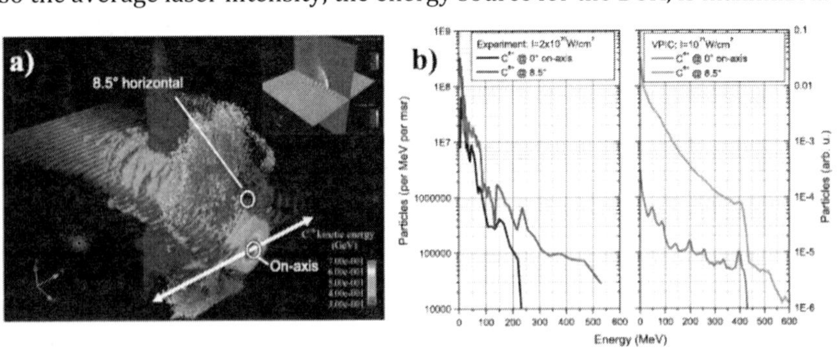

FIGURE 3. (a) 3D VPIC simulation of BOA acceleration in the interaction of a 156 fs (FWHM) laser pulse at 10^{21} W cm^{-2} with a 30 nm solid density carbon target. Shown is a 3D view near time t_2, at the end of the BOA phase. Iso-surfaces of laser pulse intensity are represented by the color coded structure on axis and is linearly polarized in the horizontal plane. Carbon

ion kinetic energy iso-surfaces are color-coded for their energy as depicted by the legend and reach energies of ~700 MeV. The highest energies are observed at a slight angle in the vertical plane. The horizontal plane shows projections of the laser field, while the vertical plane shows projections of the ion energies. The inset shows a cutaway view of the ion acceleration, with the ion energies shown color coded on the vertical plane and the accelerating electric field color-coded on the horizontal plane. The longitudinal field co-moves with the ion front, with the highest field strength coinciding with the highest ion energies. (b) C^{6+} spectra on axis and at 8.5°. From experiment (left) and from the 3D VPIC simulation (right) at $\tau = 165$ fs. The simulation was performed well in advance of the experiment with parameters slightly different from the actual shot parameters.

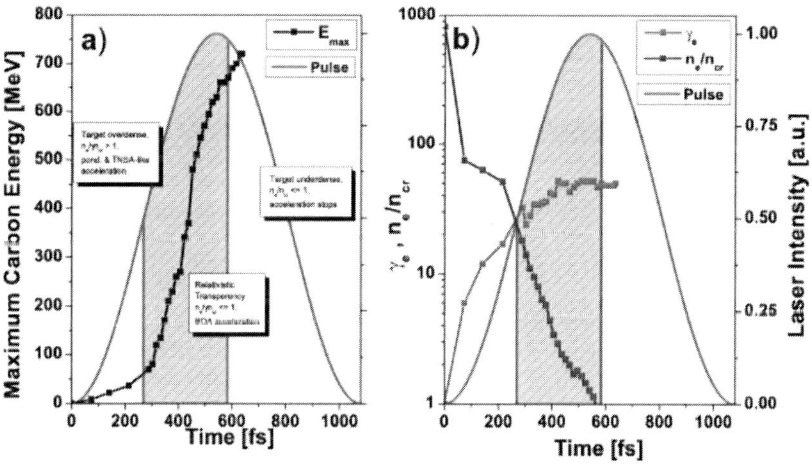

FIGURE 4. 2D VPIC simulation for Trident pulse: $E = 90$ J, $t = 540$ fs, $I_{ave} = 2 \times 10^{20}$ W cm^{-2}, $I_{peak} = 5 \times 10^{20}$ W cm^{-2}, target 58 nm DLC at $821 n_{e,0}/n_{cr}$. Three quantities are plotted over the course of the pulse: maximum ion energy (a), electron density n_e/n_{cr} (b) and the electron γ-factor (b). Over the course of the interaction, three phases can be identified: phase I: $n_e/(n_{cr}\gamma_e) \geq 1$ the target is overdense, the laser pulse heats electrons, the target expands and moderate ion acceleration to ~0.1c takes place due to ponderomotive pressure and quasi-static field build-up. The ions reach ~10–20% of their final energy; phase II: $n_e/(n_{cr}\gamma_e) \leq 1$, the target becomes relativistically transparent and the pulse starts to propagate through the target, efficiently coupling laser energy into the ions. The remaining 80–90% of the final ion energy is gained in this stage. Phase III: $n_e/n_{cr} < 1$, the target becomes classically underdense, the laser coupling efficiency to the electrons is minimal and acceleration ceases.

3. ANALYTICAL MODEL

Based on the understanding gained in the VPIC simulations and a first set of data from targets ranging from 4–60 nm, we developed a reduced analytical model that predicts the maximum ion energies for a given set of laser and target parameters. While not as complete as the PIC simulations that include all details of the highly nonlinear, relativistic physics, it captures the essential behavior and guides design of experiments and laser systems without requiring substantial computing resources. The best coupling between the laser pulse and foil electrons arises when the electron density approaches the relativistic transparency density at $t = t_1$, estimated assuming one-dimensional target expansion driven by a laser pulse $a = a_0 \sin 2(\Omega t)$, where a_0 is the dimensionless light amplitude of the laser pulse, with pulse length τ and $\Omega = \pi/2\tau$. Provided the target thickness d is much smaller than the focal spot diameter, $t_1 = [(M_i/m_e)(3N^2d^2/Z_i\Omega^2c^2a^3{}_0)]^{1/4}$, where Z_i is ion core charge, $N = n_e/n_{cr}$ is the normalized electron density, d the target thickness, c the speed of light and m_e and M_i are the electron and ion masses, as described in [58]. Alternate descriptions of relativistic transparency can be found, e.g. [59, 60]. Between t_1 and t_2, the laser drives a 3D, spherical, isotropic expansion of the target until $n_2 = n_e(t_2)/n_{cr} = 1$; expansion occurs over a time interval $t_1 - t_2 = Nd(\gamma^{1/3} - 1)/\gamma C_s \sin(\Omega\tau)$, where $C_s \cong (Z_i m_e c^2 a_0/M_i)^{1/2}$. After t_2, no more significant ion acceleration occurs. An important difference between the BOA and TNSA is the former's collective, as opposed to stochastic, electron motion. In TNSA, electrons only interact with the laser field over the time spent within a skin-depth layer of the target, which tends to wash out collective effects. In the BOA, relativistically induced transparency allows the laser field to interact directly with the bulk electron population over the primary acceleration phase. This leads to collective electron motion driven by the laser field and a coupling among the laser, electrons and ions during the acceleration period. We evaluate the ion energy gain in the BOA model by applying the analysis reported in [58], in which the energy gain between t_1 and t_2 (the predominant contribution to the ion energy) is obtained. This ion energy is approximately $E_{ion} = (2\alpha + 1)Z_i m_e c^2 f(t_1,t_2,\alpha)$, where Z_i is the ion core charge and $\bar{a}_0$ is the average normalized laser amplitude between t_1 and t_2. The function $f(t_1,t_2,\alpha) = [1 + \Omega(t_2 - t_1)]^{1/(2\alpha+1)-1}$ arises from a time integral over the interaction and contains the collective stopping parameter α indicating the strength of coupling between electrons and ions; $\alpha \sim 3$ for Trident parameters.

For a given set of laser parameters and target density, this model predicts an optimum target thickness, at which the coupling among the laser ponderomotive drive, electrons, and ions is maximized. The evaluation of the model and its comparison with both PIC simulations and Trident data are shown in figure 5(a). The data were collected in three experimental campaigns from April 09 until April 11. All shots considered have $E > 70$ J, $I_{min} > 2.1 \times 10^{20}$ W cm^{-2}. We evaluated the model (green lines) for three different peak intensities: the maximum intensity observed on a shot for the presented data set ($I_{max} \approx 6.4 \times 10^{20}$ W cm^{-2}), the minimum intensity ($I_{min} \approx 2.1 \times 10^{20}$ W cm^{-2}) and a typical intensity, which was also used in the simulations ($I_{sim} \approx 5 \times 10^{20}$ W cm^{-2}). Also, target expansion will occur due to remaining imperfections even in the cleaned pulse. We evaluated the model for expansions corresponding to 300–600 n_c, from the cold target electron density of 900 n_c. In addition to the model we plot cutoff energies from a suite of 2D VPIC simulations9 (green triangles) and a corresponding model calculation (green dotted line). Figure 5(b) shows proton cutoff energies as a function of target thickness as measured on the recent TPW experiments. The calculated curve (dashed black) used the recorded beam parameters averaged over all shots, with $E = 62$ J (energy on-target), $\tau = 160$ fs and a spot size radius of 5 μm. While the PMs significantly improve the TPW laser contrast, there are still levels of prepulse present exceeding those in the Trident system. Consequently we observe an increase in target pre-expansion to about 1/3 of the initial cold target density of $n_e = 280 n_{cr}$ (dotted black). Using our relativistic model, we find a best fit to the density at peak interaction at a value of $n_e = 90 n_{cr}$.

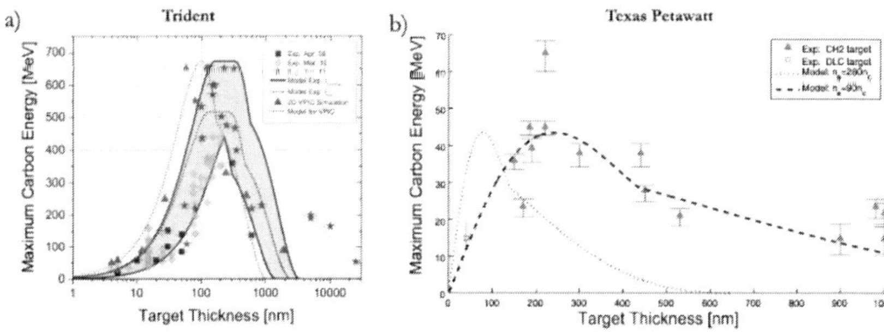

FIGURE 5. Cutoff energy versus target thickness: (a) comparison of Trident experimental data, simulations and an analytical model. The symbols are the measured maximum cutoff energies for different target thickness, the error bars are a function of target fluctuation and number of shots at a given thickness. The green triangles are the equivalent cutoff energies obtained by 2D VPIC simulations. The blue lines are the predictions of our analytical model for different laser intensities (Min, Max, typical) and $n_e = 300$–$600\ n_c$. The model predictions are in good agreement over the validity range of the model up to ~600 nm (beyond that, the model predicts much lower energies since it does not contain the physics of thick target interaction and therefore underestimates the TNSA in this regime). Also shown in green is the analytical model prediction for the 2D VPIC

simulation conditions. (b) Results from TPW experiments employing PMs to improve the on-target contrast. Data were collected by either the iWasp or Cu stacks. BOA calculated curves are shown assuming an initial density of a cold target, $n_e = 280n_{cr}$ (for CH_2), and for a best fit initial density value of $n_e = 90n_{cr}$.

We observe the same qualitative behavior in the data, the simulations and the analytic model: a clear maximum around an optimal thickness of 150 nm with a sharp decrease toward thinner targets and a gradual decrease toward thicker targets. For thinner targets (4–30 nm), the maximum energy decreases because the target becomes transparent early and expands rapidly, leading to a shorter acceleration period and lower average laser intensity during the BOA. On the other hand, much thicker targets (300–600 nm) only become transparent at the end of the pulse, when the laser intensity is low. Above 600 nm, the target remains opaque and the model and experimental data deviate, since TNSA physics is not included in the model. Figure 6(a) shows the FROG trace of the reflected pulse which we used to directly probe the induced target transparency, from a shot on a 165 nm thick DLC target. Figure 6(b) shows the analyzed temporal intensity profile (red) and temporal phase (blue). Overlaid is a representation of the input pulse (black). As explained in detail in [26], the steep drop in the phase is indicative of the onset of relativistic transparency at t_1. The vertical green line indicates the value for t_1 as predicted by our analytic model. As the density drops further, reflection decreases and finally ceases completely. The orange line shows the model prediction for t_2, where $n_e \approx n_{cr}$. Despite the large uncertainty in target intensity, we observe good quantitative agreement between the model and the observed maximum ion energies. For those shots we can assume the target was at best focus and the experimental parameters correspond closely to the model and simulation parameters. The model thus presents an upper bound on ion energy for a given set of laser and target conditions and is predictive: indeed, the model was developed based on experimental data for target thicknesses smaller 60 nm, yet it provided an accurate prediction for data, obtained later, above that thickness. This lends credibility to this model as a design tool for future experiments. Also, the VPIC simulations agree well with both the data and the model. Given the experimental uncertainties regarding both laser and target parameters (**O** ~ ± 20%), the observed deviations between experimental data and PIC simulations are surprisingly small. The modest systematic trend toward higher energies and thinner optimal target thickness in the simulations can be understood in terms of dimensionality effects in 1D and 2D simulations, which would tend to expand the target more slowly than in 3D, thus delaying t_2. Overall, the experimental data, simulations and reduced model are in good quantitative agreement and the data can be explained qualitatively and quantitatively by the BOA acceleration model. We therefore interpret our data as the first experimental demonstration of BOA acceleration.

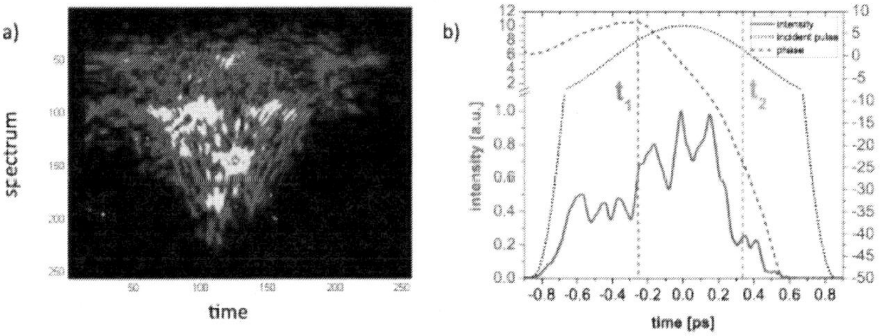

FIGURE 6. FROG measurement of a Trident pulse (85 J, 760 fs) reflected from a 165 nm CVD diamond target. (a) FROG trace. (b) Analyzed temporal intensity profile (red) and temporal phase of reflected pulse (blue). Overlaid is the input pulse (black dots). Red and orange lines indicate t_1 and t_2 calculated by the analytic model. They match the start of a steep drop in phase, indicating the onset of relativistic transparency and the end of significant reflection as the plasma becomes classically underdense, and are consistent with both the measurements.

4. CONCLUSIONS

We present here the first experimental demonstration of BOA acceleration reaching carbon energies in excess of 0.5 GeV and demonstrating the required particle energies for carbon-driven IFI. The data are in good agreement with kinetic simulations, both previously published and dedicated to this set of data. Furthermore, we developed a reduced analytical model that predicts ion cutoff energies for different target and laser parameters. This model is in good quantitative agreement with the observed data and has already made astonishingly accurate predictions. The good quantitative agreement between experimental data, simulations and analytic model gives confidence in our understanding of the BOA process and its further predictions, including scalability, high efficiency and spectral tailoring. Based on these results, we predict with confidence that GeV carbon energies for hadron therapy are achievable with today's technology. This is only the very first demonstration of this new acceleration mechanism.

Both experiment and simulation indicate that a main feature of BOA is the equal velocity distribution between different ion species. With the goal of efficient generation of proton beams, a target with a small fraction of heavier nuclei is desired. On the experimental campaigns describe here, DLC and CH_2 targets were used. Optimally one would expect that an optimal laser–target energy coupling will be achieved using a pure (cryogenic) hydrogen target. Figure 7 shows 2D VPIC simulations results for proton generation by a TPW pulse focused to 2 μm FWHM interacting with a 4 μm cryogenic hydrogen target (other parameters of this simulation are shown in figure 7). The resulting proton spectra reach an endpoint of almost 450 MeV.

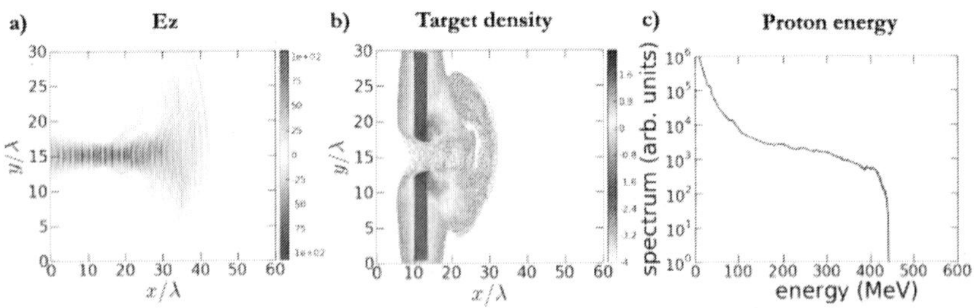

FIGURE 7. 2D VPIC simulation for TPW pulse: $E = 160$ J, $t = 150$ fs, $I = 2 \times 10^{22}$ W cm^{-2}, interacting with a 4 μm solid density pure hydrogen target at $t = 130\lambda/c$. (a) Laser field magnitude in the z direction and (b) target density color coded (c) proton spectrum extracted down the centerline.

As has been the case with other laser-driven particle acceleration mechanisms, progress in target design, both geometry and composition, as well as in laser performance will significantly improve the performance of BOA, moving laser-based acceleration one step closer to application.

ACKNOWLEDGMENTS

We gratefully acknowledge the expert support of the Trident and TPW laser teams in conducting the experiments and the support of V Liechtenstein of the Kurchatov Institute, Moscow, in the development of DLC targets. The VPIC simulations were run on the LANL Roadrunner supercomputer. This work was supported by LANL Laboratory Directed Research & Development (LDRD), the DOE Office of Fusion Energy Sciences (OFES) and by the Deutsche Forschungsgemeinschaft (DFG) through Transregio SFB TR18, the DFG Cluster of Excellence Munich-Centre for Advanced Photonics (MAP) and DFG LMU-Excellence (M Hegelich).

FOOTNOTES

1. Targets developed and supplied by Applied Diamond, Inc., Wilmington, DE 19805.
2. On this beam time, a reduced laser energy of 90 J was used instead of the usual 150 J due to optical damaged mirrors in the laser chain.
3. Pulse duration 156 fs instead of 540 fs FWHM, intensity 10^{21} W cm^{-2}.
4. Simulation parameters as specified above for the 2D case.

REFERENCES

1. Snavely R A et al 2000 Intense high-energy proton beams from petawatt-laser irradiation of solids Phys. Rev. Lett. **85** 2945–8
2. Clark E L, Krushelnick K, Zepf M, Beg F N, Tatarakis M, Machacek A, Santala M I K, Watts I, Norreys P A and Dangor A E 2000 Energetic heavy-ion and proton generation from ultraintense laser–plasma interactions with solids Phys. Rev. Lett. **85** 1654–7
3. Brambrink E, Schreiber J, Schlegel T, Audebert P, Cobble J, Fuchs J, Hegelich M and Roth M 2006 Transverse characteristics of short-pulse laser-produced ion beams: a study of the acceleration dynamics Phys. Rev. Lett. **96** 154801
4. Schollmeier M et al 2008 Controlled transport and focusing of laser-accelerated protons with miniature magnetic devices Phys. Rev. Lett. **101** 055004
5. Cowan T E et al 2004 Ultralow emittance, multi-MeV proton beams from a laser virtual-cathode plasma accelerator Phys. Rev. Lett. **92** 204801
6. Schreiber J et al 2004 Source-size measurements and charge distributions of ions accelerated from thin foils irradiated by high-intensity laser pulses Appl. Phys. **79** 1041–5
7. Hegelich M et al 2002 MeV ion jets from short-pulse-laser interaction with thin foils Phys. Rev. Lett.**89** 085002
8. Hegelich B M, Albright B J, Cobble J, Flippo K, Letzring S, Paffett M, Ruhl H, Schreiber J, Schulze R K and Fernández J C 2006 Laser acceleration of quasi-monoenergetic MeV ion beams Nature **439**441–4
9. Schwoerer H, Pfotenhauer S, Jackel O, Amthor K U, Liesfeld B, Ziegler W, Sauerbrey R, Ledingham K W D and Esirkepov T 2006 Laser-plasma acceleration of quasi-monoenergetic protons from microstructured targets Nature **439** 445–8
10. Albright B J, Yin L, Hegelich B M, Kevin J Bowers, Kwan T J T and Fernández J C 2006 Theory of laser acceleration of light-ion beams from interaction of ultrahigh-intensity lasers with layered targets Phys. Rev. Lett. **97** 115002
11. Hatchett S P et al 2000 Electron, photon and ion beams from the relativistic interaction of Petawatt laser pulses with solid targets Phys. Plasmas **7** 2076–82
12. Hegelich B M et al 2005 Spectral properties of laser-accelerated mid-z MeV u^{-1} ion beams Phys. Plasmas **12** 056314
13. Roth M et al 2001 Fast ignition by intense laser-accelerated proton beams Phys. Rev. Lett. **86** 436–9
14. Temporal M, Honrubia J J and Atzeni S 2002 Numerical study of fast ignition of ablatively imploded deuterium–tritium fusion capsules by ultra-intense proton beams Phys. Plasmas **9** 3098–107
15. Fernández J C, Albright B J, Flippo K A, Hegelich B M, Kwan T J, Schmitt M J and Yin L 2008 Progress on ion based fast ignition J. Phys.: Conf. Ser. **112** 022051
16. Tajima T, Habs D and Yan X 2009 Laser acceleration of ions for radiation therapy Rev. Accel. Sci. Technol. **2** 201–28

17. Esirkepov T, Borghesi M, Bulanov S V, Mourou G and Tajima T 2004 Highly efficient relativistic-ion generation in the laser-piston regime Phys. Rev. Lett. **92** 175003
18. Robinson A P L, Zepf M, Kar R G, Evans S and Bellei C 2008 Radiation pressure acceleration of thin foils with circularly polarized laser pulses New J. Phys. **10** 013021
19. Klimo O, Psikal J, Limpouch J and Tikhonchuk V T 2008 Monoenergetic ion beams from ultrathin foils irradiated by ultrahigh-contrast circularly polarized laser pulses Phys. Rev. ST Accel. Beams **11**031301
20. Zhang X, Shen B, Li X, Jin Z, Wang F and Wen M 2007 Efficient GeV ion generation by ultraintense circularly polarized laser pulse Phys. Plasmas **14** 123108
21. Henig A et al 2009 Radiation-pressure acceleration of ion beams driven by circularly polarized laser pulses Phys. Rev. Lett. **103** 245003
22. Qiao B, Zepf M, Borghesi M and Geissler M 2009 Stable GeV ion-beam acceleration from thin foils by circularly polarized laser pulses Phys. Rev. Lett. **102** 145002
23. Yin L, Albright B J, Hegelich B M and Fernández J C 2006 GeV laser ion acceleration from ultrathin targets: the laser break-out afterburner Laser Part. Beams **24** 291–8
24. Albright B J, Yin L, Kevin J Bowers, Hegelich B M, Flippo K A, Kwan T J T and Fernandez J C 2007 Relativistic buneman instability in the laser breakout afterburner Phys. Plasmas **14** 094502
25. Yin L, Albright B J, Hegelich B M, Bowers K J, Flippo K A, Kwan T J T and Fernandez J C 2007 Monoenergetic and GeV ion acceleration from the laser breakout afterburner using ultrathin targets Phys. Plasmas **14** 056706
26. Palaniyappan S et al 2012 Dynamics of relativistic transparency and optical shuttering in expanding overdense plasmas Nature Phys. **8** 763–9
27. Yin L, Albright B J, Bowers K J, Jung D, Fernández J C and Hegelich B M 2011 Three-dimensional dynamics of breakout afterburner ion acceleration using high-contrast short-pulse laser and nanoscale targets Phys. Rev. Lett. **107** 045003
28. Gaul E W, Martinez M, Blakeney J, Ringuette M, Hammond D, Jochmann A, Escamilla R, Borger T, Dyer G and Ditmire T 2009 Activation of a 1.1 petawatt hybrid, OPCPA-Nd:glass laser Conf. on Lasers and Electro-Optics/Int. Quantum Electronics Conf. p JWB2
29. Bahk SW, Rousseau P, Planchon T, Chvykov V, Kalintchenko G, Maksimchuk A, Mourou G and Yanovsky V 2005 Characterization of focal field formed by a large numerical aperture paraboloidal mirror and generation of ultra-high intensity (10^{22} W/cm^2) Appl. Phys. B **80** 823–32
30. Albright B J, Schmitt M J, Fernández J C, Cragg G E, Tregillis I, Yin L and Hegelich B M 2008 Studies in capsule design for mid-z ion-driven fast ignition J. Phys.: Conf. Ser. **112** 022029
31. Fernández J C, Honrubia J J, Albright B J, Flippo K A, Cort Gautier D, Hegelich B M, Schmitt M J, Temporal M and Yin L 2009 Progress and prospects of ion-driven fast ignition Nucl. Fusion **49**065004
32. Honrubia J J, Fernández J C, Temporal M, Hegelich B M and Meyer ter Vehn J 2009 Fast ignition of inertial fusion targets by laser-driven carbon beams Phys. Plasmas **16** 102701
33. Habs D, Pretzler G, Pukhov A and Meyer ter vehn J 2001 Laser acceleration of electrons and ions and intense secondary particle generation Prog. Part. Nucl. Phys. **46** 375–7
34. Cobble J A, Johnson R P, Cowan T E, Renard-Le N Galloudec and Allen M 2002 High resolution laser-driven proton radiography J. Appl. Phys. **92** 1775–9
35. Ledingham K W D et al 2004 High power laser production of short-lived isotopes for positron emission tomography J. Phys. D: Appl. Phys. **37** 2341
36. Ma W, Liechtenstein V Kh, Szerypo J, Jung D, Hilz P, Hegelich B M, Maier H J, Schreiber J and Habs D 2011 Preparation of self-supporting diamond-like carbon nanofoils with thickness less than 5 nm for laser-driven ion acceleration Nucl. Instrum. Methods Phys. Res. A **655** 53–6

37. Liechtenstein V Kh, Ivkova T M, Olshanski E D, Feigenbaum I, DiNardo R and Döbeli M 1997 Preparation and evaluation of thin diamond-like carbon foils for heavy-ion tandem accelerators and time-of-flight spectrometers Nucl. Instrum. Methods Phys. Res. A **397** 140–5
38. Batha S H et al 2008 Trident high-energy-density facility experimental capabilities and diagnostics Rev. Sci. Instrum. **79** 10F305
39. Shah R C, Johnson R P, Shimada T, Flippo K A, Fernandez J C and Hegelich B M 2009 High-temporal contrast using low-gain optical parametric amplification Opt. Lett. **34** 2273–5
40. Wang Y and Luther-Davies B 1994 Optical-parametric-amplification-based prepulse eliminator for a chirped-pulse-amplification Nd:glass laser J. Opt. Soc. Am. B **11** 1531–8
41. Antici P, Fuchs J, d'Humiéres E, Robiche J, Brambrink E, Atzeni S, Schiavi A, Sentoku Y, Audebert P and Pépin H 2009 Laser acceleration of high-energy protons in variable density plasmas New J. Phys.**11** 023038
42. Martinez M et al 2005 The Texas Petawatt Laser Proc. SPIE **5991** 59911N
43. Zielbauer B, Green J, Neely D, Roth M, Scott G, Stöhlker T, Wagner F and Bagnoud V Development of a plasma mirror setup for pulse contrast enhancement at PHELIX GSI Scientific Report 201 PNI-PP-06
44. Henig A et al 2009 Enhanced laser-driven ion acceleration in the relativistic transparency regime Phys. Rev. Lett. **103** 045002
45. Dromey B, Kar S, Zepf M and Foster P 2004 The plasma mirror? A subpicosecond optical switch for ultrahigh power lasers Rev. Sci. Instrum. **75** 645–9
46. Clarke R J et al 2008 Nuclear activation as a high dynamic range diagnostic of laser-plasma interactions Nucl. Instrum. Methods Phys. Res. A **585** 117–20
47. Jung D et al 2011 A novel high resolution ion wide angle spectrometer Rev. Sci. Instrum. **82** 043301
48 Ziegler J F, Ziegler M D and Biersack J P 2010 SRIM–the stopping and range of ions in matter (2010) Nucl. Instrum. Methods Phys. Res. B **268** 1818–23
49. Paterson I J, Clarke R J, Woolsey N C and Gregori G 2008 Image plate response for conditions relevant to laser? Plasma interaction experiments Phil. Mag. Ser. **19** 095301
50. Fleischer R L, Price P B and Walker R M 1965 Ion explosion spike mechanism for formation of charged-particle tracks in solids J. Appl. Phys. **36** 3645–52
51. Thomson J J 1911 The resonance spectra of iodine vapour and their destruction by gases of the helium group Phil. Mag. Ser. 6 **22** 469–81
52. Jung D et al 2011 Development of a high resolution and high dispersion thomson parabola Rev. Sci. Instrum. **82** 013306
53. Rusch G, Winkel E, Noll A and Heinrich W 1991 The Siegen automatic measuring system for track detectors: new developments Int. J. Radiat. Appl. Instrum. D **19** 261–6
54. Cowan T E et al 1999 High energy electrons and nuclear phenomena in petawatt laser-solid experiments AIP Conf. Proc. **472** 358–67
55. Gautier D C, Flippo K A, Letzring S A, Workman J, Shimada T, Johnson R P, Hurry T R, Gaillard S A and Hegelich B M 2008 A novel backscatter focus diagnostic for the trident 200 tw laser Rev. Sci. Instrum.**79** 10F547
56. Palaniyappan S et al 2010 Pulse shape measurements using single shot-frequency resolved optical gating for high energy (80 J) short pulse (600 fs) laser Rev. Sci. Instrum. **81** 10E103
57. Bowers K J, Albright B J, Yin L, Bergen B and Kwan T J T 2008 Ultrahigh performance three-dimensional electromagnetic relativistic kinetic plasma simulation Phys. Plasmas **15** 055703
58. Yan X, Tajima T, Hegelich M, Yin L and Habs D 2010 Theory of laser ion acceleration from a foil target of nanometer thickness Appl. Phys. B **98** 711–21
59. Albright B J et al 2010 Ultraintense laser interaction with nanoscale targets: a simple model for layer expansion and ion acceleration J. Phys.: Conf. Ser. **244** 042022

60. Mourou G A, Tajima T and Bulanov S V 2006 Optics in the relativistic regime Rev. Mod. Phys. **78** 309–71

CITATIONS

1. *Intra-operative radiation therapy with laser-accelerated carbon ions P. Mur et al 2017 Journal of Instrumentation* **12** *C03083*
2. *Heavy ion acceleration in the radiation pressure acceleration and breakout afterburner regimes G M Petrov et al 2017 Plasma Physics and Controlled Fusion* **59** *075003*
3. *A light-weight compact proton gantry design with a novel dose delivery system for broad-energetic laser-accelerated beams U Masood et al 2017 Physics in Medicine & Biology* **62** *5531*
4. *Colloquium: Nanoplasmas generated by intense radiation Kostya (Ken) Ostrikov et al 2016 Reviews of Modern Physics* **88**
5. *Laser-driven ion accelerators for tumor therapy revisited Ute Linz and Jose Alonso 2016 Physical Review Accelerators and Beams* **19**
6. *Chirped-Standing-Wave Acceleration of Ions with Intense Lasers F. Mackenroth et al 2016 Physical Review Letters* **117**
7. *Radiation pressure acceleration: The factors limiting maximum attainable ion energy S. S. Bulanov et al 2016 Physics of Plasmas* **23** *056703*
8. *Improved pulse contrast on the Texas Petawatt Laser E Gaul et al 2016 Journal of Physics: Conference Series* **717** *012092*
9. *New frontiers in nuclear physics with high-power lasers and brilliant monochromatic gamma beams S Gales et al 2016 Physica Scripta* **91** *093004*
10. *Acceleration of high charge-state target ions in high-intensity laser interactions with sub-micron targets C McGuffey et al 2016 New Journal of Physics* **18** *113032*
11. *Reduction of X-ray generation in high-intensity laser ion acceleration Bastian Aurand et al 2015 Applied Physics B* **118** *247*
12. *Calibration of BAS-TR image plate response to high energy (3-300 MeV) carbon ions D. Doria et al 2015 Review of Scientific Instruments* **86** *123302*
13. *Uniform heating of materials into the warm dense matter regime with laser-driven quasimonoenergetic ion beams W. Bang et al 2015 Physical Review E* **92**
14. *The integrated laser-driven ion accelerator system and the laser-driven ion beam radiotherapy challenge Paul R. Bolton 2015 Nuclear Instruments and Methods in Physics Research Section A: Accelerators, Spectrometers, Detectors and Associated Equipment*
15. *Visualization of expanding warm dense gold and diamond heated rapidly by laser-generated ion beams W. Bang et al 2015 Scientific Reports* **5** *14318*
16. *Laser-to-hot-electron conversion limitations in relativistic laser matter interactions due to multi-picosecond dynamics M. Schollmeier et al 2015 Physics of Plasmas* **22** *043116*
17. *Enhanced laser-driven proton acceleration from a relativistically transparent transversely nano-striped target Jingwei Wang et al 2015 Plasma Physics and Controlled Fusion* **57** *115009*
18. *Towards Laser Driven Hadron Cancer Radiotherapy: A Review of Progress Ken Ledingham et al 2014 Applied Sciences* **4** *402*
19. *Probing the quantum vacuum with ultra intense laser pulses B.M. Hegelich et al 2014 The European Physical Journal Special Topics* **223** *1093*

20. *Angle-dependent modulated spectral peaks of proton beams generated in ultrashort intense laser-solid interactions* L. N. Su et al 2014 Physics of Plasmas **21** 093111
21. *Ultrashort Pulsed Neutron Source* I. Pomerantz et al 2014 Physical Review Letters **113**
22. *Nuclear Physics with 10 PW laser beams at Extreme Light Infrastructure – Nuclear Physics (ELI-NP)* N.V. Zamfir 2014 The European Physical Journal Special Topics **223** 1221
23. *A compact solution for ion beam therapy with laser accelerated protons* U. Masood et al 2014 Applied Physics B
24. *Azimuthal asymmetry in collective electron dynamics in relativistically transparent laser–foil interactions* R J Gray et al 2014 New Journal of Physics **16** 093027
25. *Focus on high energy density physics* R Paul Drake and Peter Norreys 2014 New Journal of Physics **16** 065007

CHAPTER 5

Analysis of the Efficient High-Temperature In Situ Photoluminescence from GaN Layers during Epitaxial Growth

Christoph Prall [1,*], Daniel Erni [2,3] and Dirk Rueter [1]

[1]*Institute of Measurement Engineering and Sensor Technology, University of Applied Sciences Ruhr West, Mülheim an der Ruhr D-45407, Germany*

[2]*General and Theoretical Electrical Engineering (ATE), Faculty of Engineering, University of Duisburg-Essen, Duisburg D-47048, Germany*

[3]*CENIDE—Center for Nanointegration Duisburg-Essen, Duisburg D-47048, Germany*

ABSTRACT

Photoluminescence (PL) in GaN or InGaN layers monitored during epitaxial growth at high temperatures permits a quasi-continuous in situ characterization of opto-electronic properties. Therefore, epitaxial parameters can now be optimized at the earliest possible stage. A pulsed and high-power UV laser was required for PL excitation at high temperatures. Herein, the underlying nonlinear mechanism was studied via time-resolved PL experiments and rate equation-based modeling. A temperature-activated and saturable path for quenching over defects was identified. Beyond the saturation threshold, reasonably-intensive PL sets in. At high temperatures not only is the near band gap-PL present, but also—as a new observation—a defect-assisted PL emerges. Apart from these specific electronic transitions in high-temperature PL of GaN, a simple, but reasonably predictive model of the luminescent thin film has been set up to track down interference fringes in the PL spectra. It is worth mentioning that the spectral PL modulation (aiming at the Purcell effect) is often mixed up with ordinary Fabry–Pérot interference. A distinction has become key to properly analyze the spectral signatures of high-temperature PL in order to provide a reliable in situ characterization of GaN layers during epitaxial growth.

Keywords: characterization; metalorganic vapor phase epitaxy; nitrides; light emitting diodes

1. INTRODUCTION

Photoluminescence (PL) is a classical contact-less and well-established tool for semiconductor characterization [1]. It is, thus, highly applicable to the direct band gap semiconductor gallium nitride (GaN) and its relatives, all of which are well-known for their high PL quantum efficiency. Although the required technical components have already been available for many years, the quasi-continuous PL scanning of GaN and InGaN epitaxial layers at high temperatures—typically at 700 °C or even higher if, e.g., metal organic chemical vapor deposition (MOCVD) is considered—was, however, developed and reported quite recently [2,3,4]. We have already demonstrated that reasonably strong PL spectra from GaN and InGaN layers were obtained during growth in a quasi-commercial MOCVD reactor at temperatures around 1050 °C for GaN and 730 °C for InGaN, which provided a useful correlation with the composition of the deposited $In_xGa_{1-x}N$ layers, permitting MOCVD process optimization at the earliest possible stage. This is interesting because device optimization in the sense of targeting, e.g., a particular emission wavelength, as well as a better yield for the designated light emitting diode (LED) structure is now accessible within such an in situ PL technique. The in situ PL signal also carries other information from the epitaxial layer growth [2,3], e.g., it correlates with the layer thickness and, hence, with the growth rate (while exploring the interference pattern in the PL spectra), with the layer temperature (via redshift, broadening, and attenuation of the PL with increasing temperature), with the interface roughness (while analyzing, e.g., the modulation depth of the interference pattern in the PL spectrum), and with the (self-) absorption coefficient (given by the attenuation of the PL interference pattern toward shorter wavelengths). Interestingly, we observed in our various in situ high-temperature PL measurements [2,3,4] that the general PL behavior at high temperatures was quite similar [4] (regardless of a potential difference in peak wavelengths) for a bulk GaN substrate layer and a multi-quantum well (MQW) InGaN structure for LED devices. Therefore, our considerations apply for both the bulk GaN and also for the more complex InGaN structures, rendering such a high-temperature PL characterization/monitoring scheme quite general.

In situ PL at high temperatures also correlates with the intrinsic quality of the semiconductor lattice itself because PL emission is typically affected by electronically-effective defects, such as radiative and non-radiative recombination centers in the semiconductor [1,5]. Defect monitoring at high temperatures is a new quality in this paper and may become relevant for in situ monitoring. The demand for a non-invasive, instantaneous, and (quasi-)continuous method for monitoring epitaxial layer growth in a hot MOCVD environment with narrow optical access leaves virtually no other alternative than optical techniques in the infrared, visible, and UV spectral range. Therefore, this research exclusively refers to optically-available information at high temperatures.

In an MOCVD growth process high UV excitation densities are needed to obtain reasonable PL signals at high temperatures [2,3,4]. The excitation is provided by a repetitive (up to 20 Hz) pulsed near-UV laser with a high peak power density (in the order of MW/cm^2) providing short pulses in the order of ns, with approximately 1–10 mW average power. As previously shown, a repetitive pulsed excitation (a) promotes sufficient PL at high temperatures and (b) neither damages the material nor interferes with the epitaxial growth [3]. In contrast, a continuous UV laser with equivalent power emission (e.g., 10 mW, continuous wave, CW) is hardly capable to excite reasonably efficient PL at those high temperatures. Even when increasing the CW laser power and power density by focusing the beam to small spot sizes, the PL signal remains poor, whereas the semiconductor material gets damaged, or even fully destroyed, by the intense CW laser spot.

These findings from our previous experiments [2,3,4] indicates an onset of a non-linear PL process inside the hot semiconductor layer. Laser excitation using repetitive short UV pulses with MW/cm^2 power densities and an average power of 10 mW turns out to be far more efficient than an equivalent CW UV laser excitation of 10 mW. Therefore, the aim of this study is to both explore the mechanisms responsible for these particular experimental findings and to retrieve additional characteristic signatures from the hot GaN layer.

The obtained PL spectra regularly contain interference fringes, which could be analyzed for a thickness or growth rate determination.

Similar—but not identical—interference measurements with external light sources are already well established in thin film monitoring. However, there are subtle differences between thin film interference fringes from *external illumination* and from an *internal PL process*. These differences are frequently missed in the scientific discourse of PL spectra from thin films. This contribution describes and models these two different and photonic effects in thin films. For better clarity, the photonic effects are presented and discussed separately from the electronic transitions.

Therefore, the remainder of the paper is organized as follows: Section 1.1 introduces characteristic electronic transitions in high-temperature PL and Section 1.2 examines the characteristic photonic effects in thin luminescent layers. The experimental methods and models for electronic transitions and photonic effects are introduced in Section 2.1 and Section 2.2. The results of both topics are presented in Section 3.1 and Section 3.2, followed by conclusions in Section 4.

1.1. Considerations for Characteristic Electronic Transitions in High-Temperature PL

Referring to the above mentioned and superlinear excitation mechanisms [2,3,4] it is to note that a similar non-linearity in the GaN material or InGaN MQW material is already present at room temperature and can be readily visualized using, e.g., a sapphire wafer coated with GaN/InGaN MQW layers (representing the standard material system for LEDs) illuminated by a conventional UV fluorescence lamp (i.e., a "black light" at 350–400 nm) with a low irradiation density in the order of 10^1–10^2 μW/cm^2. At these low excitation densities, the LED wafer will not emit the characteristic emission wavelengths of the LED (e.g., blue light around 450 nm). Instead, a rather effective and pale-yellowish fluorescence (centered at approximately 570 nm, 2.2 eV) is observed, which is readily visible to the naked eye. This yellowish emission, which we term as the "yellow band" (YB) is attributed to defects in the GaN lattice [5,6].

When increasing the excitation density on the wafer, e.g., by using a CW HeCd pump laser beam at 325 nm with some mW/mm^2 or with an ordinary electronic injection current of several mA/mm^2 through the LED chip, the designated wavelength, namely, the blue light at 450 nm, clearly dominates the PL emission. This is the expected emission associated to the InGaN band gap and herein denoted as the "blue band" (BB); however, it certainly also covers the near-UV wavelength range for GaN at room temperature.

Both the YB and BB emissions emerge from competing relaxation processes, and the interplay between these processes is found to be similar for simple GaN or complex InGaN MQW [4].

A non-equilibrium carrier in a BB state can relax into a vacant YB state (Figure 1) and, subsequently, emit a yellowish light at the expense of the BB emission. Since both the YB defect density and the defects' individual recombination rates at room temperature are limited, the YB provides a quenching process for the BB that is easily saturated [5] (e.g., in a conventional LED at few mA injection current). Thus, above certain excitation thresholds, the BB gains intensity and will clearly dominate the PL emission at room temperature; however, the saturated YB is still present with considerably weaker relative intensities (as later shown in Section 3.1). The competition between BB and YB is experimentally observable in GaN bulk material and also—very similar—in more complex InGaN MQW samples, as shown later in Section 3.1. As very common in luminescence, a transfer from BB inside an InGaN quantum well to YB even outside of the quantum well (GaN) can readily occur via energy transfer, an actual escape of carriers from the QW is not required for this process. Energy transfer over a distance of several nm between two insulated luminescence centers (=high potential barrier between them) is a common, efficient, and fast process, such as non-radiative near-field dipole dipole interactions, like Förster resonance energy transfer (FRET). The energies, but not the particles are transferred. In the resonance case (the emission level fits the absorption level), the transfer can quickly occur over several nm and energy transfer readily passes potential barriers. This explains the similar behavior of GaN and InGaN MQW.

As a conjecture of this research, the YB defects not only convert the excitation energy into an emission of yellowish light with a comparably slow and saturable recombination process [5,8], but such defects (or those which closely interact) also establish an additional parallel recombination channel, which is thermally activated and non-radiative. At room temperature, this non-radiative recombination is slow and not very effective. Hence, the luminescence efficiency of high quality GaN samples reaches rather high internal quantum yields virtually at 100% [9,10,11]. However, at higher temperatures, the non-radiative recombination process drastically accelerates and efficiently overtakes—i.e., quenches—both the YB and BB emissions rendering the overall PL to be affected at higher temperatures.

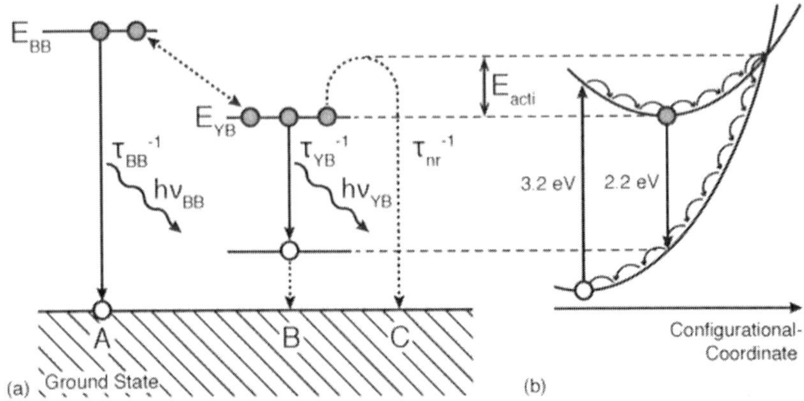

Figure 1. Schematic PL transitions: (**a**) Recombination model for GaN with the radiative BB recombination at a rate of $1/\tau_{BB}$ (transition A), radiative YB recombination at a rate of $1/\tau_{YB}$ (transition B), and the temperature activated non-radiation recombination with the rate of $1/\tau_{nr}$ (transition C). (**b**) One-dimensional configuration coordinate (CC) diagram for a YB-defect in GaN [1,5,7].

Nevertheless, when providing UV excitation at a higher rate (i.e., at higher power densities) that exceeds the thermally-quenching process, reasonably intensive BB and YB emissions are possible. The thermally-activated fast non-radiative recombination pathway over a limited number of defects is finally limited and can, therefore, be saturated with sufficient excitation power. As experimentally observed, this single mechanism turned out to be well-suited to describe the non-linear PL phenomena at high temperatures. The nonlinear behavior, as described here, may promote the irreplaceability of high (peak) power density UV excitation when sufficient PL emission at high temperatures is needed, e.g., for monitoring the growth of InGaN quantum wells inside a quasi-industrial MOCVD system via PL emission [2,3]. At room temperature (and above) only the near band gap BB and the broad YB emissions are visible. Other, more detailed, spectral features associated with, e.g., excitons or defined defect levels [5], are merely present at low temperatures, rendering them either blurred or dark at high temperatures, or even unobservable.

Since the prominent YB defects, together with closely-interacting dark defect states, are conjectured to play a governing role as a temperature-activated quenching path [7], their nature must be discussed in more detail. Although the YB luminescence and associated defects have been extensively studied, a detailed prescription towards an efficient PL at high temperatures is still lacking. In the following, the characteristic features of YB defects are briefly outlined along with the comprehensive review by Reshchikov et al. [5].

A considerable number of rather different, but mostly point-like, defects are associated with the YB-PL, such as Ga vacancies, complexes, and intentional or unintentional impurities [5]. For investigating the nature of the YB, 35 elements were implanted in GaN and most of them appear to increase the YB luminescence [5,12]. Additional experiments and calculations support the theory of gallium vacancies or impurities (such as carbon) are responsible for the YB [5,6]. The diversity of such luminescent defects in GaN and their possible transitions are considerably high. Therefore, it is remarkable that the collective behavior of all of these different defects can be represented just by one effective defect around the YB level modeled with rather few parameters, such as a single defect density and the relevant transition channels. With its broadband emission at around 2.2 eV, the apparent significant Stokes shift at excitation energies >3 eV and the temperature-independent peak emission wavelength [1,5], this defect can be associated with a typical luminescence center with strong electron-lattice coupling, whose configuration coordinate (CC) model is illustrated in Figure 1b [1,5,7]. Furthermore, the effective YB defect must be considered as an acceptor because the luminescent GaN layers are typically n-conducting, with free electron densities of approximately 10^{18} cm^{-3} at room temperature [13]. Therefore, the relevant non-equilibrium carriers are usually holes, and these interact with, more or less, deep acceptor levels above the valence band. The ground state of the YB acceptor in Figure 1a was achieved by a complete filling of the defect with electrons. In the ground state, the acceptor can capture a hole, which in *n*-GaN can be generated by a UV photon, from the valence band.

Alternatively, a UV excitation at around 3.2 eV [5] can directly lift an electron from the acceptor to the conduction band, or to suitable donor levels, leaving a hole behind. The bound hole in the acceptor represents the excited state of the YB defect. The excited state quickly relaxes via multiphonon emissions (as indicated in Figure 1b by the cascaded relaxation process) to a minimum, i.e., to a somewhat shifted coordinate of the underlying nucleus. The more the apparent shift of the coordinate of the excited state with respect to the ground state, the stronger the electron-lattice coupling of the system, which is characterized by the Huang-Rhys factor [1,5]. This factor is approximately 6.5 for the YB defect [14], which indicates a strong electron-lattice coupling. For this reason, only approximately 2.2 eV is available for the photon emission at the typical YB center wavelength, however, for an excitation energy of 3.2 eV. The excess energy is thermalized by multiple phonon emissions. The strong Stokes shift results in negligible self-absorption of the emitted light. In contrast, the short-wavelength wing of the BB luminescence usually undergoes notable self-absorption in either the GaN or InGaN material system. The YB transition back to a ground state occurs via capturing a free or bound electron (from a donor level) followed by photon emission and the elimination of the bound hole while restoring the ground state of the defect.

It should be noted that, for a luminescent center with strong electron-lattice coupling and, thus, with a significant coordinate offset between both dispersion parabolas in the schematic CC diagram (cf. Figure 1b), a crossover point between the excited and the ground exists at a certain energy level. At elevated temperatures, the electrons in the excited state can absorb a sufficient number of phonons (i.e., thermal excitation from the environment) while approaching the crossover point from which they can thermalize to the ground state without any photon emission. This is a well-known mechanism [1,5,15,16] for a thermally-activated non-radiative recombination channel of a luminescent center with high electron-lattice coupling, which is capable of providing the aforementioned quenching of the YB and BB.

The activation energy for such temperature quenching of the YB luminescence is reported [7] in the order of approximately 0.9 eV at low excitation densities, as well as at relatively moderate temperatures (<700 K) [5]. However, when reaching a high excitation power at typically 20 W/cm^2 [5,15,16], which is actually far less than applied here, a lowered activation energy of around 0.65 eV was observed in [5,15], which is similar to the values found in this study. It remains unclear how the proper quenching mechanism becomes effective: whether at high temperatures the YB defects, themselves, either quench the excitation or whether an energy transfer occurs towards certain other invisible defects providing a similar outcome for the temperature activated quenching. Currently, these two contributions are hardly distinguishable for the reported PL methods. However, the proposed PL model neither requires, nor involves, such additional transitions for explaining the above observations. In our model for rather high temperatures and high-quality samples (the designated application), just the broad BB and the broad YB remain. There is nothing more observable, as mentioned above and as previously shown [2,3,4] in high-temperature PL spectra. Just these two bands, their individual behavior, and their interaction are modeled.

Thus, to flesh out the proposed model the aforementioned three major recombination pathways have to be considered (cf. Figure 1b): the BB emission (transition A) with a certain rate over the material band gap, which is tapped and eventually quenched by the luminescent and saturable YB defects (transition B), where the YB defects are simultaneously depopulated by a temperature activated non-radiative relaxation process (transition C).

1.2. Considerations for Characteristic Photonic Effects in Thin Luminescent Layers

In addition to the microscopic electronic PL mechanism, macroscopic photonic effects have to be taken into account. The broadband YB emission (with strong stokes shift and virtually no self-absorption) from the thin GaN layer showed a typical and almost regularly observed modulation in the luminescence spectra (e.g., Section 3.1) that is commonly linked to Fabry–Pérot interference fringes. The latter can be directly assigned to the layer thickness and the index of refraction of the thin film. Interestingly, this interpretation seems to not be exactly right as the actual Fabry–Pérot oscillations (FPO) are usually retrieved from *external* white light illumination of thin films, whereas the nearly-identical modulation pattern observed in the present experiment stems from the PL as an *internal* light source intrinsic to the luminescent thin film.

It can be argued that a FPO cannot be held responsible for the modulation pattern of internal PL due to principal reasons: When considering a lossless optically-smooth thin film with a certain index of refraction, energy conservation is immediately applicable. Given an experiment with external white light illumination, the interference fringes of the thin film´s reflection spectra must find its exact counterparts to a 180° phase shift in the film´s transmission spectra. Both spectra, when superimposed, must reproduce the smooth spectrum of the original white light source, without any interference modulation. Such results are regularly achieved in sufficiently careful experiments. Hence, when looking at the modulation of the YB's luminescence spectra from thin GaN films (e.g., Section 3.1), similar interference fringes can be observed in the PL spectra emitted from the film´s front side. Inside the film an internal white light source is active and no self-absorption of the emitted white light is assumed. However, when inspecting the back side of the thin film, the modulation pattern in the emitted PL spectra looks the same, but without a 180° phase shift. This is due to symmetry because there is no difference between the front and back of a free-standing, homogeneous thin film with a built-in light source (i.e., the PL). In the lossless case this necessarily implies that the atomic emission itself has to undergo a spectral modulation, aiming at what is known as the Purcell effect [17,18,19,20,21]. Therefore, the Purcell effect presented here represents not only a conventional interference, but an imprint of the (macroscopic) thin film cavity on the spontaneous emission process itself.

In order to scrutinize the various aforementioned contributions to the PL modulation an intuitive frequency-scaled 1D equivalent circuit model of the optically-activated one-dimensional InGaN/GaN thin-film layer system is presented (cf. Section 2.2). The model is based on RF transmission lines mimicking the light propagation in the corresponding layers while using broadband voltage sources either placed externally (white light illumination) or internally (active luminescence) for the optical excitation of the layer system. The electrical equivalent circuit model, thus, allows a one-dimensional simulation of the spectral features of PL emission from the overall layer stack providing clarification of the observed interference/luminescence effects, even for arbitrarily complex layer structures.

2. MATERIALS AND METHODS

2.1. Experimental Methods and Models for Electronic Transitions in High-Temperature PL

The experimental setup for measuring both the decay time and the spectrum of the PL is similar to the setup used in previous investigations [4]. Samples (1 cm^2) of a blue InGaN LED structure on a sapphire substrate and a 3.4 μm GaN buffer on a sapphire substrate (provided by AIXTRON SE, Herzogenrath, Germany) served as test structures.

These test samples were placed on a temperature-controlled susceptor. As previously shown [2,3,4], this setup sufficiently reproduces the conditions of in situ PL monitoring in a real MOCVD system (AIX2600G3-HT, Aixtron SE, Herzogenrath, Germany). In the PL monitoring setup shown in Figure 2 the PL was excited by a frequency-tripled Nd:YAG laser at a wavelength of 355 nm (1 HP 355-300, Crylas, Berlin, Germany), with a reduced pulse energy of 88 μJ, a rated pulse length of 1.3 ns, and a repetition rate of 20 Hz. The laser light is guided through a fiber and widened (without lens optics) to a conical beam providing an illumination spot with a diameter of 1.6 mm, up to 60 mm, on the test sample that is mounted on top of the susceptor. The peak power density on the wafer surface is, thus, reduced to a maximum of 34 kW/cm^2 (8×10^{13} photons/pulse/cm^2). This is comparably low with respect to the power densities (MW/cm^2) that have been recently applied within a prior setup [2,3,4] and allows the observation of the influence of the excitation density and non-linear effects. A mode scrambler was used to provide a homogeneous beam profile. To record the decay time of the PL ranging from μs to ns, a photomultiplier tube (PMT) (H10721-01, Hamamatsu Photonics, Hamamatsu, Japan) was connected over a short coax cable (50 Ω) to a sufficiently fast oscilloscope (OSC) (DSO-X 3024A, Agilent, Santa Clara, CA, USA) with a 50 Ω terminal impedance. The overall system's decay time for a laser pulse (355 nm), resulting from the serial signal chain summing up contributions, such as the laser pulse decay time, the PMT response time, and the OSC response time, was observed to be 1.5 ± 0.3 ns, which represents the limit of temporal resolution for the observable PL decays. A combination of suitable colored glass filters (CGF) UG11 (1 mm thickness) and BG39 (1 mm) were placed in front of the laser for passing the UV excitation (at 355 nm) and strongly suppressing the other undesired emission wavelengths as the laser sub-harmonics at 532 nm, 1064 nm, and the internal diode pump laser at 808 nm. In front of the PMT (Figure 2), a stack of colored glass filters (CGF) and a notch filter (RazorEdge LP 355, Semrock, Rochester, NY, USA) distinctively suppressed the UV excitation (at 355 nm) and passed the BB PL around 440 nm for InGaN with the used CGF BG24 (2 mm), BG39 (1 mm), and the YB PL around 610 nm, which is clearly apart from the blue emission, using CGF OG590 (2 mm) and BG39 (1 mm). A metal-based neutral density (ND) filter was used to adapt the total signal levels (Figure 2). In addition to the time-resolved PL measurements of the BB and YB in the PMT, additional spectral information was obtained with a fiber-optic spectrometer (QE 65000, Ocean Optics, Dunedin, FL, USA) aiming at the spectral range of 305–1100 nm, while ignoring the temporal evolution.

For analyzing the various recombination processes associated to the PL, as shown in Figure 1, a simple, but sufficient, rate equation model was applied for the determination of the total excitation density $N(t)$ (i.e., of the non-equilibrium holes, either bound or free), with the following input parameters: initial excitation $N(0)$, the YB defect density D, the fixed radiative recombination rates (experimentally-derived at 300 K, where no influence of non-radiative quenching is present due to high PL quantum efficiency) for the BB and YB, as well as the activation energy E_{acti} (experimentally-derived from the corresponding Arrhenius plot), and the (vibrational) attempt frequency ν_{nr} (experimentally-derived from the Arrhenius plot) for thermal quenching.

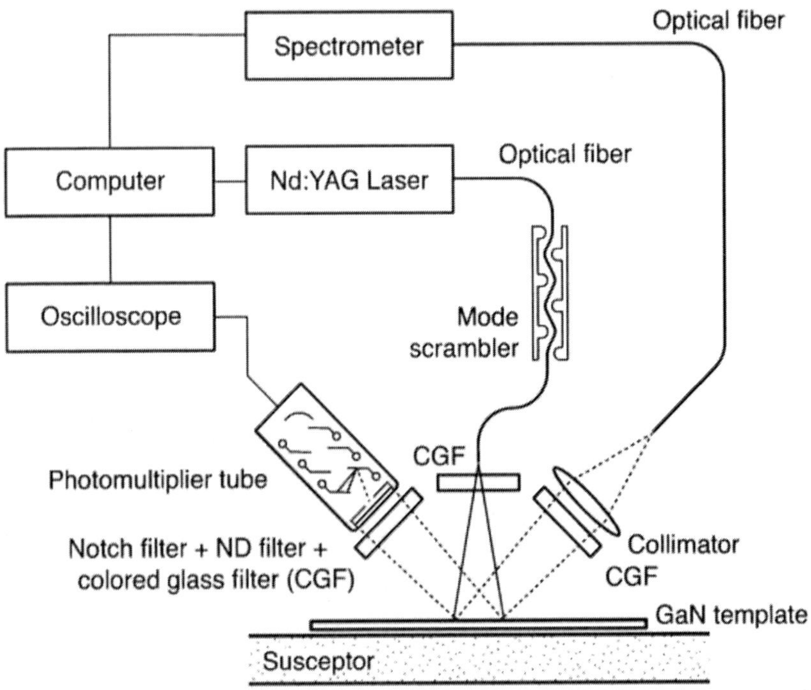

Figure 2. PL measurement setup: the 355 nm excitation light from a frequency-tripled Nd:YAG laser is guided via an optical fiber toward the hot test sample. The PL signal is filtered by colored glass filters (CGF), neutral density (ND) filters, notch filter, and then detected with a fast photomultiplier tube (PMT) and in parallel with a sensitive spectrometer.

The resulting dynamic of the excitation density is assumed to be similar to the case-distinguishing differential Equation (1) which includes the above-described mechanisms and at still moderate excitation densities, i.e., with the equilibrium majority carrier density being larger than *N*:

$$-\frac{dN}{dt} = \begin{cases} \frac{N-D}{\tau_{\mathrm{BB}}} + D \times \left(\frac{1}{\tau_{\mathrm{YB}}} + \nu_{nr} \times e^{-\frac{E_{acti}}{k_B T}}\right) & \text{for } N(t) > D \\ N \times \left(\frac{1}{\tau_{\mathrm{YB}}} + \nu_{nr} \times e^{-\frac{E_{acti}}{k_B T}}\right) & \text{for } N(t) < D \end{cases} \tag{1}$$

A thermally-activated redistribution of YB excitation states to the BB states by applying Fermi–Dirac statistics for the local equilibrium between YB excitation and BB excitation was not explicitly applied here, because no defined energy level or energy distribution of the heterogeneous YB acceptors were accessible. The dynamic behavior according to Equation (1) was numerically solved in the framework of a temporal finite-difference scheme:

1. Start the numerical calculation $N(t)$ for a given temperature T and given initial excitation $N(0)$.
2. Within a sufficiently short time step Δt, calculate the relative decrease of N: (a) a decrease of the BB population with the radiative recombination rate $1/\tau_{BB}$; (b) a decrease of the YB population with the radiative recombination rate $1/\tau_{YB}$; and (c) an additional decrease of the YB population over the thermally-activated and non-radiative path $1/\tau_{nr}$(temperature-dependent rate from the Arrhenius equation).
3. Redistribute the remaining total non-equilibrium population $N(t)$ to the YB states until saturation density D, with the excess non-equilibrium population allocated to BB states.
4. Loop back to step 2.

2.2. Experimental Methods and Models for Photonic Effects in Thin Luminescent Layers

For modeling the interference effects of PL emission within optical thin films, as observed in Section 3.1, the aforementioned frequency-scaled 1D equivalent circuit model of the layer system has been used. As shown in Figure 3 the model consists of a multi-section transmission line (TL) circuit operating in a (downscaled) frequency range of 5.4–5.8 GHz. Each individual TL represents a single layer of the processed InGaN wafer together with the free space at both sides of the wafer. The impedances (Z) of both TL_1 and TL_5 are, therefore, assigned to 377 Ω, which corresponds to the impedance of free space. The time delays for TL_1 and TL_5 were set to 2 μs, which corresponds to a physical length of approximately 11,200 wavelengths at 5.6 GHz. When translated to optical frequencies this would correspond to a distance of only 6 mm (i.e., 11,200 times 535 nm) yielding a scaling factor for the RF model of 10^5. The characteristic impedances of TL_2, TL_3, and TL_4 were adjusted in proportion to the reciprocal value of the corresponding layer's refractive index n_i. Hence, for TL_2 and TL_3—namely for the InGaN and GaN layer with $n_{2,3}$ = 2.6 [22]—the impedances were set both to 145 Ω, and for TL_4representing the sapphire substrate with n_4 = 1.7 [23], the characteristic impedance amounts to 221 Ω. The physical lengths/time delays of TL_2 and TL_3 were chosen according to the scaling factor and the optical path lengths to mimic the corresponding layer thicknesses within the LED wafer. Hence, TL_2 = 0.017 ns and TL_3 = 4.51 ns, which compares to 0.097 λ and 25.27 λ at 5.6 GHz, standing for a 20 nm-thick InGaN MQW system and a 5.2 μm GaN buffer layer, both values which are close to the dimensions of the real sample. The resistors R_1 and R_2 with 377 Ω represent perfect wave impedance matching for outgoing waves mimicking free space conditions—without reflection—for both infinite half spaces. The RF equivalent circuit model was introduced into LT-SPICE, a numerical electrical circuit simulator where two different cases were analyzed: (a) the intrinsic illumination scenario: the emission signals were picked up at the front and back side of the wafer for a built-in PL light source between the InGaN MQW system and the GaN layer; and (b) the external illumination scenario: the wafer is illuminated by an external white light source, which is located 6 mm away from the front layer surface.

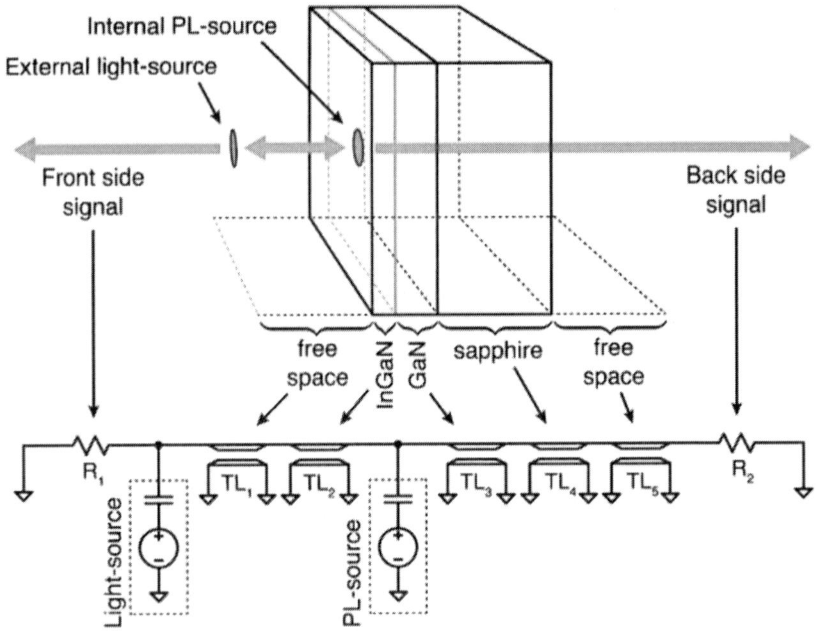

Figure 3. Frequency-scaled electrical equivalent circuit model of the optically-activated one-dimensional InGaAn/GaN thin-film layer system: the broadband AC voltage source on the left side represents the external white-light illumination of the layers and the other one on the right side mimics the generated PL inside the thin films. The optical thickness of the involved layers are represented by corresponding transmission line sections, whereas the resulting free space radiation output (i.e., back and front side signal) is retrieved at the corresponding terminal resistors.

In the scaled RF circuit model the front and back side signals were determined by the power dissipated in the terminations R_1, R_2. The broadband light sources were represented by a frequency-swept AC voltage source with 1 V amplitude. Both voltage sources were weakly coupled using a series capacitor with only 0.01 pF (i.e., 28 kΩ at 5.6 GHz). The reasoning for this intentional mismatch is provided in the framework of the intrinsic light source scenario, where the weak electrodynamic coupling of an atomic oscillator to the surrounding cavity has to be maintained. As such a dipole source is very small compared to the emitted wavelength, a weak capacitive coupling to the cavity volume is appropriate, at least in the context of classical electrodynamics. An alternative, and rather simple, argument for weak coupling is sketched along a single emission event of the atomic dipole oscillator. If an atomic luminescence center emits a single photon with an energy of 2 eV (i.e., λ = 620 nm or f = 480 THz) into free space, this would be caused by a 2 V drop of one electron. In a classical electronic picture the maximally possible discharge current is associated with an emission event lasting about one single period and may, therefore, amount to $(1.6 \times 10^{-19}\ \text{As})/(2 \times 10^{-15}\ \text{s}) = 80\ \mu\text{A}$. Thus, the characteristic coupling impedance of the single atomic emitter would be at least 2 V/80 μA = 25 kΩ.

3. RESULTS

3.1. Experimental and Numerical Results for Electronic Transitions

Figure 4 shows the PL spectra of an InGaN MQW test sample from which the non-linear interaction of both BB and YB emissions at room temperature can be deduced depending on the excitation density. A much-related interplay between BB and YB is readily observable and also well-known from simple GaN without complex InGaN MQW, e.g., [24]. Since these spectra were excited with rather short UV pulses (~1.3 ns), it is evident that the defect states already existing at room temperature may rapidly capture the excitation holes from the BB levels in the MQW (within less than 1 ns); as virtually no BB PL was detectable at low excitation densities around 1.3×10^{12} photons/puls/cm^2 (cf. Figure 4). The latter corresponds to a power density of 540 W/cm^2 within the 1.3 ns UV pulse. From a quantitative comparison of both BB and YB emission spectra a considerably high internal quantum efficiency of more than 65% was deduced for the YB transition when the internal quantum efficiency of the BB transition was set to 100% [9]. The immediate effectiveness of vacant defect states is essential for the above reasoning. Furthermore, a considerable fraction of effective defects was observable as YB PL at room temperature due to high quantum efficiency. Although the provided amount of non-equilibrium holes (~1.3×10^{12} cm^{-2}) was not sufficient to saturate the number of present defects in the GaN, the situation totally changed when the excitation density is gradual increased within only one order of magnitude. At approximately 2.5×10^{12} cm^{-2}, the BB emission set in and strongly gained intensity at higher excitation densities. Conversely, the YB emission tends to saturate and follows a distinct sub-linear behavior with increasing excitation intensities. The BB intensity, beyond the total saturation of YB, increases linearly (beyond the higher power at about 2×10^{13} photons /puls/cm^2 cf. Figure 4) and not superlinearly with excitation. In the herein-described scenario the excitation density is still small with respect to the equilibrium electron density. Therefore, the BB intensity increases linearly (beyond saturation of YB) and not in a quadratic dependence.

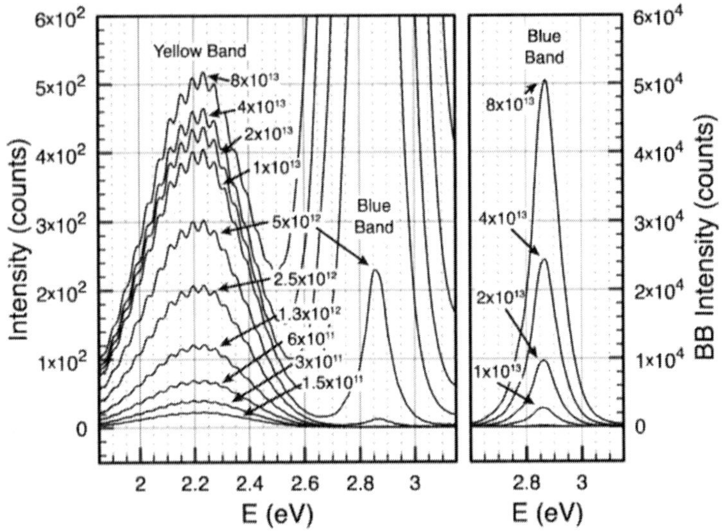

Figure 4. PL spectra from an InGaN MQW test sample at room temperature, when excited by a ns-pulsed Nd:YAG laser at 355 nm with increasing excitation densities (photons/puls/cm^2). Beyond a certain excitation density (5×10^{12} photons/puls/cm^2), the limited number of YB defects become saturated, then allowing for a fairly linear increase of the BB. Below this threshold, the luminescent defects effectively depopulate the BB levels and thereby affect the BB PL. Particularly in the YB, with its broadband emission and low self-absorption, a distinct interference modulation is observed.

Additionally, inherently evident from Figure 4 is the intimate connection between YB and BB: Before observing the BB linear behavior, the YB had to be saturated. Furthermore, (from quantitative comparison and energy conservation described before) there is virtually no other process present besides the observable YB and BB in Figure 4.

Approximately 5×10^{12}–10×10^{12} cm^{-2} of defect density were needed to be effective within the GaN thin-film sample, which, when considering a sample thickness of 3.4 μm, a penetration depth of the UV excitation of 1 μm [25], and a diffusion length of the holes up to approximately 1 μm [26] yielding an electronically-active YB defect density of approximately 5×10^{16}–10×10^{16} cm^{-3} for this sample. When looking to the time-resolved PL signal from YB and BB at room temperature (300 K) (Figure 5 and Figure 6), a relatively prolonged lifetime (Figure 6 $\tau_{YB} = 6.4 \times 10^{-6}$ s) of the YB states with respect to the BB recombination (Figure 6 $\tau_{BB} = 2.4 \times 10^{-8}$ s) was observable. Therefore, once a YB defect level was occupied, a certain time had to pass until it could depopulate another BB excitation. Thus, at room temperature, the defect-associated quenching channel in GaN emitters can be easily saturated with the power density of the excitation.

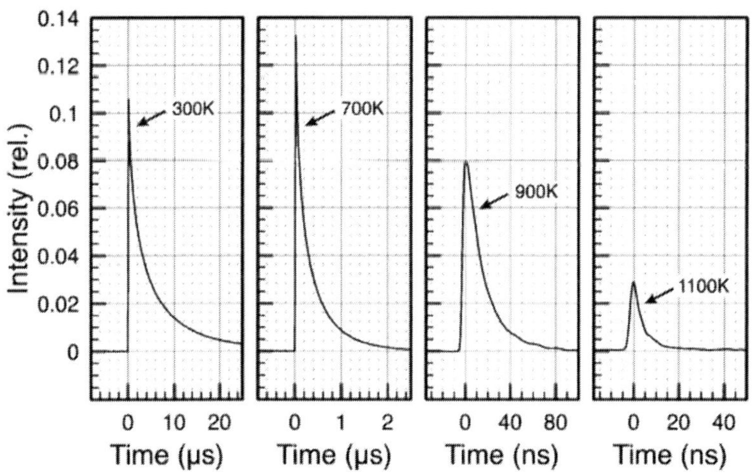

Figure 5. Time resolved PL measurements: the YB PL signals from GaN at 610 nm (and as intended here at a critically low excitation density about 5×10^{12}photons/puls/cm^2 where the BB emission is still very weak) show different temporal responses with respect to increasing temperature values 300 K, 700 K, 900 K, and 1100 K. The maximal peak intensity at 700 K might have been caused by a better coincidence of vibrational levels at the thermal activation. It is worth mentioning that it is less the peak intensity of the PL response than the decay time that is strongly affected at high temperatures.

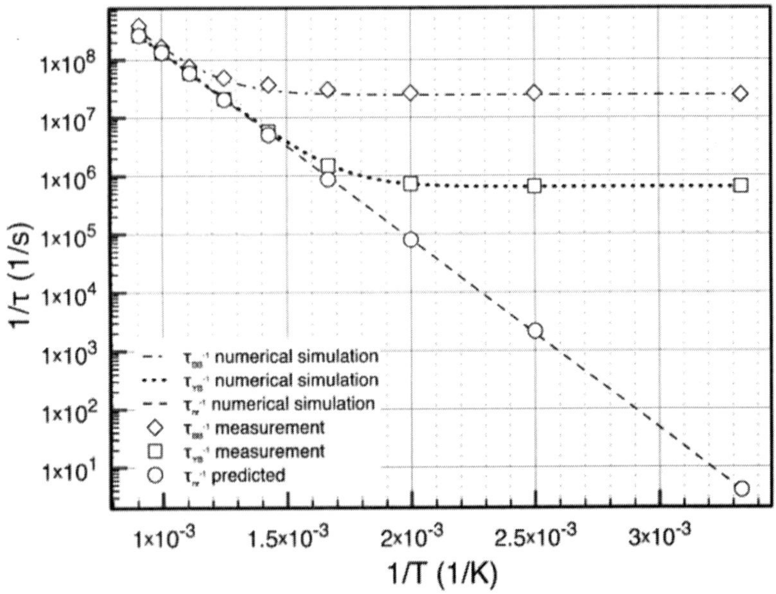

Figure 6. Arrhenius plot of the involved recombination mechanisms: the sample values depict the measured BB recombination rate ($1/\tau_{BB}$), the YB recombination rate ($1/\tau_{YB}$), and the predicted non-radiative recombination rate ($1/\tau_{nr}$) together with the dashed lines showing the simulation results of the corresponding rate equation model for both the direct and indirect cascaded transitions.

Notably, the situation changes at higher temperatures: although the YB intensity was rapidly decreasing at increasing temperature for the PL spectra without time resolution (e.g., shown in [4]), and then not visible anymore, the time-resolved analysis shows (cf. Figure 5) that the initial YB intensity was barely affected even by considerable temperatures and even at low excitation densities. Instead, only the decay time, and not the initial amplitude, was significantly lowered at higher temperatures. Hence, the defect associated YB emission becomes observable, even at rather high temperatures, thereby offering additional and potentially valuable information from an in situ PL measurement about defects in the newly-growing GaN layers. At much higher excitation densities in MOCVD, which are possible, it would certainly still be observable.

The reciprocal of the now-observable YB decay time consists of the sum of the recombination rates from YB emission $1/\tau_{YB}$ and the parallel and non-radiative recombination $1/\tau_{nr}$ (Figure 1). The Arrhenius plot of the total YB recombination rate, as shown in Figure 6, revealed a rather constant radiative recombination rate at low temperatures and an additional temperature-activated rate toward higher temperatures. The activation energy deduced from the Arrhenius equation was approximately 0.64 eV. Thus, a considerably high and predominantly non-radiative recombination rate of more than 10^8 $s^{-1}$was obtained at high temperatures above 1000 K.

The YB defects rapidly (e.g., below 1 ns, evident from Figure 4) catch the carriers from the BB level while getting saturated. At room temperature (300 K), the luminescent decay rate of the YB (transition B) is approximately $1/\tau_{YB} = 10^5 s^{-1}$, which reflects a comparably slow and easily saturable process with respect to the BB decay (transition A) with approximately $1/\tau_{BB} = 10^7$ s^{-1}. However, at high temperatures (e.g., 1100 K) the non-radiative relaxation rate over the defects (transition C) increases towards 10^9 s^{-1}. The non-equilibrium carriers can then be depopulated more efficiently from the BB level. Therefore, at this elevated temperature, approximately 10^9 (non-equilibrium) holes per second for each present defect must be provided to saturate this quenching rate before reasonably strong PL emission can be observed. This is possible with a pulsed laser in the MW range, which produces ns pulses at moderate pulse energies in the μJ to mJ range. The resulting excitation flux density at 1 MW/cm^2 is approximately 2×10^{24} photons per second, per cm^2. An analysis based on the corresponding rate equations (cf. Section 4) predicted that this is enough to saturate each defect (even for high defect densities up to 10^{19} cm^{-3}) in the μm-thin-film with as much as 10^9 non equilibrium holes per second.

Once recombined, the defect will rapidly capture another excitation carrier from the BB. Therefore, the decay rates in the BB and YB are finally governed by the temperature activated non-radiative decay. Conversely, more and more power density must be applied for overcompensating the quenching process at high temperatures; however, because time constants are rather short, a short laser pulse with moderate energy can suffice.

Figure 6 depicts an Arrhenius plot containing both the experimental and numerical modeling results of the PL's "cascaded" recombination channels where the following modeling parameters are identified: the activation energy of 0.64 eV for thermal quenching of the YB, the vibrational attempt frequency $\nu_{nr} = 2.5 \times 10^{11}$ s^{-1}, which is derived from Arrhenius plot and is associated to the resulting thermally-activated non-radiative recombination rate that ranges from irrelevant 4 s^{-1}at 300 K up to 2.6×10^8 s^{-1} at 1100 K, the excitation density (10×10^{12} cm^{-2}), and the YB defect density (5×10^{12} cm^{-2}). This rate-equation model of the transitions underlying the PL emission is, therefore, reasonably apt to quantify the efficiencies and the temperature behavior of the measured

YB and BB, as shown in Figure 6. As already mentioned, the numerical model does not include a gradual distribution of excitation between BB and YB. Notably, such a gradual distribution was already implicitly involved in the model data. For example, the measured YB decay rate $1/\tau_{YB}$ (Figure 6) already includes an additional depopulation of the YB over a BB emission, even at room temperature (see Figure 4, e.g., at 2.5×10^{12} cm^{-2}). Therefore, since the presented model is already adapted to the measured decay rates, an additional consideration by applying a Fermi–Dirac-Statistics was not performed, since defined energy levels for the YB are not readily available. The rate equations are only an approximate representation of the YB and BB dynamics. Nevertheless, in the high-temperature regime and for the high-quality samples used here, they reasonably describe the observations.

3.2. Experimental and Numerical Results for Photonic Effects in Thin Luminescent Layers

In addition to the time dependence and the temperature behavior of the PL, a numerical analysis of the PL emission from the InGaN/GaN thin-film layer system was carried out using the frequency-scaled electrical equivalent circuit model. Its comparison to experimental data is shown in Figure 7 and discussed in the following.

The broad YB emission spectrum as observed in Figure 4 shows distinct interference fringes with both coinciding maxima and minima when recorded from either the front or the back side of the emitting layer system (cf. Figure 7b). This is not surprising due to the symmetry argument for the thin film with an internal light source as earlier provided in Section 1.2. The opposite applies to the external light source illumination scenario where the spectral responses of the reflected and transmitted light support a coincidence now between the reflection maxima and transmission minima, and vice versa (cf. Figure 7a), which is reasoned by simple energy conservation (for a virtually non-absorbing layer system). These observations were fully supported by the corresponding simulations using the frequency-scaled electrical equivalent circuit model as shown in Figure 7c,d. The visible discrepancies between the measured and simulated data were mainly assigned to rough surface scattering at the back side of the used sapphire wafer.

It is impossible to explain the observed PL spectra (with in-phase fringes from front and back sides) on the basis of conventional Fabry–Pérot interference within a non-absorbing material system. In Figure 7b, a lower overall emission intensity (signals from front and backside together) is apparent at, e.g., 557 THz and a higher overall emission at

552 THz. With only passive mechanisms and interference, the overall decrease at 557 THz could only be explained with a somehow effective absorption in the thin film. However, there is virtually no absorption around 2.2 eV (560 THz) in a μm thin GaN film.

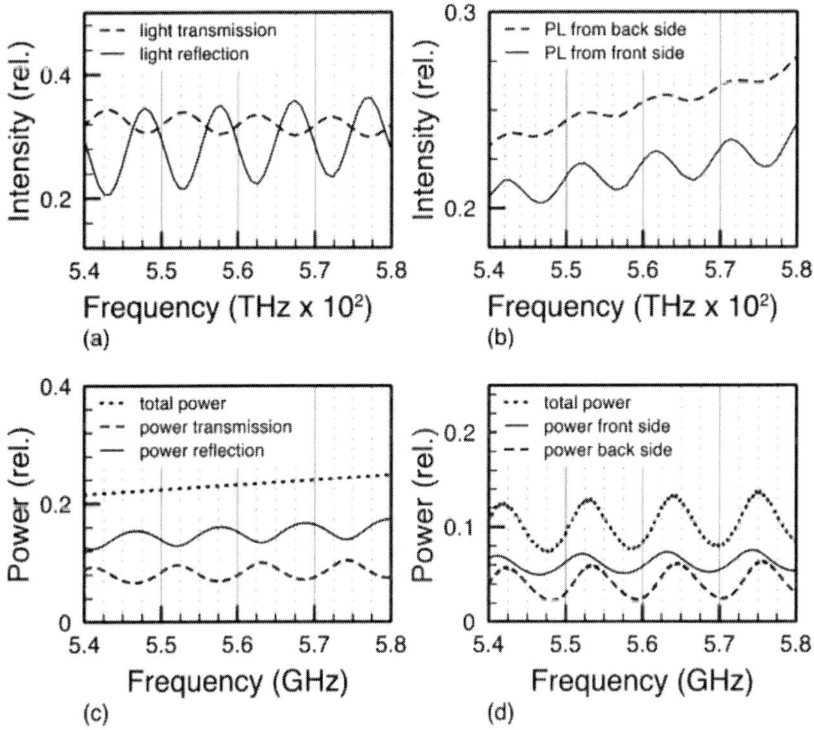

Figure 7. Measured and simulated behavior of the InGaN/GaN thin-film layer system as depicted in Figure 3: (**a**) measurement of the reflected and transmitted light from the layer system for the front-wise white light illumination; (**b**) measurement of the PL light detected at the front and back side of the emitting layer system; (**c**) simulated spectral responses of the reflected and transmitted signals for an external illumination source using the frequency-scaled electrical equivalent circuit model; and (**d**) simulated spectral responses of the signal emitted from the front and back side of the layer system for an internal source in the active layer. Please note that in the equivalent circuit modeling the total emitted power is also accessible.

Even to the contrary, notable absorption in the film suppresses the observed oscillations; see e.g., spectra in Figure 4towards 2.8 eV (BB) and also seen in many previous spectra towards the absorption edge. Thus, it cannot be passive Fabry–Pérot interference, but must be an active imprint on the emission process itself.

Thus, the power delivery of the internal source itself was spectrally modulated by the sandwiching of the resonator structure. It cannot be explained via passive interference filtering, but must incorporate a spectral modulation of the emission process itself. With this convenient electrodynamic model, an even more complex staple of layers can be readily simulated. A quantum mechanical description of a photon emission process under the influence of a more macroscopic resonator was not required. When applying an in situ PL technique during MOCVD of GaN layers, the described Purcell effect must be applied to the spectral data (symmetrical in-phase fringe effect), rather than matching the more familiar FPO (asymmetrical and energy conserving) to this kind of data.

4. CONCLUSIONS

The modeling of optical thin films with electrodynamic transmission lines as an analogue model reasonably describes spectral interference effects, reflection, and transmission of a luminescent thin film staple. The model distinguishes conventional Fabry–Pérot interference from external light sources and the Purcell effect from spontaneous luminescence emission within a thin film. The Purcell effect, and not Fabry–Pérot interference, must be adapted for these in situ PL techniques.

The electronic transition results aim at a temperature-activated and non-radiative recombination channel in the GaN material system, which can, even at high MOCVD temperatures, be saturated with a sufficiently high excitation density. Therefore, the indirectly-observed and non-radiative recombination channels are quantitatively limited and likely assigned to a certain defect concentration in the GaN material.

Fortunately, a pulsed UV laser can readily provide sufficient excitation density at still-tolerable pulse energies. Thereby, an efficient PL excitation is possible even at high temperatures, and such PL signals can be utilized for in situ monitoring of growing crystal layers during MOCVD epitaxy. In situ PL offers valuable information about the growing material [2,3]. The layer thickness and the directly-linked growth rate are accessible via the Purcell effect, the actual surface temperature as a crucial parameter in MOCVD can be derived. Additionally, the roughness, self-absorption coefficient, and the composition $In_xGa_{1-x}N$ is available, as already shown in previous contributions [2,3].

Through determining the excitation threshold for PL at high temperatures, even the concentration of electronically-active defects is observable, as described for the first time here. The threshold provides in situ information for the electronic material's quality regarding the newly-growing crystal layer at the earliest possible stage, here described for the first time with convenient in situ PL methods. With the experience of our prior work [2,3,4] we are very confident that the threshold can be practically applied in real MOCVD processes, since even less excitation power is required [2,3]. The performance of a device is typically affected by the semiconductor's defect density. The proposed PL technique, for example, can be applied for in situ monitoring and optimizing the electronic quality of growing GaN buffer layers as a substrate for InGaN MQW structures intended for LED or laser diodes. It should be noted that only high-quality samples (as designated for the application) are investigated in this study. In case of more and unintended non-radiative recombination centers, even at room temperature, the in situ PL signal at higher temperatures drastically decreases. Such a drastic decrease of the in situ PL is already a characterization criterion, itself, for quality monitoring.

The in situ investigation of more impure samples opens up an extended field for future investigations. The current manuscript only presents the basic observations and relations for high-quality samples.

However, the lifetime and the efficiency of laser diodes strongly depend on the material quality and the semiconductor's defect density, which can now be observed and corrected during growth. Furthermore, material's imperfections are held responsible for noise, reduced carrier mobility, and/or reduced breakdown thresholds in nitride-FETs for high-performance RF applications. It can be anticipated that adapted PL methods will also apply to the important and closely-related GaN-AlGaN system.

ACKNOWLEDGMENTS

The authors would like to thank the following organizations for their commitment to this study: AIXTRON SE in Herzogenrath, Germany for their donation of GaN material, LayTeC AG in Berlin, Germany for their tremendous technical support, and the ZIM funding program by the German Federal Ministry for Economic Affairs and Energy for their financial support.

AUTHOR CONTRIBUTIONS

Christoph Prall performed the experiments, analyzed the data, and derived the essential conclusions, Daniel Erni supported the modeling activities and the theoretical discussions, Dirk Rueter designed and supervised the general approach and work. All of the authors wrote the paper. All authors have read and approved the final manuscript.

CONFLICTS OF INTEREST

The authors declare no conflict of interest.

REFERENCES

1. Pelant, I.; Valenta, J. Luminescence Spectroscopy of Semiconductors; Oxford University Press: Oxford, UK, 2012.
2. Prall, C.; Haberland, K.; Kaspari, C.; Brunner, F.; Weyers, M.; Rueter, D. Process control of MOCVD growth for LEDs by in-situ photoluminescence. Proc. SPIE 2016, 9768, 97681A.
3. Prall, C.; Kaspari, C.; Brunner, F.; Haberland, K.; Weyers, M.; Rueter, D. In-situ photoluminescence measurements during MOVPE growth of GaN and InGaN MQW structures. J. Cryst. Growth 2015, 415, 1–6.
4. Prall, C.; Ruebesam, M.; Weber, C.; Reufer, M.; Rueter, D. Photoluminescence from GaN layers at high temperatures as a candidate for in situ monitoring in MOVPE. J. Cryst. Growth 2014, 397, 24–28.
5. Reshchikov, M.A.; Morkoç, H. Luminescence Properties of Defects in GaN. J. Appl. Phys. 2005, 97, 061301.
6. Li, G.; Chua, S.J.; Xu, S.J.; Wang, W.; Li, P.; Beaumont, B.; Gibart, P. Nature and Elimination of Yellow-Band Luminescence and Donor-Acceptor Emission of Undoped GaN. Appl. Phys. Lett. 1999, 74, 2821–2823.
7. Ogino, T.; Aoki, M. Mechanism of Yellow Luminescence in GaN. Jpn. J. Appl. Phys. 1980, 19, 2395–2405.
8. Korotkov, R.; Reshchikov, M.; Wessels, B. Acceptors in undoped GaN studied by transient photoluminescence. Physica B 2003, 325, 1–7.
9. Reshchikov, M.A.; Gu, X.; Nemeth, B.; Nause, J.; Morkoç, H. High quantum efficiency of photoluminescence in GaN and ZnO. MRS Proc. 2006, 892.
10. Pankove, J.I. Luminescence in GaN. J. Lumi. 1973, 7, 114–126.
11. Pankove, J.I.; Berkeyheiser, J.; Maruska, H.; Wittke, J. Luminescent properties of GaN. Solid State Commun. 1970, 8, 1051–1053.
12. Pankove, J.I.; Hutchby, J.A. Photoluminescence of ion-implanted GaN. J. Appl. Phys. 1976, 47, 5387–5390.
13. Binet, F.; Duboz, J.Y.; Off, J.; Scholz, F. High-excitation photoluminescence in GaN: Hot-carrier effects and the Mott transition. Phys. Rev. B 1999, 60, 4715–4722.
14. Ridley, B.K. A note on the origin of the yellow luminescence in GaN. J. Phys. 1998, 10, L461–L463.

15. Armitage, R.; Hong, W.; Yang, Q.; Feick, H.; Gebauer, J.; Weber, E.R.; Hautakangas, S.; Saarinen, K. Contributions from gallium vacancies and carbon-related defects to the "yellow luminescence" in GaN. Appl. Phys. Lett. 2003, 82, 3457–3459.
16. Reshchikov, M.A.; Korotkov, R.Y. Analysis of the temperature and excitation intensity dependencies of photoluminescence in undoped GaN films. Phys. Rev. B 2001, 64, 115205.
17. Purcell, E.M. Spontaneous Emission Probabilities at Radio Frequencies. Phys. Rev. 1946, 69, 674.
18. Drexhage, K. Influence of a dielectric interface on fluorescence decay time. J. Lumin. 1970, 1–2, 693–701.
19. Shen, Y.C.; Wierer, J.J.; Krames, M.R.; Ludowise, M.J.; Misra, M.S.; Ahmed, F.; Kim, A.Y.; Mueller, G.O.; Bhat, J.C.; Stockman, S.A.; et al. Optical cavity effects in InGaN/GaN quantum-well-heterostructure flip-chip light-emitting diodes. Appl. Phys. Lett. 2003, 82, 2221–2223.
20. Hwang, J.K.; Ryu, H.Y.; Lee, Y.H. Spontaneous emission rate of an electric dipole in a general microcavity. Phys. Rev. B 1999, 60, 4688–4695.
21. Ryu, H.Y.; Notomi, M. Enhancement of spontaneous emission from the resonant modes of a photonic crystal slab single-defect cavity. Opt. Lett. 2003, 28, 2390–2392. [PubMed]
22. Barker, A.S.; Ilegems, M. Infrared Lattice Vibrations and Free-Electron Dispersion in GaN. Phys. Rev. B 1973, 7, 743–750.
23. El-Refaei, H.; Yevick, D.; Betty, I. Stable and Noniterative Bidirectional Beam Propagation Method. IEEE Photonics Technol. Lett. 2000, 12, 389–391.
24. Grieshaber, W.; Schubert, E.F.; Goepfert, I.D.; Karlicek, R.F.; Schurman, M.J.; Tran, C. Competition between band gap and yellow luminescence in GaN and its relevance for optoelectronic devices. J. Appl. Phys. 1996, 80, 4615–4620.
25. Ambacher, O.; Rieger, W.; Ansmann, P.; Angerer, H.; Moustakas, T.; Stutzmann, M. Sub-bandgap Absorption of Gallium Nitride Determined by Photothermal Deflection Spectroscopy. Solid State Commun. 1996, 97, 365–370.
26. Chernyak, L.; Osinsky, A.; Nootz, G.; Schulte, A.; Jasinski, J.; Benamara, M.; Liliental-Weber, Z.; Look, D.C.; Molnar, R.J. Electron beam and optical depth profiling of quasibulk GaN. Appl. Phys. Lett. 2000, 77, 2695–2697.

CHAPTER 6

Structure, Equation of State, Diffusion and Viscosity of Warm Dense Fe Under The Conditions of A Giant Planet Core

Jiayu Dai[1,3], Yong Hou[1], Dongdong Kang[1], Huayang Sun[1], Jianhua Wu[1] and Jianmin Yuan[1,2,3]

[1] *Department of Physics, College of Science, National University of Defense Technology, Changsha 410073, People's Republic of China*

[2] *State Key Laboratory of High Performance Computing, National University of Defense Technology, Changsha 410073, People's Republic of China*

[3] *Department of Physics, College of Science, National University of Defense Technology, Changsha 410073, People's Republic of China*

ABSTRACT

Fe exists abundantly in the universe. In particular, the dynamic structures and transport properties of warm dense Fe are crucial for understanding the evolution and structures of giant planets. In this paper, we present the ionic structures, equation of states, diffusion and viscosity of Fe at two typical densities of 33.385 and 45 g cm^{-3} in the temperature range of 1–10 eV, giving the data by the first principles calculations using quantum Langevin molecular dynamics. Furthermore, the validation of the Stokes–Einstein (SE) relation in this regime is discussed, showing the importance of choosing the effective atomic diameter. The results remind us of the careful usage of the SE relation under extreme conditions.

1. INTRODUCTION

The properties of complex materials such as Fe under extreme conditions are crucial for understanding the evolution of planets (Earth, giant planets and extro-solar giant planets) [1] and stars such as the Sun [2, 3]. More interestingly, the laser-driven and Z-pinch experiments can lead to very high densities up to hundreds of g cm^{-3} and temperatures up to thousands of eV [4, 5]. In particular, temperatures of 1–10 eV with associated pressures of 10–1000 TPa are likely to exist in the interiors of massive planets [6, 7]. Recent studies have shown the possible stable structures of Fe at these pressures [6, 7], where face-centered cubic (fcc) structures are stable in the range of 7–21 TPa. However, the effect of temperature on these structures is still not known, since the dynamic structures such as melting, diffusion and viscosity can be induced by temperatures; the phases of Fe in Earth-like exoplanets are likely to be liquid. In order to understand these behaviors, since the experimental determination currently seems to be extremely expensive and difficult, accurate simulations are required for the determination of the structures and dynamics. The dynamic structures of Fe and its compounds from first principles molecular dynamics under the physical conditions of the Earth's core have been studied widely [8–10], showing the complexity of the new phenomena in high energy density physics. In particular, it is worth verifying whether the Stokes–Einstein (SE) relation at such high densities is valid, since the SE relation is one important bridge between dynamics and statistics; it has usually been adopted in previous calculations of viscosities from diffusion coefficients [8–12]. Up to now, the diffusion and viscosity can be obtained directly by molecular dynamics simulations with the Kubo–Green relation [10–13]. Moreover, we can construct statistical models according to the simulation results by introducing easy applications in hydrodynamics [14, 15]. However, molecular dynamics simulations are strongly dependent on the accuracy of many-body interatomic potentials; there are usually some limitations to statistical methods.

A cold stable fcc structure, with its inducing liquid at a high temperature, is considered an important phase in giant planets such as Jupiter and exoplanets [6, 7]. In this regime, the densities are from 33.9 to 66.7 g cm^{-3} and the temperatures are from 1 to 10 eV. For these states, the melting behaviors and the transport properties are interesting and deserve to be studied carefully. Besides, a high density will introduce many different chemical bonds since a pressure greater than 1 Mbar will change the traditional chemistry dramatically [16–20]. Furthermore, when the pressure increases to 100 Mbar, the core electron charge density can be changed significantly [16–18]. Therefore, the transport properties such as equation of states (EOS), diffusion and viscosity will change a lot since these properties are related to the electronic structures. With respect to the methods of calculating these properties accurately, first principles calculations are thus required because we did not understand their behaviors from semiclassical models, such as the average atom (AA) model [21, 22] and the Thomas–Fermi model or the orbital free method [23–25], in which the orbital behaviors or chemical bonds information cannot be described well.

Molecular dynamics combining the finite-temperature density functional theory (DFT) [26], called quantum molecular dynamics (QMD), and quantum Langevin molecular dynamics (QLMD) have been successfully used in warm and hot dense matter, especially under very high pressures [8, 10, 18–20, 27, 28]. The validations of QMD and QLMD at high densities and temperatures have been verified many times by comparing with experiments and interestingly reasonable results are found. In particular, QMD is considered as one of the most promising methods and has been successfully used to predict the transport properties of warm dense matter (WDM) [29, 31]. The biggest argument against QMD applications is that it limits the number of particles because of its computational cost at high temperature. However, QMD and QLMD can be efficient for the calculations of WDM under the conditions of a giant planet core; they can give relatively accurate transport properties using a few hundred particles [29–32]. In this paper, we adopted QLMD to simulate the transport properties of dense Fe at 45 g cm^{-3} from 1 to 10 eV. Except for considering the density of 33.385 g cm^{-3} at 100 eV along the Hugoniot curve [18], we also calculate the dynamical properties at 33.385 g cm^{-3} up to 10 eV. Furthermore, the approximation of the SE relation is discussed, showing the debatable validation under such dense regimes.

2. METHODS

For the complexity of Fe at high density, both QMD with the velocity rescaling method and QLMD simulations are tested first. In fact, at low temperatures, the electron–ion collision induced friction is small since the friction is proportional to the temperature. Therefore, QMD and QLMD are equivalent for the equilibrium properties at low temperatures. Eventually, the advantage of QLMD is its higher computational efficiency compared to traditional QMD [33]. The electronic structures are solved based on finite temperature DFT implemented in the Quantum ESPRESSO package [34]. Considering the small atomic sizes at high density, the time step of 0.5 fs is used in order to keep the correct trajectories of ions. The pseudopotential (PP) is one of the key points in QMD and QLMD simulations. Here, we construct a new PP with 16 electrons in the valence and 0.9 atomic units for the radius cutoff within the generalized-gradient approximation [35], which can promise correctness with respect to the conditions we are studying. Using this PP, we can reproduce the bulk modulus properties of Fe as previous results calculated by full-electron calculations and experiments. More importantly [18, 36], at high pressure, the pressure and band structures at the density of 33.9 and 48.23 g cm^{-3} are in good agreement with the pressure in [7] using the same PP [18]. Furthermore, the same PP has been successfully used to calculate the EOS of Fe in a hot dense regime within a wide range of densities and temperatures [18], indicating the validation of PP in this work. Besides, 2000 time steps with a large convergent tolerance of 1.0×10^{-4} in electronic structure calculations are adopted in order to reach the thermalization; 10 ps time lengths with a small convergent tolerance of 1.0×10^{-6} [33] are simulated to obtain the transport properties and thermal average. In order to accelerate the calculations, the Gamma point is only used for the representation of the Brillouin zone. To compare with the results from semiclassical methods, averaged atom molecular dynamics (AAMD) [22, 37] is performed for all cases. When two atoms are close enough, their electronic distributions overlap, inducing interactions between them. In order to describe these interactions, we should establish the interatomic potential. In the AAMD method, the temperature-dependent pair potential, which can describe the contributions of overlapping electronic distributions, is constructed from the AA electronic calculations, including the ionic correlations within the framework of the AA model. Within this framework, the ions are moving on this pair potential, giving the ionic trajectories and ionic configurations at different times. From AAMD, we can basically use a large number of particles and give the convergent results corresponding to the size effects. The comparison between QLMD and AAMD has been reported before [19, 33], showing their agreement at high temperatures. The details of AAMD can be found in [37].

The self-diffusion coefficient D can be obtained from two methods. One is from the trajectories by the mean-square displacement DR

$$D^R = \frac{1}{6t}\langle|\mathbf{R}_i(t) - \mathbf{R}_i(0)|^2\rangle, \tag{1}$$

where the $\mathbf{R}i$ is the position of the ith particle. The other method is from the velocity autocorrelation function DV

$$D^V = \frac{1}{3}\int_0^\infty \langle \mathbf{V}_i(t)\,\mathbf{V}_i(0)\rangle\,\mathrm{d}t, \tag{2}$$

where $\mathbf{V}i$ is the velocity of the ith particle. Besides, the viscosity η can be obtained from the autocorrelation function of the off-diagonal component of the stress tensor, the so-called Green–Kubo equation [10, 38]

$$\eta = \lim_{t\to\infty} \bar{\eta}(t) = \lim_{t\to\infty} \frac{V}{k_B T} \int_0^t \langle P_{12}(0)\, P_{12}(t') \rangle \, dt', \tag{3}$$

where P_{12} represents the averaged result for the five independent off-diagonal components of the stress tensor *Pxy*, *Pyz*, *Pzx*, (*Pxx* – *Pyy*)/2 and (*Pyy* – *Pzz*)/2. Using this method, the self-diffusion coefficient and viscosity of warm and hot dense matter have been reported widely [8–12, 29, 39–42] within the framework of QMD, showing its validation in these extreme conditions.

On the other side, the viscosity can also be calculated directly from the SE relation

$$D\eta = k_B T/(2\pi a), \tag{4}$$

where a is an effective atomic diameter. This relation is statistically obtained from the Brownian motion of a macroscopic particle in liquid, but it is only an approximation for the atoms. If the validation of the SE relation can be verified, the size of particles and the transport behaviors can be understood well [43]. In fact, for a Brownian particle, the SE relation is equivalent to equations (2) and (3). In particular, for dense matter, the diameters of atoms are very small so the validation of the SE relation should be examined very carefully.

3. CONVERGENCE TESTS

For calculating the transport properties from first principles a number of atoms should be tested, since diffusion and viscosity are strongly dependent on the system sizes. For this purpose, in QLMD simulations, we tested face-centered cubic (fcc) structures with 32, 108 and 256 atoms and a body-centered cubic (bcc) structure with 54 atoms in a supercell at 45 g cm^{-3}. The pressure and diffusion versus temperature relations with different numbers of atoms are shown in figure 1. It is concluded that the pressure is not incredibly sensitive to the system sizes. In particular, the pressures are closer for different sizes when the temperatures are higher. The small difference at low temperatures may be caused by the long ranged orders with crystal structures. It is worth noting that the initial configurations of bcc and fcc structures are important for calculating diffusion in the solid phases, but the differences from different initial configurations disappear gradually with the increasing temperatures in the liquid phases (here above 5 eV).

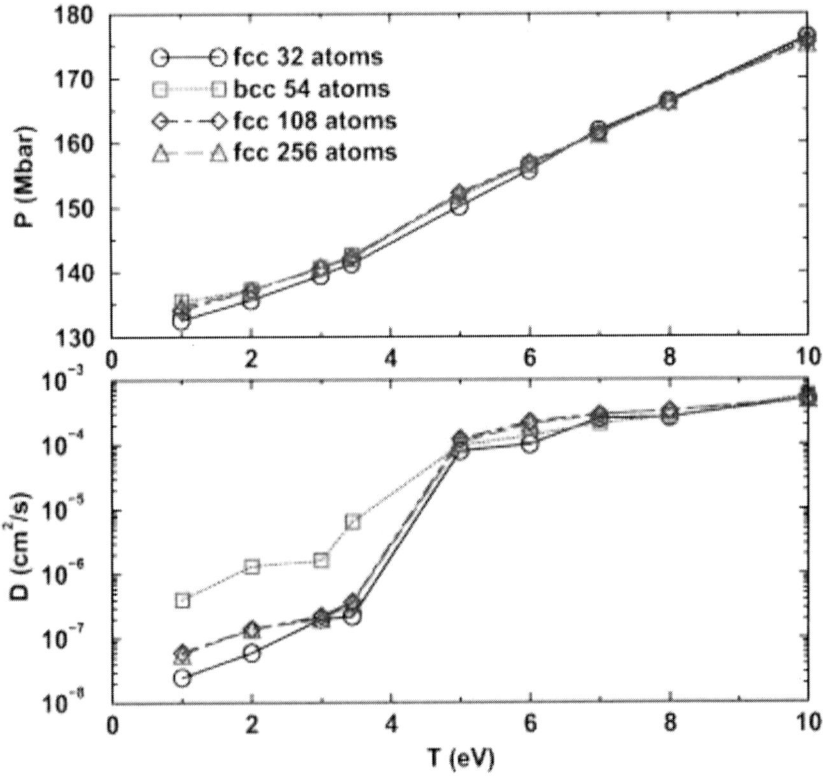

FIGURE 1. Size effect on the pressure (upper panel) and diffusion coefficient (lower panel) of Fe at 45 g cm^{-3}.

According to the tests, it can be stated that our systems are in fact convergent for the calculation of pressure and diffusion coefficients when we use 108 atoms with initial fcc structures. It should be noted that when the system becomes liquid (after the jump in the diffusion coefficient), the results of 54 atoms, 108 atoms and 256 atoms are consistent. The difference at low temperatures would be from the different initial structures of bcc or fcc crystals. In fact, the small differences from different sizes can also come from the statistical errors for small sizes. Basically, this error can be overcome by longer time simulations.

For the viscosity, the results of the convergent tests are shown in figure 2. It is only useful for the liquids here since the viscosity is very large for solids. Here, the liquid phases are shown above 5 eV in our simulations. By comparison, we can know that the viscosity is incredibly sensitive to the sizes. For a small size of 32 atoms, the statistical errors are really large, which is not likely to be compensated by increasing the simulation time.

When the number of atoms is increased up to 108 and 256, the viscosities at different temperatures seem convergent. Therefore, we can safely use the 108 atoms for the calculations of pressure, diffusion and viscosity in our cases. In fact, about 100 or 200 atoms are usually used in QMD simulations for the transport properties [8, 10, 12, 29–31], which has been shown to be convergent within reasonable errors.

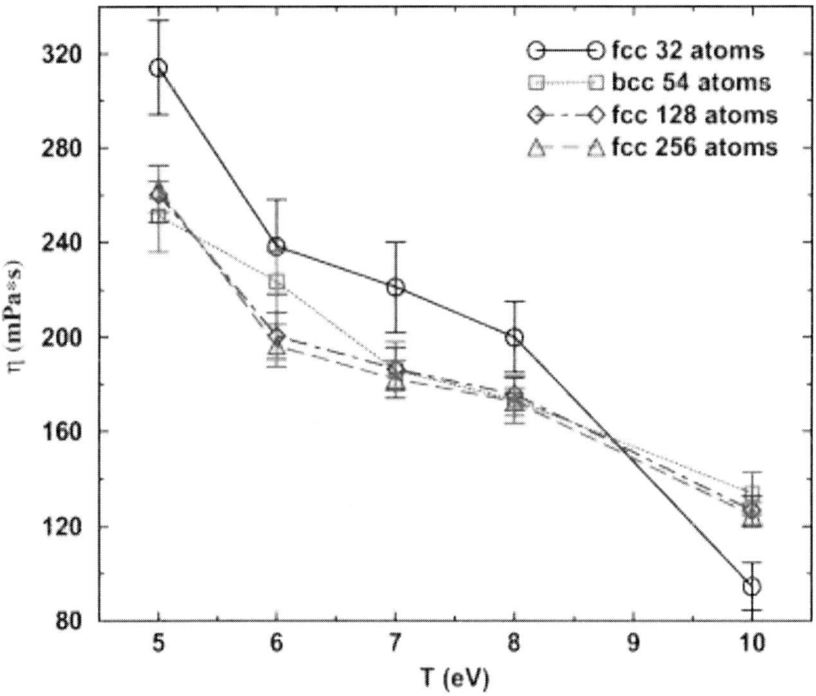

FIGURE 2. Size effect on the viscosity of Fe at 45 g cm^{-3} from QLMD simulations.

For the effect of the particle number, we use a large number of particles up to 5324 in AAMD calculations to verify the convergence, as shown in figure 3. The simulation time is up to 100 ps in each case in order to obtain the correct correlations of particles in these semiclassical calculations. It can be shown that although the number of particles can affect the convergence of diffusion and viscosity, all the differences are within 10% or smaller. When the number reaches 4000, the diffusion and viscosity are almost convergent. In order to compare the results, we use 4000 particles in AAMD to calculate the transport properties below. It is worth noting that AAMD adopted pair potentials, which should be more sensitive to the size and debatable for the viscosity since the unsymmetrical shear of the system can not be described well by pair potentials.

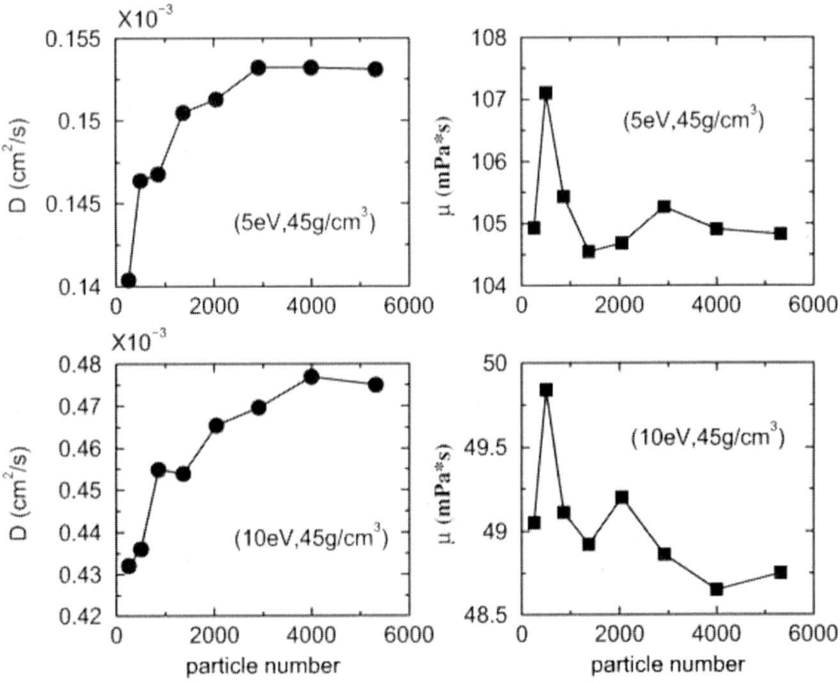

FIGURE 3. Size effect on the viscosity of Fe at 45 g cm^{-3} from AAMD simulations.

For the density of 33.385 g cm^{-3}, we used 54 atoms with initial bcc structures to calculate their transport properties. The simulation details are also the same as those of 45 g cm^{-3}. Basically, the solid phase of Fe at 33.385 g cm^{-3} may be hexagonal close-packed (hcp). However, determining the lattice constants of hcp at high temperatures is very difficult; the dynamic properties such as the structure and melting behavior are very similar to the bcc structures [7, 45]. Therefore, we used bcc structures as the initial configurations. In fact, from the tests of Fe at 45 g cm^{-3}, it is now known that the transport properties are independent of the initial configurations when Fe melts. Furthermore, the effects of larger system sizes on calculated viscosities have been previously studied for models such as hard-sphere and Lennard–Jones liquids [44], showing meaningful values for μ even with only 32 atoms. Thus, the initial configurations and sizes in this work can promise the reasonable accuracy of our simulations.

4. STRUCTURAL DYNAMICS

The dynamic structures for dense matter are basic when attempting to understand the melting behaviors and the dynamics in a planet's core. However, dense liquid Fe has rarely been studied before due to the lack of effective methods. Dense Fe can hold the solid phase or ordered structures at very high temperatures, form clusters and chemical bonds assisted by core electrons [18] and keep a high melting temperature [45–47]. Here, the structures are investigated by looking at the radial distribution functions (RDF) $g(r)$, as shown in figures 4 and 5.

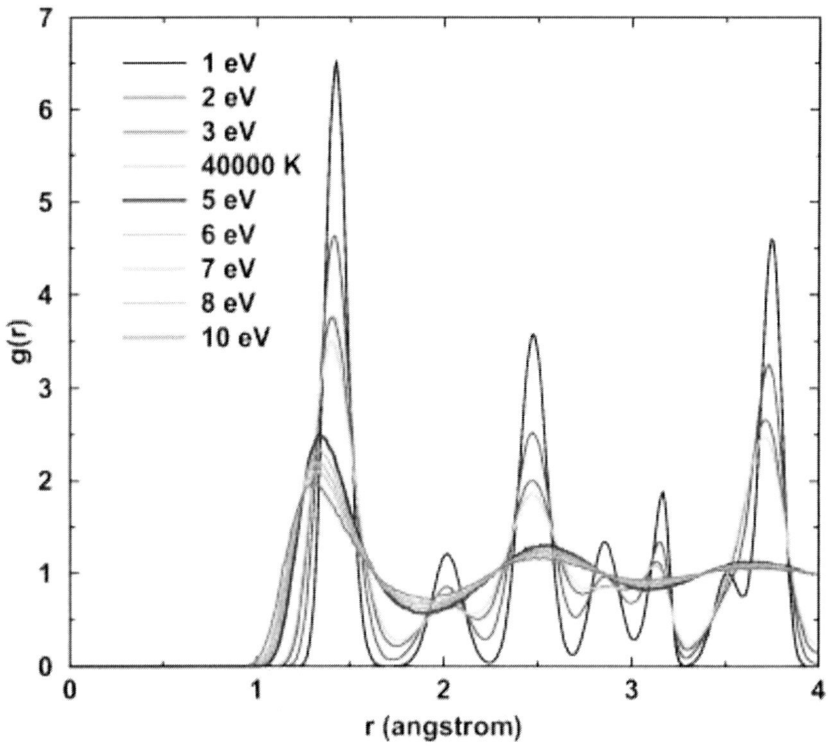

FIGURE 4. RDF versus temperature for Fe at 45 g cm^{-3}.

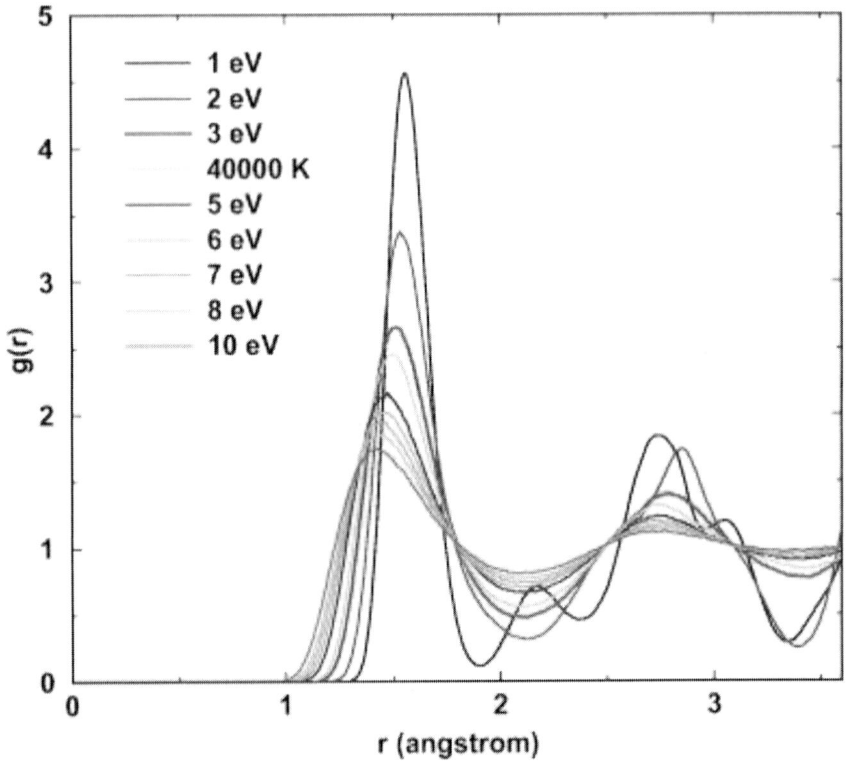

FIGURE 5. RDF versus temperature for Fe at 33.385 g cm^{-3}.

From the RDF, we can know that when the temperature increases up to 5 eV, Fe exhibits liquid behaviors at 45 g cm^{-3} and above 3 eV at 33.385 g cm^{-3}. We calculated the details of the dynamic structures near 5 and 3 eV for the two densities respectively, using the two phase approach [45] with 216 atoms and 108 atoms to obtain the melting temperature. The melting temperatures are found to be around 57 000 and 30 000 K, within the errors of 200 K in our simulations. The same behaviors can also be found in the data of diffusion coefficients and viscosities, which are shown in figures 1and 2 and will be discussed in detail later. In this stage, the Fe at 45 g cm^{-3} and 40 000 K is also in the solid phase in the Fe phase diagram [7].

We should note that the first peaks of the RDF move toward to the zero point, indicating that the effective distance between ions decreases with the increasing temperatures. Therefore, we may use the positions of the first peaks in the RDF as the effective atomic diameters when discussing the SE relation later. If we use only the averaged ionic radius, the temperature effect cannot be included since the radius is only dependent on the type of elements and system densities.

5. TRANSPORT PROPERTIES

After the convergence tests, we can obtain the transport properties including EOS, diffusion coefficients and viscosities for dense Fe. The summaries of the data are shown in tables 1 and 2. In the tables, pressures and diffusion coefficients from AAMD are also shown. When the solid phases are not broken, the diffusion coefficients are very small and the viscosities are very large (therefore they are not convergent in the Green–Kubo equation).

Table 1. Summary of the results of dense Fe at different temperatures at 45 g cm⁻³. *DR* and *DV*are respectively defined in equations (1) and (2). D_{AAMD} and μ_{AAMD} are the diffusion coefficient and viscosity from AAMD calculations, respectively.

T (eV)	P (Mbar)	D^R (cm² s⁻¹)	D^V (cm² s⁻¹)	μ (mPa s)	P_{AAMD} (Mbar)	D_{AAMD} (cm² s⁻¹)	μ_{AAMD} (mPa s)
1	134.128	0.61×10^{-7}	0.60×10^{-7}	–	245.13	2.39×10^{-8}	–
2	137.148	1.40×10^{-7}	1.61×10^{-7}	–	250.25	9.55×10^{-8}	–
3	140.594	2.24×10^{-7}	2.68×10^{-7}	–	255.73	2.07×10^{-7}	–
3.447	142.295	3.72×10^{-7}	3.82×10^{-7}	–	258.11	2.91×10^{-7}	–
5	152.218	1.21×10^{-4}	1.23×10^{-4}	62.07 ± 6	270.77	1.54×10^{-4}	104.29
6	156.825	2.22×10^{-4}	2.11×10^{-4}	44.11 ± 5	275.11	2.11×10^{-4}	83.38
7	161.296	3.15×10^{-4}	3.05×10^{-4}	41.29 ± 4	279.43	2.73×10^{-4}	69.47
8	166.081	3.26×10^{-4}	3.37×10^{-4}	38.13 ± 4	283.65	3.39×10^{-4}	60.34
10	175.421	4.86×10^{-4}	4.83×10^{-4}	27.72 ± 3	292.46	4.77×10^{-4}	48.64

For a density of 45 g cm⁻³, from table 1, the diffusion coefficients change rapidly between the temperatures of 40 000 K (3.447 eV) and 5 eV, and reach the orders of liquids diffusion at 5 eV. Also, the viscosities above 5 eV go to convergence when we integrate the pressure autocorrelations in equation (3), indicating the existence of liquid phases. However, AAMD gives the liquid Fe at 4 eV (not shown here), showing its limitations from pair potentials in very strongly coupling regimes. Furthermore, we can see that the diffusion coefficients from equations (1) and (2) are equal, which shows the validation and convergence of our simulations. For the pressure, AAMD gives much larger pressures than those of QLMD simulations, which

Table 2. Summary of the results of dense Fe at different temperatures at 33.385 g cm-3.

T (eV)	P (Mbar)	D^R (cm^2 s^{-1})	D^V (cm^2 s^{-1})	μ (mPa s)	P_{AAMD} (Mbar)	D_{AAMD} (cm^2 s^{-1})	μ_{AAMD} (mPa s)
1	60.547	3.50×10^{-7}	3.21×10^{-7}	–	79.52	2.11×10^{-6}	–
2	63.471	2.45×10^{-6}	2.19×10^{-6}	–	84.05	2.69×10^{-5}	–
3	67.739	6.70×10^{-5}	6.71×10^{-5}	47.1996 ± 5	86.99	1.49×10^{-4}	29.43
3.447	69.786	1.16×10^{-4}	1.14×10^{-4}	40.3466 ± 5	88.36	1.84×10^{-4}	27.32
5	75.282	2.98×10^{-4}	2.85×10^{-4}	24.217 ± 3	92.57	3.28×10^{-4}	22.39
6	79.727	3.67×10^{-4}	3.74×10^{-4}	19.8484 ± 2	95.33	4.15×10^{-4}	20.41
7	82.248	5.99×10^{-4}	5.91×10^{-4}	18.1716 ± 2	98.30	5.08×10^{-4}	19.00
8	85.825	6.08×10^{-4}	5.98×10^{-4}	17.1372 ± 2	101.11	6.13×10^{-4}	17.83
10	93.098	7.51×10^{-4}	7.69×10^{-4}	15.7396 ± 2	107.30	8.01×10^{-4}	16.28

is caused by the electronic structure calculations from Thomas–Fermi methods. It is very interesting that when the temperatures are above 5 eV, the diffusion coefficients from QLMD and AAMD are very close. One reason might be that the diffusion coefficients are strongly dependent on the nearest neighbor distributions. Pair potentials can deal with the nearest neighbor interactions between ions with good accuracy. This can be verified using the comparisons of RDF from QLMD and AAMD, as shown in figure 6, where the positions of the first peaks of RDF are almost overlapped. Therefore, the nearest neighbor interactions should be reasonable in AAMD. However, AAMD cannot deal with many-body interactions, suggesting that the shear viscosities from AAMD are incorrect here, since dense Fe liquid holds a lot of medium ordered structures; therefore, the melting behavior is not correct within the AAMD framework.

For a density of 33.385 g cm^{-3}, the behaviors are almost the same. The melting temperature locates below 3 eV, according to the RDF and diffusion coefficients. It is noted that the pressures from AAMD are much closer to those from QLMD compared with the results of 45 g cm^{-3}. This means that the electronic structures from the AA model are reasonable here but give more free electrons at 45 g cm^{-3}. Furthermore, the melting temperature from AAMD is also lower than that from QLMD; the diffusion coefficients are in the same order when Fe melts above 3 eV.

The small sizes in the calculations can introduce statistical errors. Especially for the viscosities, the errors are within 10%. Therefore, calculations with more than 10 000 atoms at least and accurate many-body potentials are needed basically. We need to construct more accurate semiclassical potentials for obtaining more accurate diffusion coefficients and viscosities.

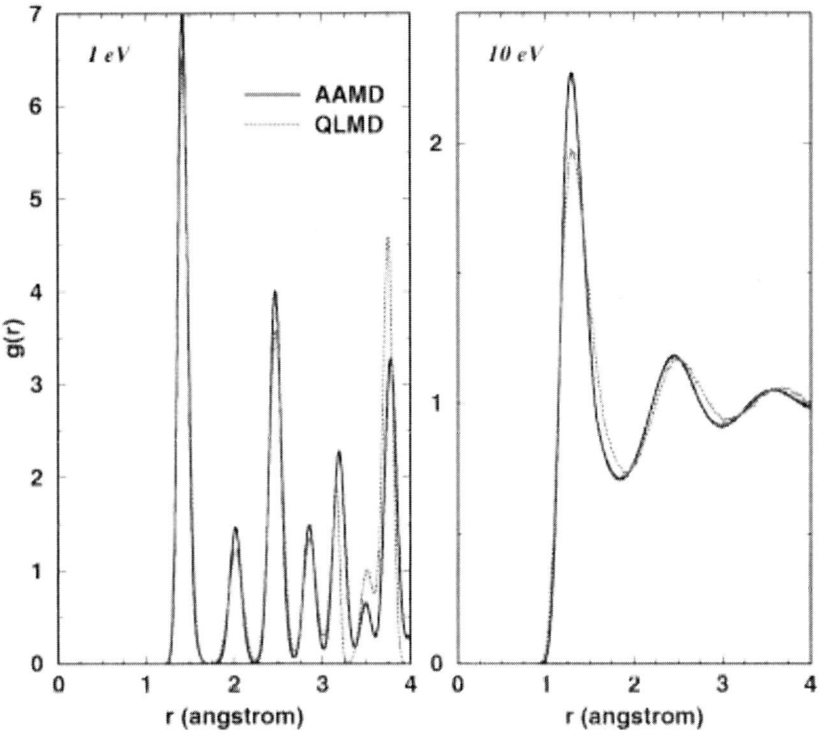

FIGURE 6. RDF comparison between AAMD and QLMD for Fe at 45 g cm^{-3}.

6. STOKES–EINSTEIN RELATION

With respect to the simulation time and system sizes, we find that the diffusion coefficients can reach convergence much more easily than the viscosities. Therefore, if we can obtain viscosities from calculated diffusions, a lot of computational resources can be saved. Therefore, the validation of the SE relation is important and meaningful. In order to test the validation, we calculate viscosities from different definitions of effective atomic diameters: averaged ionic distance at a specific density $r_i = V^{(1/3)}$ (where V is the averaged volume of one atom at a specific density), the position of the first peaks in RDF (r_f) and the effective radius

(r_E) from effective coordination number (ECN) definition. ECN can reveal the topology of the structures partly and can also give the local information around one atom. The definition of ECN is as follows:

$$\mathrm{ECN} = \frac{1}{N}\sum_{i=1}^{N} \mathrm{ECN}_i = \frac{1}{N}\sum_{i=1}^{N}\sum_{j\neq i} \exp\left[1 - \left(\frac{d_{ij}}{d_{av}^i}\right)^6\right], \quad (5)$$

where dij is the distance between atoms i and j; N is the total number of atoms in the system; di_{av} and its average value d_{av} are defined as

$$d_{av}^i = \frac{\sum_j d_{ij} \exp\left[1 - \left(\frac{d_{ij}}{d_{av}^i}\right)^6\right]}{\sum_j \exp\left[1 - \left(\frac{d_{ij}}{d_{av}^i}\right)^6\right]}, \quad d_{av} = \frac{1}{N}\sum_{i=1}^{N} d_{av}^i. \quad (6)$$

With this definition, the effective diameters $r_E = d_{av}$, which includes the temperature and density effects.

The results of different effective atomic diameters and the calculated viscosities from the SE relation are shown in tables 3 and 4. Comparing the results with the calculated viscosities from the Green–Kubo equation, we find that the viscosities are very different at low temperatures. With the increasing temperature, the SE relation seems to be more and more validated. Moreover, the choice of the effective atomic diameters plays an important role in the validation of the SE relation, where definitions of the positions of the first peaks of RDF and diameters from ECN calculations are more appropriate. Considering the calculation uncertainties and statistical errors, the SE relations can be considered validated only when the choice of the effective diameters are reasonable.

In order to avoid using the SE relation, some models have been constructed. For example, Murrilo reported the construction of the YVM based on molecular dynamics simulations [14], which can be used in a warm dense regime. We verify the validation of the YVM model here. First of all, we give the average ionization degree ($\overline{Z}$) from the AA model with electronic energy-level broadening [22] and then show the viscosities (μ_{YVM}) from the YVM model, as shown in tables 3 and 4. It can be shown that the YVM model improves the accuracy of the SE relation, especially for cases at high temperatures. This suggests that the Yukawa potential might be reasonable for states at such high densities and at relatively high temperatures.

What is the possible reason for the debatable validation of the SE relation in this regime? The basic physics is whether the ions exhibit as Brownian particles. For the specific case in this work, heavy Fe ions move in the liquid-like electron sea and randomly collide with many free-like electrons.

Table 3. Summary of the effective diameters and their corresponding viscosities for dense Fe at different temperatures at 45 g cm^{-3}. $\overline{Z}$ is the average ionization degree from AA calculations. μ_i, μ_f and μ_E are the viscosities corresponding to different definitions of effective atomic diameters of ions r_i, r_f and r_E, respectively. μ_{YVM} is the viscosity from the Yukawa viscosity model (YVM) (see details in [14]).

T (eV)	r_i (Å)	μ_i (mPa s)	r_f (Å)	μ_f (mPa s)	r_E (Å)	μ_E (mPa s)	$\overline{Z}$	μ_{YVM} (mPa s)
5	1.275	83.361	1.335	79.612	1.338	79.428	13.011	43.99
6	1.275	54.002	1.330	51.767	1.323	52.052	13.009	37.97
7	1.275	46.527	1.325	44.770	1.309	45.302	13.009	33.77
8	1.275	49.051	1.320	47.377	1.297	48.219	13.002	30.71
10	1.275	41.166	1.295	40.529	1.276	41.127	12.993	26.64

Table 4. Summary of the effective diameters and their corresponding viscosities for dense Fe at different temperatures at 33.385 g cm^{-3}.

T (eV)	r_i (Å)	μ_i (mPa s)	r_f (Å)	μ_f (mPa s)	r_E (Å)	μ_E (mPa s)	$\overline{Z}$	μ_{YVM} (mPa s)
3	1.408	81.002	1.515	75.301	1.488	76.667	9.344	28.65
3.447	1.408	53.831	1.505	50.375	1.473	51.459	9.343	25.81
5	1.408	30.300	1.475	28.931	1.437	29.687	9.332	20.23
6	1.408	29.633	1.465	28.488	1.419	29.410	9.326	18.41
7	1.408	21.147	1.455	20.469	1.404	21.214	9.323	17.28
8	1.408	23.847	1.450	23.162	1.388	24.197	9.312	16.56
10	1.408	24.122	1.425	23.840	1.364	24.912	9.296	15.91

However, the other dominant factor for ionic motions is the interactions between ions, i.e. the strong coupling of ions. Every ion moves in a specific force field determined by both electrons and ions; this force field is not constant. The ions in dense matter have a memory effect, i.e. the ionic positions at this time play an important role in the ionic movements the next time. In other words, if the correlation time of ions is larger than the observable time, the motion of the ions should not have a Markov process. In order to analyze the correlation time of the systems, the velocity autocorrelation functions (VAF) of ions at 5 and 10 eV are shown in figure 7. Here, we adopt the simplest definition of the correlation time (τ_c)

to compare the correlation time at different temperatures. Here, τ_c is defined as the time for which the VAF(t) decreases to $1/e$ [48]. With this definition, τ_c is about 4.1 and 2.7 fs for the temperatures of 5 and 10 eV, respectively. Therefore, with the increasing temperature, the behaviors of ions have a more and more Brownian-like motion. In fact, if the system is ideal gas, the correlation time will be zero, indicating uncorrelated behaviors between two near time steps. That is to say, when the temperature is high enough, the random collisions between particles are dominant and the statistical behaviors can work well. If the correlation time of the system is zero, we can recover the SE relation directly from equations (2) and (3) [49]. We can then conclude that the SE relation is strongly dependent on the coupling parameters of ions $\Gamma = Z^{*2}/(k_B Ta)$, with T the system temperature, k_B the Boltzmann constant, a the mean ionic sphere radius defined as $a = (3/(4\pi n_i))^{1/3}$, Z^* the average ionization degree and n_i the ionic number density. If Γ is small enough, the SE relation can be valid. Furthermore, one possible reason that QMD or QLMD can obtain reasonable transport properties within a relatively short time (within 10 ps) is the short correlation time of particles, as shown in figure 7.

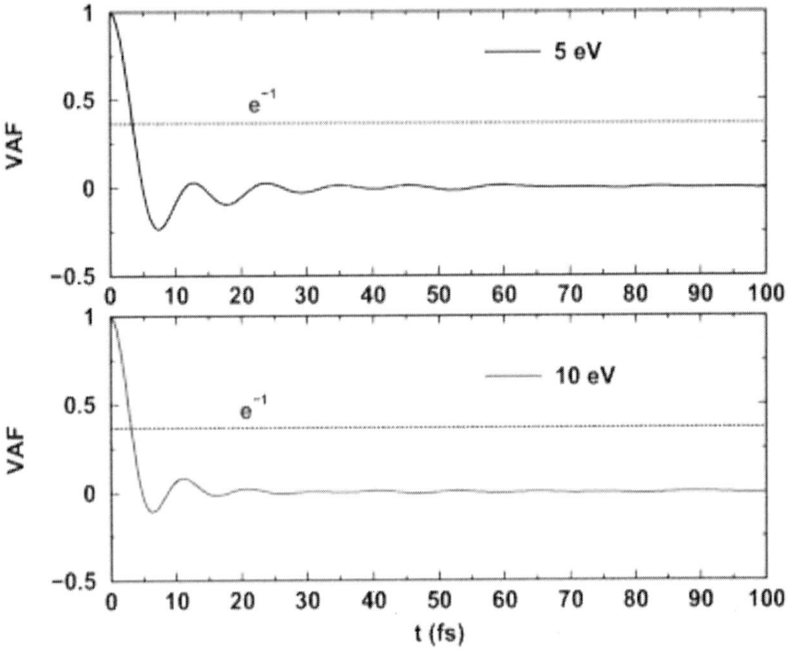

FIGURE 7. Velocity autocorrelation function (VAF) of Fe ions at the temperatures of 5 eV (upper panel) and 10 eV (lower panel) from QLMD calculations for Fe at 45 g cm^{-3}.

7. CONCLUSION

In conclusion, the dynamic structures and transport properties including EOS, diffusion and viscosities are calculated using the QLMD method within the framework of first principles. The dynamic structures provide information about RDF, melting behaviors, diffusions and viscosities. The validation of the SE relation is also discussed, showing the importance of the methods of choosing the effective atomic diameters. This work studies the dense liquid Fe existing in giant planets such as Jupiter and exoplanets, indicating the necessity of first principles calculations or constructing accurate many-body interactions. Furthermore, the results are crucial for understanding laser-shock induced compression experiments, reminding us of the complexity of dense matter. In this field, a great deal of new physics needs to be studied to a greater depth.

ACKNOWLEDGMENTS

This work was supported by the National NSFC under grant numbers 11104351, 11274383, 60921062, 11005153 and 11104350. Calculations are carried out at the Research Center of Supercomputing Application, NUDT.

REFERENCES

1. Swift D C et al 2012 Mass-radius relationships for exoplanets Astrophys. J. **744** 59
2. Jin F T et al Radiative opacity of iron studied using a detailed level accounting model Astrophys. J. **693**597–609
3. Gao C and Zeng J L 2008 Spectrally resolved and Rosseland and Planck mean opacities of iron plasmas at temperatures above 100 eV: a systematic study Phys. Rev. E **78** 046407
4. Moses E I et al 2009 The National Ignition Facility: Ushering in a new age for high energy density science Phys. Plasmas **16** 041006
5. Jeanloz R et al 2007 Achieving high-density states through shock-wave loading of precompressed samples Proc. Natl Acad. Sci. USA **104** 9172–7
6. Pickard C J and Needs R J 2009 Stable phases of iron at terapascal pressures J. Phys.: Condens. Matter **21** 452205
7. Stixrude L 2012 Structure of iron to 1 Gbar and 40 000 K Phys. Rev. Lett. **108** 055505
8. Wijs Gilles A et al 1998 The viscosity of liquid iron at the physical conditions of the Earth's core Nature**392** 805
9. Alfè D, Gillan M J and Price D G 1999 The melting curve of iron at the pressures of the Earth's core from ab initio calculations Nature **401** 462
10. Alfè D and Gillan M J 1998 First-principles calculation of transport coefficients Phys. Rev. Lett. **81**5161

11. Alfè D and Gillan M J 1998 First-principles simulations of liquid Fe–S under Earth's core conditions Phys. Rev. B **58** 8248
12. Vočadlo L, Alfè D, Price G D and Gillan M J 2000 Transport coefficients of liquids from first principles Phys. Earth Planet. Inter. **120** 145
13. Saigo T and Hamaguchi S 2002 Shear viscosity of strongly coupled Yukawa systems Phys. Plasmas **9**1210–6
14. Murillo M S 2008 Viscosity estimates of liquid metals and warm dense matter using the Yukawa reference system High Energy Density Phys. **4** 49–57
15. Daligault J 2012 Diffusion in ionic mixtures across coupling regimes Phys. Rev. Lett. **108** 225004
16. Ma Y M et al 2009 Transparent dense sodium Nature **458** 182
17. Pickard C J and Needs R J 2010 Aluminium at terapascal pressures. Nature Mater. **9** 624–27
18. Dai J Y et al 2011 Dynamic ionic clusters with flowing electron bubbles from warm to hot dense iron along the Hugoniot curve Phys. Rev. Lett. **109** 175701
19. Dai J Y, Hou Y and Yuan J M 2010 Unified first principles description from warm dense matter to ideal ionized gas plasma: electron–ion induced friction Phys. Rev. Lett. **104** 245001
20. Dai J Y, Hou Y and Yuan J M 2010 Quantum Langevin molecular dynamics determination of the solar-interior equation of state Astrophys. J. **721** 1158–63
21. Yuan J M 2002 Self-consistent average-atom scheme for electronic structure of hot and dense plasmas of mixture Phys. Rev. E **66** 047401
22. Hou Y, Jin F T and Yuan J M 2006 Influence of the electronic energy level broadening on the ionization of atoms in hot and dense plasmas: an average atom model demonstration Phys. Plasmas **13**093301 and references therein
23. Mazevet S, Lambert F, Bottin F, Zérah G and Clérouin J 2007 Ab initio molecular dynamics simulations of dense boron plasmas up to the semiclassical Thomas–Fermi regime Phys. Rev. E **75**056404
24. Lambert F, Clérouin J and Zérah G 2006 Very-high-temperature molecular dynamics Phys. Rev. E **73**016403
25. Danel J-F, Kazandjian L and Zérah G 2012 Numerical convergence of the self-diffusion coefficient and viscosity obtained with Thomas–Fermi–Dirac molecular dynamics Phys. Rev. E **85** 066701
26. Mermin N D 1965 Thermal properties of the inhomogeneous electron gas Phys. Rev. **137** A1441–3
27. Graziani F R et al 2012 Large-scale molecular dynamics simulations of dense plasmas: the Cimarron project High Energy Density Phys. **8** 105–31
28. Dai J Y, Hou Y and Yuan J M 2011 Influence of ordered structures on electrical conductivity and XANES from warm to hot dense matter High Energy Density Phys. **7** 84–7
29. Kress J D, Cohen James S, Horner D A, Lambert F and Collins L A 2010 Viscosity and mutual diffusion of deuterium–tritium mixtures in the warm-dense-matter regime Phys. Rev. E **82** 036404
30. Horner D A, Lambert F, Kress J D and Collins L A 2009 Transport properties of lithium hydride from quantum molecular dynamics and orbital-free molecular dynamics Phys. Rev. B **80** 024305
31. Lambert F and Recoules V 2012 Plastic ablator and hydrodynamic instabilities: a first-principles set of microscopic coefficients Phys. Rev. E **86** 026405
32. Kress J D et al 2011 Quantum molecular dynamics simulations of transport properties in liquid and dense-plasma plutonium Phys. Rev. E **83** 026404
33. Dai J Y and Yuan J M 2009 Large scale efficient Langevin dynamics and why it works Europhys. Lett.**88** 20001
34. Giannozzi P et al 2009 Quantum ESPRESSO: a modular and open-source software project for quantum simulations of materials J. Phys.: Condens. Matter **21** 395502
35. Perdew J P, Burke K and Ernzerhof M 1996 Generalized gradient approximation made simple Phys. Rev. Lett. **77** 3865–8

36. Zhang H L et al 2010 Static equation of state of bcc iron Phys. Rev. B **82** 132409
37. Hou Y and Yuan J M 2009 Alternative ion–ion pair-potential model applied to molecular dynamics simulations of hot and dense plasmas: Al and Fe as examples Phys. Rev. E **79** 016402
38. Allen M P and Tildesley T J 1987 Computer Simulation of Liquids (Oxford: Clarendon)
39. Collins L, Kwon I, Kress J, Troullier N and Lynch D 1995 Quantum molecular dynamics simulations of hot, dense hydrogen Phys. Rev. E **52** 6202–19
40. Mithen J P, Daligault J and Gregori G 2012 Molecular dynamics simulations for the shear viscosity of the one-component plasma Contrib. Plasma Phys. **52** 58–61
41. Lorenzen W, Holst B and Redmer R 2009 Demixing of hydrogen and helium at megabar pressures Phys. Rev. Lett. **102** 115701
42. Plagemann K U et al 2012 Dynamic structure factor in warm dense beryllium New J. Phys. **14** 055020
43. Cappelezzo M, Capellari C A, Pezzin S H and Coelho L A F 2007 Stokes–Einstein relation for pure simple fluids J. Chem. Phys. **126** 224516
44. Schoen M and Hoheisel C 1985 The shear viscosity of a Lennard–Jones fluid calculated by equilibrium molecular dynamics Mol. Phys. **56** 653–72
45. Morard G, Bouchet J, Valencia D, Mazevet S and Guyot F 2011 The melting curve of iron at extreme pressures: implications for planetary cores High Energy Density Phys. **7** 141–4
46. Nguyen Jeffrey H and Holmes Neil C 2004 Melting of iron at the physical conditions of the Earth's core Nature **427** 339–42
47. Alfè D 2009 Temperature of the inner-core boundary of the earth: melting of iron at high pressure from first-principles coexistence simulations Phys. Rev. B **79** 060101
48. Kesselring T A, Franzese G, Buldyrev S V, Herrmann H J and Stanley H E 2012 Nanoscale dynamics of phase flipping in water near its hypothesized liquid–liquid critical point Sci. Rep. **2** 474
49. Frenkel D and Smit B 1996 Understanding Molecular Simulation (San Diego, CA: Academic)

CITATIONS

1. *Average-atom model for two-temperature states and ionic transport properties of aluminum in the warm dense matter regime Yong Hou et al 2017 High Energy Density Physics **22** 21*
2. *Melting properties of Pt and its transport coefficients in liquid states under high pressures Pan-Pan Wang et al 2016 International Journal of Modern Physics B **30** 1550250*
3. *Equations of state and transport properties of mixtures in the warm dense regime Yong Hou et al 2015 Physics of Plasmas **22** 022711*
4. *Properties of Liquid Nickel along Melting Lines under High Pressure Cao Qi-Long et al 2015 Chinese Physics Letters **32** 086201*
5. *Quantum molecular dynamics study on the structures and dc conductivity of warm dense silane Huayang Sun et al 2014 Physical Review E **89***
6. *Nuclear quantum dynamics in dense hydrogen Dongdong Kang et al 2014 Scientific Reports **4***
7. *Transport coefficients and entropy-scaling law in liquid iron up to Earth-core pressures Qi-Long Cao et al 2014 The Journal of Chemical Physics **140** 114505*
8. *Focus on high energy density physics R Paul Drake and Peter Norreys 2014 New Journal of Physics **16** 065007*

CHAPTER 7

5D World-Universe Model. Multicomponent Dark Matter

Vladimir S. Netchitailo

Biolase Inc., Irvine, USA

ABSTRACT

5D World-Universe Model (WUM) is based on the decisive role of the Medium of the World composed of massive particles: protons, electrons, photons, neutrinos, and Dark Matter Particles (DMP). The model forecasts the masses of DMP, discusses the possibility of all macroobject cores consisting of DMP (galaxy clusters, galaxies, star clusters, extrasolar systems, and planets), and explains the diffuse cosmic gamma-ray background radiation as the sum of contributions of multicomponent dark matter annihilation. The signatures of DMP annihilation with expected masses of 1.3 TeV, 9.6 GeV, 70 MeV, 340 keV, and 3.7 keV, are found in spectra of the diffuse gamma-ray background and the emission of various macroobjects in the World. The correlation between different emission lines in spectra of macroobjects is connected to their structure, which depends on the composition of the cores and surrounding shells made up of DMP. Consequently, the diversity of Very High Energy (VHE) gamma-ray sources in the World has a clear explanation.

Keywords: 5D World-Universe Model, Medium of the World, Dark Matter Particles, Cores of Macroobjects, Gamma-Ray Background Radiation, Pioneer Anomaly

1. INTRODUCTION

We can't solve problems by using the same kind of thinking we used when we created them.

Albert Einstein

In the World-Universe Model (WUM), we introduce the basic unit of mass m_0 that equals to

$$m_0 = \frac{h}{ac} = 70.025267 \text{ MeV}/c^2$$

where h is Planck constant, c is the electrodynamic constant, $a = 2\pi a_0$, and a_0 is the classical electron radius. m_0 plays a key role when the masses of Dark Matter Particles (DMP) are discussed in the next Section.

The Fine-structure constant (FSC) α is a fundamental physical constant that has several physical interpretations. α is the rest mass of an electron m_e measured in terms of basic unit m_0. FSC plays a central role in WUM.

According to WUM, all stable particles are created in the 3-sphere World due to the surface energy of the 4-ball Nucleus of the World provided by the 4-dimensional Universe. The World consists of the Medium (protons, electrons, photons, neutrinos, and DMP) and Macroobjects (Galaxy clusters, Galaxies, Star clusters, Extrasolar systems, planets, etc.) made of these particles. There is no empty space or dark energy in WUM. The role of the Intergalactic plasma consisting of protons, electrons, and photons as part of the Medium of the World is analyzed in [1] .

This paper discusses the Multicomponent Dark Matter and its decisive role in the Medium and Macroobjects of the World. DMP includes three Majorana fermions (Neutralinos, WIMPs, and Sterile neutrinos) with spin of 1/2 and two spin-0 bosons (named DIRACs and ELOPs in the World-Universe Model), as detailed below. Multicomponent dark matter models consisting of both bosonic and fermionic components were analyzed in literature (for example, see [2] - [10] and references therein).

2. DARK MATTER PARTICLES

Dark Matter (DM) is among the most important open problems in both cosmology and particle physics. There are three prominent hypotheses on nonbaryonic DM, namely Hot Dark Matter (HDM), Warm Dark Matter (WDM), and Cold Dark Matter (CDM).

A neutralino with mass m_N in $100 \Leftrightarrow 10000\ \mathrm{GeV}/c^2$ range is the leading CDM candidate. Light DMP that are heavier than WDM and HDM but lighter than neutralinos is DM candidates too. Subsequently, we will refer to the light DMP as WIMPs. Their mass m_{WIMP} falls into $1 \Leftrightarrow 10\ \mathrm{GeV}/c^2$ range. It is known that a sterile neutrino with mass m_{ν_s} in $1 \Leftrightarrow 10\ \mathrm{keV}/c^2$ range is a good WDM candidate. In our opinion, a tauonic neutrino is a good HDM candidate.

In addition to fermions discussed above, we offer another type of DMP—spin-0 bosons, consisting of two fermions each. There exist two types of DM bosons which we called DIRACs and ELOPS.

DIRACs are magnetic dipoles with mass m_0, consisting of two Dirac monopoles with mass $\frac{m_0}{2}$ and charge $\mu = \frac{e}{2\alpha}$, where e is an electron charge. Dissociated DIRACs can only exist at nuclear densities or at high tem-

peratures. In our opinion, Dirac monopoles are the smallest building blocks of constituent quarks and hadrons (mesons and baryons).

The second spin-0 boson is the ELOP (named by analogy to an ELectron-nortisOP dipole). ELOP weighs $\frac{2}{3}m_e$ and consists of two preons with mass $m_{pr} = \frac{1}{3}m_e$ and charge $e_{pr} = \frac{1}{3}e$. ELOPs break into two preons at nuclear densities or at high temperatures. In particle physics, preons are postulated to be "point-like" particles, conceived to be subcomponents of quarks and leptons [11] .

We did not take into account the binding energies of DIRACs and ELOPs, and thus the values of the masses of monopoles and preons are approximate. They have negligible electrostatic and electromagnetic charges because the separation between charges is very small. The signatures of these bosons' annihilation in gamma-ray spectra will be discussed in Section 6.

WUM postulates that masses of DMP are proportional to multiplied by different exponents of α and can be expressed with the following formulae:

CDM particles (neutralinos and WIMPs):

$$m_N = \alpha^{-2} m_0 = 1.3149950 \text{ TeV/c}^2 \tag{2.1}$$

$$m_{\text{WIMP}} = \alpha^{-1} m_0 = 9.5959823 \text{ GeV/c}^2 \tag{2.2}$$

DIRACs:

$$m_{\text{DIRAC}} = 2\alpha^0 \frac{m_0}{2} = 70.025267 \text{ MeV/c}^2 \tag{2.3}$$

ELOPs:

$$m_{\text{ELOP}} = 2\alpha^1 \frac{m_0}{3} = 340.66606 \text{ keV/c}^2 \tag{2.4}$$

WDM particles (sterile neutrinos):

$$m_{\nu_s} = \alpha^2 m_0 = 3.7289402 \text{ keV/c}^2 \tag{2.5}$$

These values fall into the ranges estimated in literature. The roles of those particles in macroobject cores built up from fermionic dark matter, in gamma-ray spectra of the diffuse gamma-ray background, and the emission of various macroobjects in the World will be discussed in Sections 3, 4 and 6 respectively.

Our Model holds that the energy densities of all types of DMP are proportional to the proton energy density ρ_p in the World's Medium [1] :

$$\rho_p = \frac{2\pi^2\alpha}{3}\rho_{cr} \tag{2.6}$$

where ρ_{cr} is a critical energy density of the World:

$$\rho_{cr} = 3\rho_0 \times Q^{-1} \tag{2.7}$$

$$\rho_0 = \frac{hc}{a^4} \tag{2.8}$$

ρ_0 is a basic unit of energy density and a dimensionless time-varying quantity Q equals to the ratio of the size of the World R at cosmological time τ to the Worlds' size a at the beginning:

$$Q = \frac{R}{a} \tag{2.9}$$

In all, there are 5 different types of DMP. Then the total energy density of DM is

$$\rho_{DM} = 5\rho_p = 0.24007327\rho_{cr} \tag{2.10}$$

which is close to the measured DM energy density: $\rho_{DM} \cong 0.268\rho_{cr}$ [12]. Note that one of outstanding puzzles in particle physics and cosmology relates to so-called cosmic coincidence: the ratio of dark matter density in the World to baryonic matter density in the Medium of the World $\cong 5$ [10] [13].

Neutralinos, WIMPs, and sterile neutrinos are Majorana fermions, which partake in the annihilation interaction with strength equals to α^{-2}, α^{-1}, and α^2 respectively (see Section 3). The signatures of DMP annihilation with expected masses of 1.3 TeV, 9.6 GeV, 70 MeV, 340 keV, and 3.7 keV are found in spectra of the diffuse gamma-ray background and the emission of various macroobjects in the World (see Section 6).

3. MACROOBJECT CORES BUILT UP FROM FERMIONIC DARK MATTER

In this section, we discuss the possibility of all macroobject cores consisting of DMP introduced in Section 2. The first phase of stellar evolution in the history of the World may be dark stars, powered by Dark Matter heating rather than fusion. Neutralinos and WIMPs, which are their own antiparticles, can annihilate and provide an important heat source for the stars and planets in the World.

In our view, all macroobjects of the World (including galaxy clusters, galaxies, star clusters, extrasolar systems, and planets) possess the following properties:

Ÿ Macroobject cores are made up of DMP;

Ÿ Macroobjects consist of all particles under consideration, in the same proportion as they exist in the World's Medium;

Ÿ Macroobjects contain other particles, including DM and baryonic matter, in shells surrounding the cores.

Taking into account the main principle of the World-Universe Model (all physical parameters can be expressed in terms of α, Q, small integer numbers, and π) we modify the published theory of Fermionic Compact Stars (FCS) developed by G. Narain, et al. [14] as follows. We take a scaling solution for a free Fermi gas consisting of fermions with mass m_f in accordance with following equations:

Maximum mass:

$$M_{\max} = A_1 M_F \tag{3.1}$$

Minimum radius:

$$R_{\min} = A_2 R_F \tag{3.2}$$

Maximum density:

$$\rho_{\max} = A_3 \rho_0 \tag{3.3}$$

where

$$M_F = \frac{M_P^3}{m_f^2}; \quad R_F = \frac{M_P}{m_f}\frac{L_{Cf}}{2\pi}; \quad \rho_0 = \frac{hc}{a^4} \tag{3.4}$$

and M_P is Planck mass, L_{Cf} is a Compton length of the fermion. A_1, A_2, and A_3 are parameters. Let us choose π as the value of A_2 (instead of $A_2 = 3.367$ taken by G. Narain, et al. [14]). Then diameter of FCS is proportional to the fermion Compton length L_{Cf}. We use $\pi/6$ as the value of A_1 (instead of $A_1 = 0.384$ taken by G. Narain, et al. [14]). Then A_3 will equal to

$$A_3 = \left(\frac{m_f}{m_0}\right)^4 \quad (3.5)$$

Table 1 summarizes the parameter values for FCS made up of various fermions:

Table 1. Parameter values of Compact Stars built up from fermionic dark matter particles.

Fermion	Fermion relative mass m_f/m_0	Macroobject relative mass M_{max}/M_0	Macroobject relative radius R_{min}/L_g	Macroobject relative density ρ_{max}/ρ_0
Sterile neutrino	α^2	α^{-4}	α^{-4}	α^8
Preon	$3^{-1}\alpha^2$	$3^2\alpha^{-2}$	$3^2\alpha^{-2}$	$3^{-4}\alpha^4$
Electron-proton (white dwarf)	$\alpha^2 . \beta$	β^{-2}	$(\alpha\beta)^{-1}$	$\alpha^3\beta$
Monopole	2^{-1}	2^2	2^2	2^{-4}
WIMP	α^{-1}	α^2	α^2	α^{-4}
Neutralino	α^{-2}	α^4	α^4	α^{-8}
Interacting WIMPs	α^{-1}	β^{-2}	β^{-2}	β^4
Interacting neutralinos	α^{-2}	β^{-2}	β^{-2}	β^4
Neutron (star)	$\approx\beta$	β^{-2}	β^{-2}	β^4

where

$$M_0 = \frac{4\pi m_0}{3} \times Q^{3/2} \quad (3.6)$$

$$L_g = a \times Q^{1/2} \quad (3.7)$$

$$\beta = \frac{m_p}{m_0} \quad (3.8)$$

and m_p is the mass of a proton. A maximum density of neutron stars equals to the nuclear density:

$$\rho_{\max} = \beta^4 \rho_0 \tag{3.9}$$

which is the maximum possible density of any macroobject in the World.

A Compact Star made up of heavier particles—WIMPs and neutralinos—could in principle have a much higher density. In order for such a star to remain stable and not exceed the nuclear density, WIMPs and neutralinos must partake in an annihilation interaction whose strength equals to α^{-1} and α^{-2} respectively.

Scaling solution for interacting WIMPs can also be described with Equations (3.1), (3.2), (3.3) and the following values of A_1, A_2 and A_3:

$$A_{1\max} = \frac{\pi}{6}(\alpha\beta)^{-2} \tag{3.10}$$

$$A_{2\min} = \pi(\alpha\beta)^{-2} \tag{3.11}$$

$$A_{3\max} = \beta^4 \tag{3.12}$$

The maximum mass and minimum radius increase about two orders of magnitude each and the maximum density equals to the nuclear density. Note that parameters of a FCS made up of strongly interacting WIMPs are identical to those of neutron stars.

In accordance with the paper by G. Narain, et al. [14] , the most attractive feature of the strongly interacting Fermi gas of WIMPs is practically constant value of FCS minimum radius in the large range of masses M_{WIMP} from

$$M_{\text{WIMPmax}} = \frac{\pi}{6}(\alpha\beta)^{-2} M_F = \frac{1}{\beta^2} M_0 \tag{3.13}$$

down to

$$M_{\text{WIMPmin}} = \alpha^4 M_{\text{WIMPmax}} \tag{3.14}$$

M_{WIMPmin} is more than eight orders of magnitude smaller than M_{WIMPmax}. It makes strongly interacting WIMPs good candidates for stellar and planetary cores of extrasolar systems with Red stars (see Section 4).

When the mass of a FCS made up of WIMPs is much smaller than the maximum mass, the scaling solution yields the following equation for parameters A_1 and A_2:

$$A_1 A_2^3 = \pi^4 \tag{3.15}$$

Compare $\pi^4 \cong 97.4$ with the value of 91 used by G. Narain, et al. [14] .

Minimum mass and maximum radiuses take on the following values:

$$A_{1\min} = \frac{\pi}{6}\sqrt{6}(\alpha\beta)^2 \tag{3.16}$$

$$A_{2\max} = \pi\sqrt[6]{6}(\alpha\beta)^{-2/3} \tag{3.17}$$

It follows that the range of FCS masses $(A_{1\min} \Leftrightarrow A_{1\max})$ spans about three orders of magnitude, and the range of FCS core radii $(A_{2\min} \Leftrightarrow A_{2\max})$—one order of magnitude. It makes WIMPs good candidates for brown dwarf cores too (see Section 4).

Scaling solution for interacting neutralinos can be described with the same Equations (3.1), (3.2), (3.3) and the following values of A_1^*, A_2^* and A_3^*:

$$A_{1\max}^* = \frac{\pi}{6}(\alpha^2\beta)^{-2} \tag{3.18}$$

$$A_{2\min}^* = \pi(\alpha^2\beta)^{-2} \tag{3.19}$$

$$A_{3\max}^* = \beta^4 \tag{3.20}$$

In this case, the maximum mass and minimum radius increase about four orders of magnitude each and the maximum density equals to the nuclear density. Note that parameters of a FCS made up of strongly interacting neutralinos are identical to those of neutron stars.

Practically constant value of FCS minimum radius takes place in the huge range of masses M_N from

$$M_{N\max} = \frac{\pi}{6}(\alpha\beta)^{-2}\alpha^2 M_F = \frac{1}{\beta^2}M_0 \qquad (3.21)$$

down to

$$M_{N\min} = \alpha^8 M_{N\max} \qquad (3.22)$$

$M_{N\min}$ is more than seventeen orders of magnitude smaller than $M_{N\max}$. It makes strongly interacting neutralinos good candidates for stellar and planetary cores of extrasolar systems with Main-sequence stars (see Section 4).

When the mass of a FCS made up of neutralinos is much smaller than the maximum mass, the scaling solution yields the following equation for parameters A_1^* and A_2^*:

$$A_1^* A_2^{*3} = \pi^4 \qquad (3.23)$$

Minimum mass and maximum radiuses take on the following values:

$$A_{1\min}^* = \frac{\pi}{6}\sqrt{6}\left(\alpha^2\beta\right)^2 \qquad (3.24)$$

$$A_{2\max}^* = \pi\sqrt[6]{6}\left(\alpha^2\beta\right)^{-2/3} \qquad (3.25)$$

It means that the range of FCS masses $\left(A_{1\min}^* \Leftrightarrow A_{1\max}^*\right)$ is about twelve orders of magnitude, and the range of FCS core radiuses $\left(A_{2\min}^* \Leftrightarrow A_{2\max}^*\right)$ is about four orders of magnitude. The numerical values for FCS masses and radii will be given in Section 4.

Fermionic Compact Stars have the following properties:

Ÿ The maximum potential of interaction $U_{\max}$ between any particle or macroobject and FCS made up of any fermions

$$U_{\max} = \frac{GM_{\max}}{R_{\min}} = \frac{c^2}{6} \qquad (3.26)$$

does not depend on the nature of fermions;

Ÿ The minimum radius of FCS made of any fermion

$$R_{\min} = 3R_{SH} \tag{3.27}$$

equals to three Schwarzschild radii and does not depend on the nature of the fermion;

Ÿ FCS density does not depend on $M_{\max}$ and $R_{\min}$ and does not change in time while $M_{\max} \propto \tau^{3/2}$ and $R_{\min} \propto \tau^{1/2}$.

4. MACROOBJECTS OF THE WORLD

According to WUM, all macroobjects of the World (galaxies, stars, planets) possess cores consisting of DMP. The theory of fermion compact stars made up of DMP is well developed. Scaling solutions are derived for a free and an interacting Fermi gas in Section 3. Table 2 describes the numerical values for masses and radii of FCS made up of different fermions:

The calculated parameters of FCS show that

Ÿ White Dwarf Shells (WDS) around the nuclei made of strongly interacting WIMPs or neutralinos compose cores of stars in extrasolar systems;

Ÿ Shells of dissociated DIRACs to monopoles around the nuclei made of strongly interacting WIMPs or neutralinos form cores of globular clusters;

Ÿ Shells of dissociated ELOPs to preons around the nuclei made of strongly interacting WIMPs or neutralinos constitute cores of galaxies;

Table 2. Numerical values for parameters of Compact Stars built up from fermionic dark matter particles.

Fermion	Fermion mass m_f, MeV/c^2	Macroobject mass $M_{\max}$, kg	Macroobject radius $R_{\max}$, m	Macroobject density $\rho_{\max}$, kg/m^3
Sterile neutrino	3.73×10^{-3}	1.2×10^{41}	5.4×10^{14}	1.8×10^{-4}
Preon	≥ 0.17	5.9×10^{37}	2.6×10^{11}	7.8×10^{2}
Monopole	≥ 35	1.4×10^{33}	6.2×10^{6}	1.4×10^{12}
Interacting WIMPs	9596	1.9×10^{30}	8.6×10^{3}	7.2×10^{17}
Interacting neutralinos	1315×10^{3}	1.9×10^{30}	8.6×10^{3}	7.2×10^{17}
Electron; proton (white dwarf)	0.511; 938.3	1.9×10^{30}	1.6×10^{7}	1.2×10^{8}
Neutron (star)	939.6	1.9×10^{30}	8.6×10^{3}	7.2×10^{17}

Ÿ Shells of sterile neutrinos around the nuclei made of strongly interacting WIMPs or neutralinos make up cores of galaxy clusters.

Although there are no free Dirac's monopoles and preons in the World, they can arise in the cores of FCS as the result of DIRACs and ELOPs gravitational collapse with density increasing up to the nuclear density $\left(\sim 10^{17}\ \mathrm{kg/m^3}\right)$ and/or at high temperatures, with subsequent dissociation of dipoles to monopoles and preons.

4.1. Galaxies and Galaxy Clusters

A number of non-traditional models explaining the supermassive dark objects observed in galaxies and galaxy clusters, formed by self-gravitating non-baryonic matter composed of fermions and bosons, are widely discussed in literature [2] - [10] .

Dark matter can be, in principle, achieved also through extended theories of gravity. It has been shown, for example, that in the framework of R^2gravity and in the linearized approach, it is possible to obtain spherically symmetric and stationary galaxy states which can be interpreted like an approximated solution of the Dark Matter problem [15] [16] .

According to WUM, the heaviest macroobjects include a high-density preon plasma shell around their cores:

Ÿ Macroobjects with a cold preon shell emit strong radio waves. Such objects are good candidates for the compact astronomical radio sources at centers of galaxies like Sagittarius A* in the Milky Way Galaxy;

Ÿ Red Giants are macroobjects with hot preon shells;

Ÿ Blazars are members of a larger group of active galaxies that host active galactic nuclei (AGN). They are macroobjects with hot preon and sterile neutrinos shells;

Ÿ Quasars are the most energetic and distant members of AGN. They are macroobjects with very hot preon and sterile neutrinos shells;

Ÿ Seyfert galaxies are one of the two largest groups of AGN, along with quasars. They have quasar-like nuclei, but unlike quasars, their host galaxies are clearly detectable. Seyfert galaxies account for about 10% of all galaxies.

Note that the temperature of the preon and sterile neutrinos shells depends on the composition of the macroobject core. Macroobjects whose cores are made up of WIMPs and preons remain cold. Macroobjects with cores made up of WIMPs and WDS produce hot preon and sterile neutrino shells.

Macroobjects whose cores consist of neutralinos and WDS have very hot preon and sterile neutrino shells.

The mass of an AGN is about 7-11 orders of magnitude larger than the mass of the Sun. The radius of an AGN is about 4 - 7 orders of magnitude larger than the radius of WDS (see Table 2). The area of the closed spherical surface around the AGN is 8 - 14 orders of magnitude greater than the surface area of WDS. Luminosity of the AGN is then 8 - 14 orders of magnitude higher than the luminosity of the largest star. This take on AGNs explains the fact that the most luminous quasars radiate at a rate that can exceed the output of average galaxies, equivalent to two trillion suns.

To summarize, macroobjects of the World have cores made up of DM particles. The cores are surrounded by shells made up of DM and baryonic matter. Every macroobject consists of all particles under consideration that are present in the same proportion as they exist in the World's Medium. No compact stars are made up solely of fermionic DMP, for instance.

4.2. Extrasolar Systems

There are two primary types of stars: main-sequence stars and red stars. They differ in their surface temperatures and radii:

Ÿ Red stars have cool surface temperatures: 3500 ⇔ 4500 K for Hypergiants, Supergiants, Giants, lower for Red dwarfs (2300 ⇔ 3800 K), and significantly lower for Brown dwarfs (300 ⇔ 1000 K). These stars have enormous range of radii: from $1650R_{Sun}$ for Hypergiants down to $0.08R_{Sun}$ for Red dwarfs, and lower still for Brown dwarfs.

Ÿ Main-sequence stars have surface temperatures in the range of 3000 ⇔ 45,500 K, and radii in the range from $35R_{Sun}$ for the most massive known star R136a1 down to $0.1R_{Sun}$ for least heavy stars.

In our opinion, the difference between main-sequence stars and red stars lies in composition of stellar cores. Main-sequence stars cores are made up of neutralinos, while red star cores consist of WIMPs. As we have shown in Section 3, in both cases the cores' maximum mass and minimum radius equals to that of a neutron star. The fermions, however, have drastically different interaction strength of annihilation: α^{-1} in case of WIMPs and α^{-2} in case of neutralinos.

The Core temperature is therefore much higher in main-sequence stars whose cores are made up of neutralinos. Ignition of proton-proton chain reaction with the interaction strength equal to $\beta \approx 13.4$ developing in the surrounding WDS happens much more efficiently in these stars.

The developed star model explains the very low power production density produced by fusion inside of the Sun. Wikipedia humorously notes that the power output of the Sun more nearly approximates reptile metabolism than a thermonuclear bomb. In our Model, the core made up of strongly interacting neutralinos is the supplier of proton-electron pairs into WDS and igniter of the proton-proton chain reaction developing in the surrounding WDS with small interaction strength $\beta \cong 13.4$.

New neutralinos freely penetrate through the entire stellar envelope, get absorbed into the core and support neutralino annihilation and proton fusion in the WDS. An important consequence for Solar system, and in fact for all other stars in the World, is that they will never burn their "fuel" out. On the contrary, stars accumulate more fuel with time, and output more power.

Enormous radii of Hypergiants (up to $1650R_{\rm Sun} \cong 10^{12}$ m) and huge luminosity of giant stars can be explained by an additional shell of preons—particles whose charge equals to $\frac{1}{3}e$. They compose hot high-density plasma

with surface temperature in the range of 3500 ⇔ 4500 K. The minimum radius of preon shell $R_{\rm min} \cong 2.6\times10^{11}$ m (see Table 2).

Brown dwarfs are sub-stellar objects whose masses range from 13 to 80 Jupiter masses. In our opinion, Brown dwarfs differ from red stars in that the density of their cores is smaller than nuclear density. Consequently, WIMPs annihilation takes place less efficiently.

4.3. Extrasolar System Formation

The Nebular Hypothesis is the most widely accepted model of planetary formation. It holds that 4.6 Billion years ago, the Solar System was formed during a gravitation collapse of a giant molecular cloud, some light years across. The most significant criticism of the hypothesis is its inability to explain the Sun's relative lack of angular momentum when compared to the planets [17] .

According to WUM, extrasolar systems arise from clouds of all particles under consideration with mass M_{Cl}. As a result of gravitational instability, gravitational collapse takes place and one third of M_{Cl} is concentrating at the center of the cloud, increasing the density of the core up to the nuclear density.

The heaviest particles—neutralinos or WIMPs are the first in this stream of matter. When their density achieves the nuclear density, self-annihilation process ignites. As the result, the Stellar Nucleus (SN) grows up to 10^4 for neutralinos and 10^2 times for WIMPs taking additional mass of neutralinos and WIMPs from oncoming stream.

The next heaviest particles—protons, joined by electrons—will follow neutralinos or WIMPs during the gravitational collapse, and form the White Dwarf Shell (WDS) around the SN made of strongly interacting WIMPs or neutralinos.

Expansion of the hot Stellar Core (SC), consisting of SN with WDS, is progressing. Drops of the SC are ejected from the equatorial bulges of an overspinning SC (outward centrifugal forces exceed the inward gravitational force) and give birth to the cores of planets.

The following facts support the creation picture of extrasolar systems outlined above:

Ÿ The analysis of a mass-radius ratio for compact stars made of strongly interacting fermions shows that the radius remains approximately constant for a wide range of compact stars masses;

Ÿ The analysis of a mass-radius ratio for the lowest mass white dwarfs shows the same behavior—radius does not depend on mass. It happens because at the low mass end the Coulomb pressure (which is characterized by constant density $\propto M/r^3$ and thus $r \propto M^{1/3}$) starts to compensate the degeneracy: $r \propto M^{-1/3}$. The two effects nearly cancel each other out, so $r \propto M^0$—no dependency at all;

Recent analysis of the Solar and Heliospheric Observatory (SOHO) mission data favors a faster rotation rate in the solar core (below 0.2 solar radius) than in the rest of the radiative zone [18] ;

Ÿ By analyzing the minute changes in travel times and wave shapes for earthquake doublets, the researchers of [19] concluded that the Earth's inner core is rotating faster than its surface by about 0.3 - 0.5 degrees per year;

Ÿ The authors of [20] found that Earth's inner core, made up of solid iron, "superrotates" in an eastward direction—meaning it spins faster than the rest of the planet—while the outer core, comprising mainly molten iron, spins westwards at a slower pace.

In our opinion, the Earth's inner core is made up of neutralinos, while the outer core is the WDS. The cores of the Sun and the planets comprising the Solar System are not rotating with the same speed as their surfaces. When analyzing the angular momentum distribution of the entire Solar System, one must consider these additional angular momentums. Moreover, the remainder of the original particle cloud weighing $\frac{2}{3}M_{Cl}$ may possess additional angular momentum.

As discussed above, the minimum radius of the hot neutralinos and WIMPs core $R_{\min} \cong 8.6\ \text{km}$, and it remains essentially constant whether the core belongs to a star or to a planet. The masses of planets formed around red stars and main-sequence stars differ:

Ÿ The smallest possible mass of planets formed around red stars is 8 orders of magnitude smaller than maximum star mass M_0;

Ÿ Planets formed around main-sequence stars may be 17 orders of magnitude lighter than the maximum star mass.

Consequently, all round objects in hydrostatic equilibrium, down to Mimas in Solar system, contain hot neutralinos cores with WDS and should be considered planets. Planets can arise only around main sequence and red stars. Due to the less violent nature of their formation, brown dwarfs do not create planets.

4.4. Pioneer Anomaly

According to WUM, the macroobject energy E_{MO} enclosed in surface S_{MO} is proportional to the area of that surface:

$$E_{MO} = \sigma_0 S_{MO} \qquad (4.4.1)$$

where σ_0 is the basic unit of surface energy density: $\sigma_0 = \rho_0 a$. It is natural to define surface S_{MO} as the boundary between macroobject and surrounding environment. In case of our Solar system, such a surface is named Heliosphere. We will refer to such a surface as Macroobject Boundary (MOB). According to the developed Model, Macroobjects have cores made up of fermionic DMP possessing minimum radii $R_{\min}$ described in Table 1 and Table 2. In case of extrasolar systems, the cores are made up of interacting neutralinos or WIMPs surrounded with White Dwarf Shells (WDS).

The cores are surrounded by the transitional region. In this region, the density decreases rapidly to the point of the zero level of the fractal structure [21] characterized by radius R_f and energy density ρ_f that satisfy the following equation for $r \geq R_f$:

$$\rho(r) = \frac{\rho_f R_f}{r} \tag{4.4.2}$$

According to Yu. Baryshev: For a structure with fractal dimension D = 2 the constant $\rho_f R_f$ may be actually viewed as a new fundamental physical constant [21] . In our Model, it is natural to connect this constant with the constant σ_0:

$$\rho_f R_f = 4\sigma_0 \tag{4.4.3}$$

The value of 4 above follows from the ratio for all Macroobjects of the World: 1/3 of the total energy is in the central macroobject (for example, star in extrasolar system) and 2/3 of the total energy is in the fractal structure around it. Taking the radius of a Macroobject Boundary R_{MOB} we find the macroobject energy:

$$E_{MO} = 4\pi R_{MOB}^2 \sigma_0 \tag{4.4.4}$$

The energy in the fractal structure E_{FS} at $R_{MOB} \gg R_f$ is:

$$E_{FS} = \int_{R_f}^{R_{MOB}} \frac{4\sigma_0}{r} \times 4\pi r^2 dr \approx 8\pi R_{MOB}^2 \sigma_0 \tag{4.4.5}$$

and the total energy E_{tot} equals to: $E_{tot} = 12\pi R_{MOB}^2 \sigma_0$.

It allows us to explain the so-called “Pioneer anomaly”. Wikipedia describes this effect the following way: The Pioneer anomaly is the observed deviation from predicted accelerations of the Pioneer 10 and Pioneer 11 spacecraft after they passed about 20 astronomical units $(3\times10^9 \text{ km}; 2\times10^9 \text{ mi})$ on their trajectories out of the Solar System. An unexplained force appeared to cause an approximately constant sunward acceleration of $a_P = 8.74 \pm 1.33\times10^{-10} \text{ m/s}^2$ for both spacecraft. The magnitude of the Pioneer effect a_P is numerically quite close to the product of the speed of light c and the Hubble constant H_0 hinting at cosmological connection.

Let us calculate a deceleration a_P at the distance $r_P \gg R_f$ due to the additional mass of the fractal structure $M_{FS}(r_P) \propto r_P^2$ with the following equation for the gravitational parameter G [1] :

$$G = \frac{c^4}{8\pi\sigma_0 R_0} \tag{4.4.6}$$

$$a_P = \frac{GM_{FS}}{r_P^2} = \frac{c^4}{8\pi\sigma_0 R_0}\times\frac{8\pi\sigma_0}{c^2} = \frac{c^2}{R_0} = cH_0 = 6.68\times10^{-10} \text{ m/s}^2 \tag{4.4.7}$$

which is in good agreement with the experimentally measured value (R_0 and H_0 are the values of the World's size R and Hubble's parameter H at the current time t). It is important to notice that the calculated deceleration does not depend on r_P and equals to cH_0 for all objects around the macroobject at the distance $r > R_f$.

Mass of the fractal structure around Sun M_V at distances $R_V \gg R_f$ is

$$M_V = 8\pi R_V^2 \sigma_0 / c^2 \tag{4.4.8}$$

At distance $R_V = 1.8\times10^{13}$ m away from the Sun (approximate distance to Voyager 1 [22]),

$$M_V \cong 3.3\times10^{27} \text{ kg} \tag{4.4.9}$$

that is $\sim 0.15\% M_{Sun}$. Note that the distances traveled by Voyagers are much smaller than the radius of the MOB: $R_V \ll R_{MOB} \sim 10^{15}$ m.

5. X RAYS AND GAMMA RAYS

All "elementary" particles of the World are fermions and they possess masses. Bosons such as photons, X- quants, and Gamma-quants are composite particles and consist of two fermions. Gamma rays are usually distinguished from X rays by their origin: X rays are emitted by electrons outside the nucleus, while gamma rays are emitted by the nucleus. A better way to distinguish the two, in our opinion, is the type of fermions composing the core of X-quants and Gamma-quants.

Super-soft X rays possess energies in the 0.09 2.5 keV range, whereas soft Gamma rays have energies in the 10 5000 keV range. We assume that X-quants are composed of two interacting neutrinos. New Physics with the dineutrinos in the Rare Decay $B \to K\nu\bar{\nu}$ is actively discussed in literature in recent years (for example, see [23] [24]).

Soft Gamma-quants are composed of two sterile neutrinos (3.7 keV each). Hard and super-hard Gamma-quants may be composed of two preons ($\gtrsim$ 0.17 MeV each), which are ELOPs in our Model, two Dirac monopoles ($\gtrsim$ 35 MeV each) which are, in fact, DIRACs.

We propose that Super-soft gamma rays (<10 keV) can arise as the result of sterile neutrino annihilation in the low energy case. Two or three super-soft gamma-quants with the energy < 3.7 keV are created. Similarly,

Ÿ ELOP annihilation produces hard gamma rays with energies < 340 keV;

Ÿ DIRAC annihilation produces hard gamma rays with energies < 70 MeV;

Ÿ WIMP annihilation produces super-hard gamma rays with energies < 9.6 GeV;

Ÿ Neutralino annihilation produces super-hard gamma rays with energies < 1.3 TeV.

Diffuse cosmic gamma-ray background is the sum of the contributions of the multicomponent self-interacting dark matter annihilation.

6. DARK MATTER SIGNATURES IN GAMMA-RAY SPECTRA

Large number of papers has been published in the field of X-ray and gamma-ray astronomy. The X-ray and gamma-ray background from $\lesssim$ 0.1 keV to $\gtrsim$ 10 TeV has been studied using high spectral and spatial resolution data from different spectrometers. Numerous papers were dedicated to Dark Matter searches with astroparticle data (see reviews [25] - [34] and references therein). Dark Matter annihilation is proportional to the square of the DM density and is especially efficient in places of highest concentration of dark matter, such as compact stars built up from fermionic dark matter particles (see Section 3).

The models of DM annihilation and decay for various types of macroobjects (galaxy clusters, blazars, quasars, Seyfert galaxies) are well-developed. Physicists working in the field X-ray and gamma-ray astronomy attempt to determine masses of DM particles that would fit the experimental results with the developed models.

Recall that no macroobjects are made up of just a single type of DM particles, since other DM particles as well as baryonic matter are present in the shells. It follows that macroobjects cannot irradiate gamma rays in a single spectral range. On the contrary, they irradiate gamma-quants in different spectral ranges with ratios of fluxes depending on structure of a given macroobject.

WUM forecasts existence of DM particles with 1.3 TeV, 9.6 GeV, 70 MeV, 340 keV, and 3.7 keV masses. We will look for signs of annihilation of these particles in the observed gamma-ray spectra. We connect gamma-ray spectra with the structure of macroobjects (core and shells composition).

C. Boehm, P. Fayet, and J. Silk have this to say about Light and Heavy Dark Matter Particles:

It has recently (2003) been pointed out that the 511 keV emission line detected by Integral/SPI from the bulge of our galaxy could be explained by annihilations of light Dark Matter particles into e^+e^-. If such a signature is confirmed, then one might expect a conflict with the interpretation of very high energy gamma rays if they also turn out to be due to Dark Matter annihilations.

They proposed a way to reconcile the low and high energy signatures, even if both of them turn out to be due to Dark Matter annihilations. One would be a heavy fermion, for example, the lightest neutralino (>100 GeV) and the other one a possibly light spin-0 particle (~100 MeV). Both of them would be neutral and also stable as a result of two discrete symmetries (say R and M-parities) [9] [35] .

According to our Model, the two couples of co-annihilating DMP are: a heavy fermion—neutralino with mass 1.3 TeV and a light spin-0 boson—DIRAC with mass 70 MeV; a heavy fermion—WIMP with mass 9.6 GeV and a light spin-0 boson?ELOP with mass 340 keV.

6.1. Neutralino 1.3 TeV

J. Holder has this to say about TeV Gamma-ray Astronomy: In leptonic scenarios, a population of electrons is accelerated to TeV energies, typically through Fermi acceleration by shocks in the AGN jet. These electrons then cool by radiating X-ray synchrotron photons. TeV emission results from Inverse Compton interactions of the electrons with either their self-generated synchrotron photons, or an external photon field. The strong correlation between X-ray and TeV emission which is often observed provides evidence for a common origin such as this, although counter examples do exist [36] .

In our opinion, the TeV blazar emission should be classified as extremely-hard X rays and not gamma rays, since by definition: X rays are emitted by electrons outside the nucleus, while gamma rays are emitted by the nucleus.

R.C.G. Chaves, et al. have found that a significant fraction of the Galactic VHE (Very High Energy) gamma-ray sources (from the observed approximately 100 VHE γ-ray sources [38] - [42]) do not appear to have obvious counterparts at other wavelengths [37] .

This correlation between keV emission and TeV emission can be easily explained by the annihilation of the sterile neutrinos (3.7 keV) in the shell around the core of AGN made of neutralinos (1.3 TeV). Lack of the counterpart in gamma-ray spectra means the absence of sterile neutrino shell.

A detailed global analysis on the interpretation of the latest data of PAMELA, Fermi-LAT, AMS-02, H.E.S.S, and other collaborations in terms of dark matter annihilation and decay in various propagation models showed that for the Fermi-LAT and H.E.S.S. data favor DMP mass is $m_\chi \approx 1.3$ TeV [43] - [46] . The obtained data in [47] - [55] require DM mass m_χ to be around 1 to 1.5 TeV which is in good agreement with the predicted mass of a neutralino (1.3 TeV). Pulsars are the most natural candidates for such sources [41] .

The presence of spectral break at 1.3 TeV in VHE spectra was measured for different blazars [56] - [58] . Some nearby sources, e.g. Vela, Cygnus Loop, and Monogem Supernova Remnant (SNR) have unique signatures in the electron energy spectrum in the TeV region: broken power-law at ~1.3 TeV [59] . The DM interpretations of the $e^{\pm}$ excesses observed by PAMELA, Fermi and ATIC suggest the DMP mass of 1.3 TeV [60] .

As we mentioned above, pulsars are the most natural candidates for such VHE gamma-ray sources. According to WUM, FCS made up of strongly interacting neutralinos and WIMPs have maximum mass and minimum size which is exactly equal to parameters of neutron stars (see Table 1 and Table 2). It follows that pulsars might be in fact rotating Neutralino stars and WIMP stars with different shells around them.

The cores of such pulsars may also be made up of the mixture of neutralinos (1.3 TeV) and WIMPs (9.6 GeV) surrounded by shells composed of the other DMP: DIRACs (70 MeV), ELOPs (340 keV), and sterile neutrinos (3.7 keV). Annihilation of those DMP can give rise to any combination of gamma-ray lines. Thus the diversity of VHE gamma-ray sources in the World has a clear explanation in frames of the World-Universe Model.

In our opinion, results obtained by the CALET program are the closest to the ultimate discovery of the first confirmed dark matter particle—neutralino with mass 1.3 TeV [59] .

6.2. WIMP 9.6 GeV

Dan Hooper summarized and discussed the body of evidence which has accumulated in favor of dark matter in the form of approximately 10 GeV particles, including the spectrum and angular distribution of gamma rays from the Galactic Center, the synchrotron emission from the Milky Way's radio filaments, the diffuse synchrotron emission from the Inner Galaxy (the "WMAP Haze") and low-energy signals from the direct detection experiments DAMA/LIBRA, CoGeNT and CRESST-II. Dan Hooper finds that gamma-ray signal observed from the Galactic Center is consistent with 7 - 12 GeV dark matter particles annihilating mostly to leptons [61] [62] .

Based on EGRET observations, P. Sreekumar, et al. attribute the high-energy gamma ray emissions to blazars: Most of the measured spectra of individual blazars only extend to several GeV and none extend above 10 GeV, simply because the intensity is too weak to have a significant number of photons to measure [63] . WUM proposes that cores of blazars are composed of annihilating WIMPs (9.6 GeV), explaining why no observed radiation extends above 10 GeV. The results of gamma-ray emission between 100

MeV to 10 GeV detected from 18 globular clusters in our Galaxy are also in a good correlation with the predicted mass of WIMPs [64] [65] .

The DAMA/LIBRA, CoGeNT, CRESST-II, CDMS-II collaborations conduct direct detections of DMP by nuclear recoils due to the elastic scattering of DMP. An 8.6 GeV DMP is deemed most probable [66] .

Based on its core assumptions, WUM analytically predicts WIMPs to possess the mass of 9.6 GeV. A large number of experimental results seem to converge to a number in the neighborhood of 10 GeV, providing additional support to WUM.

6.3. DIRAC 70 MeV

S. D. Hunter, et al. discuss a peak at 67.5 MeV: Below about 100 MeV, gamma rays produced via electron bremsstrahlung are the dominant component of the observed spectrum, whereas, above about 100 MeV, the gamma-rays from π^0 decay, which form the broad "pion bump" centered at 67.5 MeV, are the dominant component of the spectrum. The "pion bump", clearly visible in this spectrum, is the only spectral feature in the diffuse gamma ray emission in the EGRET energy range [67] .

70 MeV peak in EGRET data was discussed by Golubkov and Khlopov [68] . They explained this peak by the decay of π^0 -mesons, produced in nuclear reactions. B. Wolfe, et al. said that gamma rays at 70 MeV are notably detectable by GLAST and EGRET [69] . R. Yamazaki, et al. attribute the 70 MeV peak in the emission spectrum from an old supernova remnant (SNR) to π^0 -decay [70] [71] .

Note that whenever the 70 MeV peak appears in gamma-ray spectra, it is always attributed to pion decay. We claim that π^0 decay produces a 67.5 MeV peak, while DIRAC annihilation is responsible for 70 MeV peak. To find out the source of the observed broad peak about 70 MeV, we suggest utilization of exponentially cutoff power-law for analysis of experimental data for gamma-ray energies < 70 MeV. A better fit of experimental data will be evidence of DIRACs annihilation.

In our opinion, the DIRAC may indeed be the so-called U boson, target of intense search by the scientific community. Note that the mass of DIRAC proposed by WUM—0.07 GeV/c^2—falls into the mass range of U boson: $M_U = 0.02 - 0.1\ \mathrm{GeV/c^2}$ [72] - [77] .

6.4. ELOP 340 keV

An ELOP is a spin-0 boson with 340 keV mass. Existence of DMP of similar masses $(m_\chi < 0.42 \text{ MeV})$ has been discussed by Y. Rasera, et al. [78] . The experimental 100 - 400 keV "bump" [79] is in good agreement with the theoretical analysis in [78] and with annihilating ELOPs with mass 340 keV proposed in our Model.

D. E. Gruber, et al. describes a wide gamma-ray extragalactic background spectrum between 1 keV and 10 GeV: Above 60 keV selected data sets included the HEAO 1 A-4 (LED and MED), balloon, COMPTEL, and EGRET data. The fit required the sum of three power laws [80] .

According to our Model, the fit of the total diffuse spectrum in the range between 3 keV and 10 GeV should be performed based on three exponentially cutoff power-laws with injection spectral $J(E) \propto E^{-\gamma} \exp\{-E/E_{cut}\}$ with the spectral index γ and E_{cut} being the cutoff energy of the source spectra. For values of E_{cut} , we should use

Ÿ 9.6 GeV (annihilating WIMPs) in the 9.6 GeV - 70 MeV range;

Ÿ 70 MeV (annihilating DIRACs) in the 70 MeV - 340 keV range;

Ÿ 340 keV (annihilating ELOPs) in the 340 keV - 3.7 keV range.

The fit in the range between 9.6 GeV and 1.3 TeV should be done with $E_{cut} = 1.3 \text{ TeV}$, which equals to the mass of a neutralino.

6.5. Sterile Neutrino 3.7 keV

The very first signature of the emission around 3.7 keV was found in 1967 by P. Gorenstein, R. Giacconi, and H. Gursky. They analyzed the counting rate in the 2 - 5 keV range and found that the sources GX-10.7, +9.1, +13.5, and +16.7 are qualitatively different from Sco X-1, Cyg X-1 or Cyg X-2 in that the highest number of net counts is recorded in the bin centered at 3.75 keV [81] .

An important result was obtained by S. Safi-Harb and H. Ogelman in 1997. They reported that the observations of the X-ray lobes of the large Galactic source W50 [are] associated with the two-sided jets source SS 433. A broken power-law model gives the best fit. The power-law indices are 1.9 and 3.6, with the break occurring at 3.7 keV [82] .

T. Itoh analyzed the broad-band (3.0 - 50 keV) spectra of NGC 4388 and found line-like residual around 3.7 keV at the high confidence level [83] .

A. Bykov, et al. investigated the nature of the extended hard X-ray source XMMU J061804.3 + 222732 and its surroundings using XMM-Newton, Chandra, and Spitzer observations. The X-ray emission consists of a number of bright clumps embedded in an extended structured non-thermal X-ray nebula larger than 30″ in size. Some clumps show evidence for line emission at ~1.9 keV and ~3.7 keV at the 99% confidence level. A feature at 3.7 keV was found in the X-ray spectrum of Src 3 at the 99% confidence level [84] .

In our opinion, the line emission ~3.7 keV corresponds to the annihilation of sterile neutrinos and the line ~1.9 keV corresponds to their decay.

R. Fukuoka, et al. observed the South End of the Radio Arc and found the line-like residual at ~3.7 keV with ~3σ significance [85] . In 2012, A. Moretti, et al. measured the diffuse gamma-ray emission at the deepest level and with the best accuracy available today. An emission line around 3.7 keV is clearly visible in the obtained spectrum [86] .

6.6. Conclusions

Ÿ Emission lines of 1.3 TeV, 9.6 GeV, 70 MeV, 340 keV, and 3.7 keV, can be found in spectra of the diffuse gamma-ray background radiation and various macroobjects of the World in different combinations depending on their structure.

Ÿ The diffuse cosmic gamma-ray background radiation in the <1.3 TeV range is the sum of the contributions of multicomponent dark matter annihilation.

Ÿ The total cosmic-ray radiation consists of gamma-ray background radiation plus X-ray radiation from the different highly ionized chemical elements in the hot areas of the World and is due to various electron processes such as synchrotron radiation, electron bremsstrahlung, and inverse Compton scattering.

ACKNOWLEDGEMENTS

I am very grateful to referees for the important critical remarks. Many thanks to Felix Lev for our frequent discussions of history and philosophy of Physics. Special thanks to my son Ilya Netchitailo who questioned every aspect of the paper and helped shape it to its present form.

CITE THIS PAPER

Vladimir S. Netchitailo (2015) 5D World-Universe Model. Multicomponent Dark Matter. *Journal of High Energy Physics, Gravitation and Cosmology*,**01**,55-71.

REFERENCES

1. Netchitailo, V.S. (2015) 5D World-Universe Model. Space-Time-Energy. Journal of High Energy Physics, Gravitation and Cosmology, 1, 25.
2. Arrenberg, S., et al. (2013) Complementarity of Dark Matter Experiments.
3. Heeck, J. and Zhang, H. (2013) Exotic Charges, Multicomponent Dark Matter and Light Sterile Neutrinos.
4. Aoki, M., et al. (2012) Multi-Component Dark Matter Systems and Their Observation Prospects.
5. Kusenko, A., Loewenstein, M. and Yanagida, T. (2013) Moduli Dark Matter and the Search for Its Decay Line Using Suzaku X-Ray Telescope. Physical Review D, 87, Article ID: 043508.
6. Feldman, D., Liu, Z., Nath, P. and Peim, G. (2010) Multicomponent Dark Matter in Supersymmetric Hidden Sector Extensions.
7. Feng, J.L. (2010) Dark Matter Candidates from Particle Physics and Methods of Detection.
8. Zurek, K.M. (2009) Multi-Component Dark Matter.
9. Boehm, C., Fayet, P. and Silk, J. (2003) Light and Heavy Dark Matter Particles.
10. Feng, W.Z., Mazumdar, A. and Nath, P. (2013) Baryogenesis from Dark Matter.
11. D'Souza, I.A. and Kalman, C.S. (1992) Preons: Models of Leptons, Quarks and Gauge Bosons as Composite Objects. World Scientific, Singapore.
12. NASA's Planck Project Office (2013) Planck Mission Brings Universe into Sharp Focus.https://www.nasa.gov/mission_pages/planck/news/planck20130321.html#.VZ4k5_lViko
13. Feng, W.Z., Nath, P. and Peim, G. (2012) Cosmic Coincidence and Asymmetric Dark Matter in a Stueckelberg Extension.
14. Narain, G., Schaffner-Bielich, J. and Mishustin, I.N. (2006) Compact Stars Made of Fermionic Dark Matter.
15. Corda, C., Cuesta, H.J.M. and Gomez, R.L. (2012) High-Energy Scalarons in R2 Gravity as a Model for Dark Matter in Galaxies. Astroparticle Physics, 35, 362-370.
16. Corda, C. (2009) Interferometric Detection of Gravitational Waves: The Definitive Test for General Relativity. International Journal of Modern Physics D, 18, 2275-2282.
17. Woolfson, M.M. (1984) The Evolution of Rotation in the Early History of the Solar System. Philosophical Transactions of the Royal Society A, 313, 5-18.
18. García, R.A., Turck-Chieze, S., Jimenez-Reyes, S.J., et al. (2007) Tracking Solar Gravity Modes: The Dynamics of the Solar Core. Science, 316, 1591-1593.
19. Zhang, J., Song, X.D., Li, Y.C., et al. (2005) Inner Core Differential Motion Confirmed by Earthquake Waveform Doublets. Science, 309, 1357-1360.
20. Livermore, P.W., Hollerbach, R. and Jackson, A. (2013) Electromagnetically Driven Westward Drift and Inner-Core Superrotation in Earth's Core. Proceedings of the National Academy of Sciences, 110, 15914-15918.
21. Baryshev, Y.V. (2008) Field Fractal Cosmological Model as an Example of Practical Cosmology Approach.

22. Agle, D.C. and Brown, D. (2012) Data from NASA's Voyager 1 Point to Interstellar Future. http://www.nasa.gov/mission_pages/voyager/voyager20120614.html
23. Altmannshofer, W., Buras, A.J., Straub, D.M., et al. (2009) New Strategies for New Physics Search in Decays.
24. Del Amo Sanchez, P., et al., The BABAR Collaboration (2011) Search for the Rare Decay .
25. Strigari, L.E. (2012) Galactic Searches for Dark Matter.
26. Bechtol, K. (2011) The Extragalactic Gamma-Ray Background. A Census of High Energy Phenomena in the Universe.http://astro.fnal.gov/events /Seminars /Slides/Bechtol%20120611.pdf
27. Buckley, J., Byrum, K., Dingus, B., et al. (2008) The Status and Future of Ground-Based TeV Gamma-Ray Astronomy. A White Paper Prepared for the Division of Astrophysics of the American Physical Society.
28. Jeltema, T. (2012) Observational Cosmology and Astroparticle Physics.http:// physics.ucsc.edu/~joel/12Phys205/Feb6-Jeltema.pdf
29. Aharonian, F.A. (2004) Very High Energy Cosmic Gamma Radiation. A Crucial Window on the Extreme Universe. http://www.worldscientific.com /worldscibooks/10.1142/4657http://dx.doi.org/10.1142/4657
30. Totani, T. (2009) The Cosmic Gamma-Ray Background Radiation. AGNs, and More?http://www-conf.kek.jp/past/HEAP09/ppt/1day/Totani_HEAP09.pdf
31. Johnson, R.P. and Mukherjee, R. (2009) GeV Telescopes: Results and Prospects for Fermi. New Journal of Physics, 11, Article ID: 055008.
32. Giovannelli, F. and Sabau-Graziati, L. (2012) Multifrequency Behavior of High Energy Cosmic Sources. A Review. Memorie Della Societa Astronomica Italiana, 83, 17.
33. Essig, R., Kuflik, E., McDermott, S.D., et al. (2013) Constraining Light Dark Matter with Diffuse X-Ray and Gamma-Ray Observations.
34. Porter, T.A., Johnson, R.P. and Graham, P.W. (2011) Dark Matter Searches with Astroparticle Data.
35. Boehm, C., Hooper, D., Silk, J., et al. (2003) MeV Dark Matter: Has It Been Detected?
36. Holder, J. (2012) TeV Gamma-Ray Astronomy: A Summary.
37. Chaves, R.C.G., for the H.E.S.S. Collaboration (2009) Extending the H.E.S.S. Galactic Plane Survey
38. Tibolla, O., Chaves, R.C.G., de Jager, O., et al. (2009) New Unidentified H.E.S.S. Galactic Sources.
39. Hoppe, S., de Ona-Wilhemi, E., Khélifi, B., et al. (2009) Detection of Very-High-Energy Gamma-Ray Emission from the Vicinity of PSR B1706-44 with H.E.S.S.
40. Tam, P.H.T., Wagner, S.J., Tibolla, O., et al. (2009) A Search for VHE Counterparts of Galactic Fermi Bright Sources and MeV to TeV Spectral Characterization.
41. Tibolla, O., Chaves, R.C.G., Domainko, W., et al. (2009) New Unidentified Galactic H.E.S.S. Sources.
42. Tam, P.H.T., Wagner, S., Tibolla, O. and Chaves, R. (2010) A Search for VHE Counterparts of Galactic Fermi Sources.
43. Aleksic, J., Ansoldi, S., Antonelli, L.A., et al. (2013) Optimized Dark Matter Searches in Deep Observations of Segue 1 with MAGIC.
44. 44. Moralejo, A. (2013) http://projects.ift.uam-csic.es/multidark/images/ moralejoalcala.pdf.
45. Abramowski, A., Acero, F., Aharonian, F., et al. (2013) Search for Photon Line-Like Signatures from Dark Matter Annihilations with H.E.S.S.
46. Jin, H.B., Wu, Y.L. and Zhou, Y.F. (2013) Implications of the First AMS-02 Measurement for Dark Matter Annihilation and Decay.
47. Abdo, A.A., et al., Fermi/LAT Collaboration (2009) Measurement of the Cosmic Ray e+ + e- Spectrum from 20 GeV to 1 TeV with the Fermi Large Area Telescope.
48. Adriani, O., Barbarino, G.C., Bazilevskaya, G.A., et al. (2011) The Cosmic-Ray Electron Flux Measured by the PAMELA Experiment between 1 and 625 GeV.

49. He, X.G. (2009) A Brief Review on Dark Matter Annihilation Explanation for e± Excesses in Cosmic Ray.
50. Cholis, I. and Goodenough, L. (2010) Consequences of a Dark Disk for the Fermi and PAMELA Signals in Theories with a Sommerfeld Enhancement.
51. Morselli, A. (2011) Indirect Detection of Dark Matter, Current Status and Recent Results. Progress in Particle and Nuclear Physics, 66, 208-215.
52. Abazajian, K.N. and Harding, J.P. (2011) Constraints on WIMP and Sommerfeld-Enhanced Dark Matter Annihilation from HESS Observations of the Galactic Center.
53. Kawanaka, N., Ioka, K., Ohira, Y., et al. (2010) TeV Electron Spectrum for Probing Cosmic-Ray Escape from a Supernova Remnant.
54. Aharonian, F.A., Akhperjanian, A.G., de Almeida, U.B., et al. (2008) Energy Spectrum of Cosmic-Ray Electrons at TeV Energies. Physical Review Letters, 101, Article ID: 261104.
55. Ibarra, A., et al. (2010) Extragalactic Diffuse Gamma-Rays from Dark Matter Decay. http://c alet.phys.lsu.edu/Science/DGR.php
56. Orr, M. and Krennrich, F. (2011) Constraining the Extragalactic Background Light in the Near-Mid IR with the Cherenkov Telescope Array (CTA). 32nd International Cosmic Ray Conference, Beijing.http://www.ihep.ac.cn/english/conference /icrc2011/paper/proc/v8/v8_1156.pdf
57. Orr, M., Krennrich, F. and Dwek, E. (2011) Strong New Constraints on the Extragalactic Background Light in the Near- to Mid-IR.
58. Madhavan, A. (2013) The VHE γ-Ray Spectra of Several Hard-Spectrum Blazars from Long-Term Observations with the VERITAS Telescope Array. PhD Thesis, Iowa State University, Ames.
59. Torii, S., for the CALET Collaboration (2014) The CALorimetric Electron Telescope (CALET): A High Energy Cosmic-Ray Observatory on the International Space Station.http://www.crlab.wise.sci.waseda.ac.jp/eng/wp-content/uploads/downloads/2014/09/VHEPU2014-CALET_final.pdf
60. Papuccia, M. and Strumia, A. (2009) Robust Implications on Dark Matter from the First FERMI Sky Gamma Map.
61. Hooper, D. (2012) The Empirical Case For 10 GeV Dark Matter.
62. Hooper, D. and Goodenough, L. (2010) Dark Matter Annihilation in the Galactic Center as Seen by the Fermi Gamma Ray Space Telescope.
63. Sreekumar, P., Bertsch, D.L., Dingus, B.L., et al. (1998) EGRET Observations of the Extragalactic Gamma Ray Emission. The Astrophysical Journal, 494, 523-534.
64. Abdo, A.A., et al. (1997) A Population of Gamma-Ray Emitting Globular Clusters Seen with the Fermi Large Area Telescope.
65. Tam, P.H.T., Kong, A.K.H., Hui, C.Y., et al. (1997) Gamma-Ray Emission from Globular Clusters.
66. Frandsen, M.T., Kahlhoefer, F., McCabe, C., et al. (2013) The Unbearable Lightness of Being: CDMS versus XENON.
67. Hunter, S.D., Bertsch, D.L., Catelli, J.R., et al. (1997) EGRET Observations of the Diffuse Gamma-Ray Emission from the Galactic Plane. The Astrophysical Journal, 481, 205-240.
68. Golubkov, Y.A. and Khlopov, M.Y. (2000) Antiprotons Annihilation in the Galaxy as a Source of Diffuse Gamma Background.
69. Wolfe, B., Melia, F., Crocker, R.M., et al. (2008) Neutrinos and Gamma Rays from Galaxy Clusters.
70. Yamazaki, R., Kohri, K., Bamba, A., et al. (2006) TeV Gamma-Rays from Old Supernova Remnants.
71. Nakamori, T. (2012) Fermi Observations of Galactic Sources.
72. Agakishiev, G., Balanda, A., Belver, D., et al. (2013) Searching a Dark Photon with HADES.
73. Merkel, H., Achenbach, P., Gayoso, C.A., et al., A1 Collaboration (2011) Search for Light Gauge Bosons of the Dark Sector at the Mainz Microtron. Physical Review Letters, 106, Article ID: 251802.

74. Abrahamyan, S., Ahmed, Z., Allada, K., et al., APEX Collaboration (2011) Search for a New Gauge Boson in Electron-Nucleus Fixed-Target Scattering by the APEX Experiment. Physical Review Letters, 107, Article ID: 191804.
75. Drees, R.M., Waltham, C., Bernasconi, T., et al., SINDRUM I Collaboration (1992) Measurement of the π0 Electromagnetic Transition form Factor. Physical Review D, 45, 1439-1447.
76. Adlarson, P., et al., WASA-at-COSY Collaboration (2013) Search for a Dark Photon in the π0→e+e-γ Decay. Physics Letters B, 726, 187-193.
77. Babuski, D., Badoni, D., Balwierz-Pytko, I., et al., KLOE-2 Collaboration (2013) Limit on the Production of a Light Vector Gauge Boson in φ Meson Decays with the KLOE Detector. Physics Letters B, 720, 111-115.
78. Rasera, Y., Teyssier, R., Sizun, P., et al. (2006) Soft Gamma-Ray Background and Light Dark Matter Annihilation.
79. Zdziarski, A.A. (1996) Contributions of AGNs and SNe Ia to the Cosmic X-Ray and Gamma-Ray Backgrounds. Monthly Notices of the Royal Astronomical Society, 281, L9-L13.
80. Gruber, D.E., Matteson, J.L. and Peterson, L.E. (1999) The Spectrum of Diffuse Cosmic Hard X-Rays Measured with HEAO-1.
81. Gorenstein, P., Giacconi, R. and Gursky, H. (1967) The Spectra of Several X-Ray Sources in Cygnus and Scorpio. The Astrophysical Journal, 150, L85.
82. Safi-Harb, S. and Ogelman, H. (1997) ROSAT and ASCA Observations of W50 Associated with the Peculiar Source SS 433. The Astrophysical Journal, 483, 868-881.
83. Itoh, T. (2007) Suzaku Studies of Time Variable X-Ray Spectra of Edge-On Active Galactic Nuclei. PhD Thesis, University of Tokyo, Tokyo. http://ww.astro.isas.jaxa.jp /suzaku/bibliography/phd/titoh_dron_print080220.pdf
84. Bykov, A.M., Krassilchtchikov, A.M., Uvarov, Y.A., et al. (2009) Isolated X-Ray—Infrared Sources in the Region of Interaction of the Supernova Remnant IC 443 with a Molecular Cloud.
85. Fukuoka, R., Koyama, K., Ryu, S.G., et al. (2008) Suzaku Observation Adjacent to the South End of the Radio Arc.
86. Morretti, A., Vattakunnel, S., Tozzi, P., et al. (2012) Spectrum of the Unresolved Cosmic X-Ray Background: What Is Unresolved 50 Years after Its Discovery.

CHAPTER 8

Crystallography Under Extreme Conditions: State of the Art and Perspectives

Legrand Vincent[1]

[1] LUNAM Université - Université de Nantes - Ecole Centrale Nantes, Institut de Recherche en Génie Civil et Mécanique (GeM), France

1. INTRODUCTION

Sir William Henry Bragg and his son William Lawrence Bragg were the pioneers of crystallography (Bragg, W. H., 1912, 1913a, 1915a, 1915b; Bragg, W. L., 1920). In 1913, they published several articles, notably *The reflection of X-rays by crystals* (Bragg, W. H., 1913b) and *The structure of diamond* (Bragg, W. H. & Bragg, W. L., 1913) were they wrote: "We have applied the new methods of investigation involving the use of X-rays to the case of the diamond, and have arrived at a result which seems of considerable interest. The structure is extremely simple". Two years later, they were jointly awarded the Nobel Prize in Physics for their works in the analysis of crystal structure by means of X-rays.

A century after the first crystallographic experiment, new computing facilities, modern technologies and new diffraction sources (synchrotron, neutron sources...) offer a large range of possibilities and opportunities for crystallographers to probe matter. Crystallography appears nowadays as a new science.

Performing structural analyses at ambient conditions or at low temperature (i.e. above 100 K using nitrogen jet-stream) is very common and popular in laboratories to obtain the structure of powder and single-crystal materials. To understand the mechanisms governing the behaviour of materials it is essential to well know the close relations between the structure and the physical and chemical properties. First, X-ray diffraction is a unique tool to obtain routinely a detailed description of atomic structure and thermal vibrations by analysing the diffracted intensities *Ihkl* of the crystallographic *hkl*reflections:

$$I_{hkl} = S.C_{hkl} \cdot \left|F_{hkl}\right|^2 \tag{1}$$

$$F_{hkl} = \sum_{j=1}^{n} f_j.T_j.\exp(2i\pi(h.x_j + k.y_j + l.z_j)) \tag{2}$$

where S is a scale factor, C_{hkl} is an experimental corrections term (including absorption, extinction, Lorentz-polarization correction...), F_{hkl} is the structure factor of a *hkl* reflection which depends of f_j the form factor of the atom j with (x_j, y_j, z_j) coordinates in the cell and of T_j the Debye-Waller factor given in the case of isotropic harmonic vibrations by:

$$T_j = \exp\left(-B_j \frac{\sin^2\theta}{\lambda^2}\right) = \exp\left(-8\pi^2 \left\langle U_j^2 \right\rangle \frac{\sin^2\theta}{\lambda^2}\right) \tag{3}$$

where $\left\langle U_j^2 \right\rangle$ is the mean square atomic displacement of the atom j.

If one wants to increase the data quality, the best choice, without consideration about crystal quality or particular experimental conditions, is to perform a low temperature measurement, usually at 110 K, to reduce the atomic displacement parameters which affect the value of the atomic structure factor. In those experimental conditions, if high resolution X-ray diffraction measurement is performed up to large momentum transfers ($\sin\theta/\lambda > 1$ Å-1) the electron density distribution of a molecule can be determined thanks to the Hansen-Coppens pseudo-atomic multipolar expansion (Hansen & Coppens, 1978):

$$\rho(\vec{r}) = \rho_{core}(r) + P_{val}\kappa^3 \rho_{val}(\kappa r) + \sum_{l} \kappa'^3 R_{nl}(\kappa' r) \sum_{m=0}^{l} P_{lm\pm} y_{lm\pm}(\theta,\varphi) \tag{4}$$

where *ρcore(r)* and *ρval(r)* are spherically averaged core and valence electron densities calculated from Clementi Hartree-Fock wave functions for ground state isolated atoms (Clementi & Roetti, 1974). κ and κ' are contraction-expansion parameters and *Pval* is the atomic valence shell electron population. The deformation of the valence electron shell is projected on real spherical harmonics $ylm\pm(\theta,\varphi)$ times Slater-type radial functions *Rnl(r)*. *Plm±* are the multipole population parameters.

In the later case, the accuracy of the results is highly dependant of the experimental conditions (Destro et al., 2004; Zhurov et al., 2008) and particularly of the crystal quality and the measurement temperature which must be chosen to reduce at the maximum the thermal vibrations of the atoms. These *usual experiments* at static low temperature give basic structural properties, and their variations as a function of temperature can reveal particular behaviour of matter as phase transition for example with changes in structural, electronic, optical and/or magnetic properties. But, in a general consideration, performing crystallographic measurements under external perturbations is of prime importance. Nowadays, experiments at different temperatures or hydrostatic pressures can be done almost routinely and exotic sample environments are more and more used to explore materials properties. If we just consider for the moment experiments involving temperature or pressure variations, a Web of Science search for “high pressure”, “low temperature” and “high temperature” in the field of X-ray and neutron diffraction between the years 2005 and 2012 gave about 3500 experiments performed in 2005 but nearly 5000 in 2011 (figure 1). Two comments can be done observing figure 1. First, a large amount of the extreme conditions (temperature and pressure) experiments are carried out using X-ray diffraction compared to neutron diffraction. This difference is of course due to the large number of X-ray diffractometers available in laboratories, which are more and more outstanding in terms of source power and detector efficacy, combined with numerous sample environments especially designed for laboratory equipments. But this is also due to the building of several 3rd generation of synchrotron sources which offer the possibility to conduct very quick measurements on a very small quantity of matter, allowing the measurement of materials under non ambient conditions, particularly in the domains of chemistry and biology sciences (neutron diffraction experiment demands more matter quantity, in general at least some mm^3). The second observed aspect on figure 1 concerns the evolution of the cumulative number (neutron and X-ray) of publications during the considered period: about 300 supplementary published papers per year appear between 2005 and 2009, but in the years 2009-2011, a distinct decrease of this number is noticed. This effect is directly related to the number of synchrotron radiation facilities over the world. One can count about 69 particle accelerators and accelerator laboratories in 2005 and about 76 in 2012 (including about 40 synchrotrons in 2011), with most of the new facilities operative in 2006-2008.

As said before, crystallography under extreme conditions is now more and more used and plays a key role as it offers helpful understanding of the physical, chemical and mechanical properties of the solid state. The term *extreme conditions* was first used to define non ambient thermodynamical conditions of pressure or temperature. It is also now employed when out-of-equilibrium conditions are applied such as light irradiation, external magnetic or electric fields, specific chemical environments (e.g. under liquid or gas flux...) or applied strain. In situ measurements can also be considered as extreme conditions, for examples in the cases of time-resolved experiments (picosecond diffraction...) or chemical kinetic reactions (dehydrogenation reaction, diffusion process, decomposition pathway...). The present challenge is to combine two or more extreme conditions to explore new states of matter and new material properties, taking advantage of last generations of high brilliance sources (synchrotron and neutron sources) (figure 2).

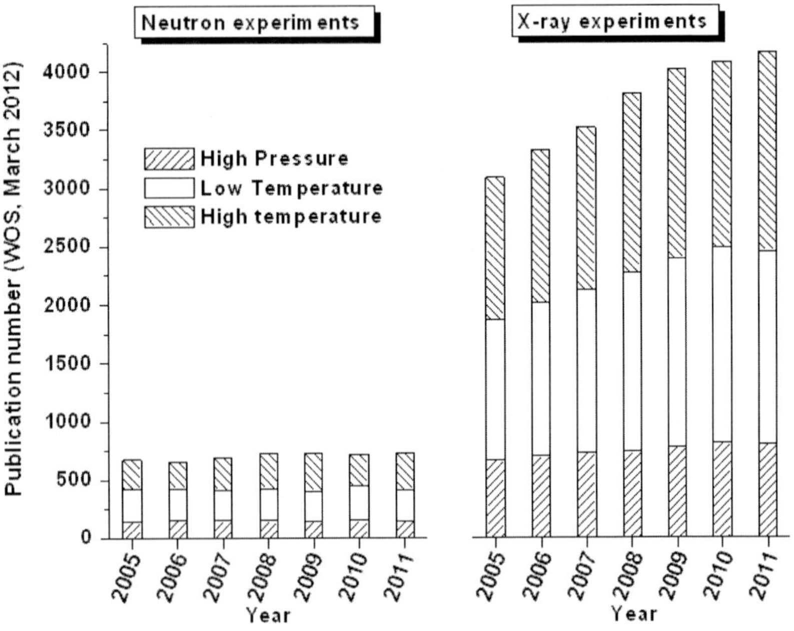

FIGURE 1. Publication number related to neutron or X-ray diffraction experiments under extreme conditions of temperature and pressure during the period 2005-2011 (Web of Science, March 2012).

When extreme conditions are applied to a material, various changes occur. They may involve sample state variations (gaseous, liquid and solid phases), phase transitions (magnetic and structural), electronic structure modifications (the chemical bonds can change from covalent to ionic or metallic) and atomic bond lengths variations (which induce variations of the atomic vibrations, of the coordination numbers, of the diffusion processes…). A large number of the material properties are modified under extreme conditions, leading sometimes to metastable states, which open new investigation opportunities for the crystallographer, but also for all scientist who wants to go deeper in the understanding of the structure-properties relationships in order to design future materials.

FIGURE 2. The neutron source of the Institut Laue-Langevin (ILL) and the European Synchrotron Radiation Facility (ESRF) in Grenoble, France (credit: V. Legrand).

This chapter is reviewing different aspects of the use of crystallography under extreme conditions to investigate the nature, the mechanisms and the dynamics of out-of-equilibrium phase transitions as well as transformations driven by external applied conditions. The part 2. is focus on the advantages of using Large scale facilities around the world and what are now the extreme sample environments available to explore unknown properties of materials. Finally, an overview of perspectives and future developments in diffraction instruments and sample environments are presented in part 3. showing that modern crystallography is in perpetual evolution and that the present century is certainly the one of *crystallography under extreme conditions.*

2. LARGE SCALE FACILITIES AND EXTREME SAMPLE ENVIRONMENTS

2.1. State-of-The-Art Sample Environments at Laboratories

As noticed before, there are more than 5000 publications per year dealing with extreme conditions of temperature and pressure. Nevertheless, those conditions illustrate only 2 of the numberless non ambient thermodynamical and exotic conditions which can be used to probe the close links between structure and properties of materials. Changing the pressure or the temperature in a crystallographic experiment is performed quite routinely in laboratories. But if one wants for examples to apply electric or magnetic fields, to explore superconductivity properties below 1 K, to perform very fast measurements of transient species or quick chemical reactions, or to understand phase transitions under combined sample environments, the diffractometer performances become overriding and in particular the maximum power of the source, the quality of the beam collimators and the efficacy of the detector. This is very important to obtain exploitable data as the sample environment in those later examples is very bulky and could dramatically affect the beam intensity. Moreover, it is very rare for a laboratory to have sample equipments to perform sub-kelvin, time-resolved or high-magnetic field measurements as these sample environments are very expensive, maintenance demanding and hard to exploit on conventional diffractometers. For a long time, extreme conditions measurements were the privilege of expert laboratories in cryogenic, high fields or high pressures with ability and technical resources to build and to use such devices. Those techniques are nowadays distributed in a larger number of laboratories and, for some of them, take place to semi-industrial devices merchandised by high technology industries. Since ten or so years, apparatus able to be used on conventional diffractometers were punctually developed to respond to the requirement of probing new properties of materials. We can cite some of them. First, to probe the structural and the electronic properties of materials under light irradiation at low temperature, *photocrystallographic* experiments (Coppens et al., 1998;

Fomitchev et al., 2000) were developed first to study transient species of transition metal nitrosyl complexes. Coppens et al. (1998) wrote that "the study of photo-induced processes in crystals is a frontier area of crystallographic research, which requires development of novel experimental and computational methods". This is particularly true as the live-time of transient species is in general not related to the time of the experiment required to collect accurate data for structural refinement or experimental electron density modelling. The success of kind experiments consists in in situ laser irradiation of a cryogenically cooled crystal, with a N_2 or He gas-flow system for less restricted access than closed cryostat (White et al., 1994), mounted on a diffractometer equipped with a CCD detector for fast data collection (Graafsma et al., 1997; Muchmore, 1999). More recently, Legrand (2005) has developed a crystallographic experimental methodology approach (Legrand et al., 2005; 2007a; 2007b) and shown that it is possible to measure with accuracy photo-induced metastable states and to refine its experimental electron density below 35 K using conventional X-ray sources (Legrand et al., 2006; Pillet et al., 2008).

We can also mention an other development to do pressure measurements using particular pressure cell, up to 3.0 GPa and for T > 9 K (Guionneau et al., 2004) (figure 3). This last device was developed to probe phase transitions and structure-properties relationships in molecular conductor (Chasseau et al., 1997) and spin transition complexes (Guionneau et al., 2001). Doing in situ laboratory pressure investigations is still difficult as the pressure cell induce several perturbations affecting the quality of the data collection. The originality of this high pressure cell is that the X-ray beam goes through the gasket and not through the diamonds, offering a wide diffraction angle of 342° rotation. In this case, Guionneau et al. (2004) indicate that using a CCD area detector to collect Bragg peaks improves data and reduces the measurement time.

To finish this non-exhaustive description, we can briefly mentioned two device developments in the field of very low temperature measurements. A mini-goniometer for X-ray diffraction studies down to 4 K on a four-circle diffractometer equipped with a CCD area detector (Fertey et al., 2007). The authors used a helium-bath orange cryostat (Brochier, 1977) and added a remarkable evolution: a magnetically coupled two-rotation-axis mini-goniometer permanently installed in the sample chamber of the helium-bath cryostat. This original apparatus was tested with success on organic charge-transfer compounds (Garcia et al., 2005; Garcia et al., 2007). An other low temperature system was developed to analyse Jahn-Teller distortion of $TmVO_4$ using conventional X-ray diffractometer between 0.22 and 3.5 K (Suzuki et al., 2002). The authors used a diffractometer equipped with a counter-monochromator and a scintillation counter, and an X-ray generator with rotating Cu Anode. To perform the ultra-low temperature measurements, a modified ^3He-^4He dilution refrigerator was mounted on the two-fold axis goniometer. In this later case, the study was also a success

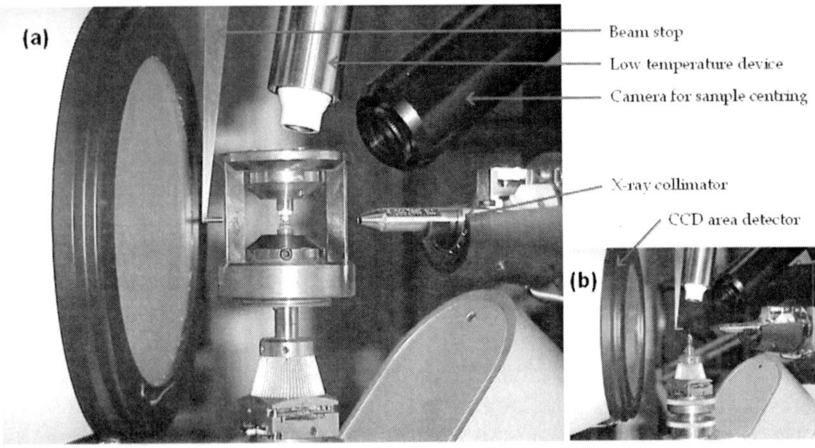

FIGURE 3. The high pressure cell developed by **Guionneau et al. (2004)**: (a) view of the high pressure cell mounted on a 4-circle diffractometer equipped with low temperature device and CCD area detector; (b) comparison of the spatial occupied volume when a conventional goniometer head is used instead of the pressure cell. (Reprinted by permission from IOP Publishing Ltd).

and the Jahn-Teller effect was fully investigated down to 0.22 K. In their paper, Suzuki et al. mentioned what was noticed above in this section concerning the use of bulky devices: "the x-ray beam generated by a Cu target, which passed through 4 walls of Be 2 mm thick and 4 aluminized mylar walls was reduced to approximately 1/100". This beam attenuation is generally inevitable when doing extreme conditions experiments and to guaranty good data collection, it must be used first class equipments on the diffractometer. Even if a lot of progresses concerning the performances of new diffractometers were obtained last years, extreme conditions measurements are sometimes impossible to perform at laboratories. Then, the best to do is to use Large scale facilities which benefits are now more and more highlighted. The term *Large scale facilities* stands for synchrotron research institutes and neutron facilities (spallation sources and reactors).

2.2. Why Using Large Scale Facilities In Extreme Crystallography?

The synchrotron working principle was shown during the Second World War by Oliphant (1943, University of Birmingham, UK). Three years later, it was shown for the first time (Goward & Barnes, 1946; Elder et al., 1947) that it is possible to built synchrotrons providing a satisfactory means of producing high energy electrons and X-rays. Nearly 70 years later, several synchrotrons of 3rdgeneration are operational throughout the world and are of first necessity for all the scientists communities (Bilderback et al., 2005). The X-ray beam, produces by a synchrotron, is a light beam (electromagnetic wave) which interacts with electrons surrounding the nuclei of heavy elements (Born, 1933, 1935; Born & Schrödinger, 1935). The actual phenomenal success of the synchrotron for performing extreme conditions crystallographic measurements comes from numerous undeniable advantages and particularly it's very intense (brilliance) and focussing beam and it's time structure allowing the measurement of small or ultra-dilute samples at different time scales. Thus, it is possible to study transient species and fast chemical reactions or to perform biological and geological studies using bulky sample environment allowing temperature and pressure changes without affecting the data.

Neutron diffraction also plays a key role in actual crystallographic studies under extreme conditions. The first nuclear reactor was built in 1942 by Enrico Fermi and Leó Szilárd (University of Chicago, USA) (Fermi, 1947). In October 2011, the International Atomic Energy Agency (IAEA) database indexed 241 operational research reactors in the world. Generally, experiments using neutrons to explore the properties of a material are based not only on the intensity of the beam (to exploit this criteria, synchrotron radiation would be preferable) but on the very nature of neutrons. It is a particle beam, allowing interactions with the nuclei and the magnetic moment of unpaired electrons in the sample, scattered by all elements also the light ones like hydrogen and its isotopes, with a deep penetration depth enables bulk studies of materials (Fermi, 1930, 1934, 1940; Szilárd, 1935). As neutrons can penetrate deeper in matter (unless exceptions like cadmium, samarium, europium or gadolinium elements for examples), sample environments requiring to put the sample inside them (e. g. pressure cells, furnaces, gas handling inside a cryostat...) can be used even if further walls, generally in aluminium, must be crossed by neutrons before to hit the sample. This neutron property is very appreciable and neutron facilities now include very well equipped services and highly qualified staff to carry out invaluable experiments as the Services for Advanced Neutron Environment (SANE) at the Institut Laue-Langevin in France, where the Orange Cryostat (Brochier, 1977) and the Cryopad (Tasset, 1989), among others equipments, were invented and built.

Synchrotron and neutron facilities are complementary techniques which allow to analyse structural details of materials at the atomic scale. Microscopic properties are then known and the challenge is to derive the macroscopic ones. To achieve that goal, investigations must concern both surface and bulk phenomena of materials in both dynamic and steady states. Large scale facilities can give fine details on the nuclear structure of materials, on the physical and chemical properties, on the electronic and magnetic structures, they can give crucial information about surface and interface states. All materials are concerned: metals, alloys, polymers, composites, colloids, glass, cement, biological macromolecules (viruses, enzymes, proteins...), minerals, food products, superconductors... The world-wide scientific community can observe since about ten years state of the art measurements in the above mentioned fields.

One field where X-ray and neutron measurements are really complementary is the domain of spin-transition molecular magnetic materials which is highly studied under *exotic environments* since 20 years (Gütlich & Goodwin, 2004; Bousseksou et al., 2011). These materials are especially interesting owing to their bistability properties which give them promising applications as data storage elements, thermal switches or display devices. These materials consist of a d^4 to d^7 transition metal coordinated to specific containing aromatic ligands whose ligand field is intermediate between weak and strong. Correlatively, the electron configuration of the central metal ion can be driven by several kinds of external perturbations such as pressure, intense magnetic field or light excitation. In the case of iron(II) complexes, the spin conversion is related to the electronic configuration of the ion changing from $t_{2g}{}^6e_g{}^0$ in the low spin (LS) state to $t_{2g}{}^4e_g{}^2$ in the high spin (HS) state. In addition to this electron redistribution, spin transition complexes present drastic structural variations between both phases, principally observed in the iron coordination sphere: an increase of the iron octahedron distortion and of the Fe-ligand bond length, by typically 0.2 Å, are observed for the LS → HS transition, generating a decrease of the molecular volume by several Å^3. Thus, owing to their high electronic and structural contrast, spin transition materials are good candidates for crystallographic studies under extreme conditions especially using Large scale facilities. Pressure dependence of the lattice parameters of several iron(II) complexes were derived from neutron powder diffraction (Legrand et al., 2008) and single-crystal X-ray synchrotron (Legrand, private communication) measurements at ambient and low temperatures. These studies highlight very promising results in the field of spin crossover phenomena under constraint, where crystallographic investigations under pressure are still very rare, especially because experimental conditions are difficult and sometimes not favourable. Concerning investigations under high magnetic field, Goujon et al. (2003) reported the photo-induced magnetization density of the iron(II) spin crossover compound $Fe(ptz)_6(BF_4)_2$ using an experimental setup allowing combined application of light irradiation (λ = 473 nm) at low temperature (2 K) and high magnetic field (5 T) (figure 4a). Thanks to this polarized neutron diffraction measurement, the temperature dependence of the magnetization was well described and the thermal relaxation of the photo-induced state well observed.

Moreover, electronic and structural properties of iron(II) spin transition complexes were studied under light irradiation and low temperature (< 15 K) using neutron Laue (Goujon et al., 2006) and X-ray synchrotron (Legrand et al., 2007; Pillet et al., 2008) diffraction (figure 4b). These latter authors notably refined for the first time the experimental electron density of a photo-induced metastable state in a spin transition compound. Finally, probing photo-induced phase transitions in molecular materials can also be performed using ultra-fast measurements at a picosecond time scale (Coppens et al., 2004; Collet et al. 2012a; 2012b).

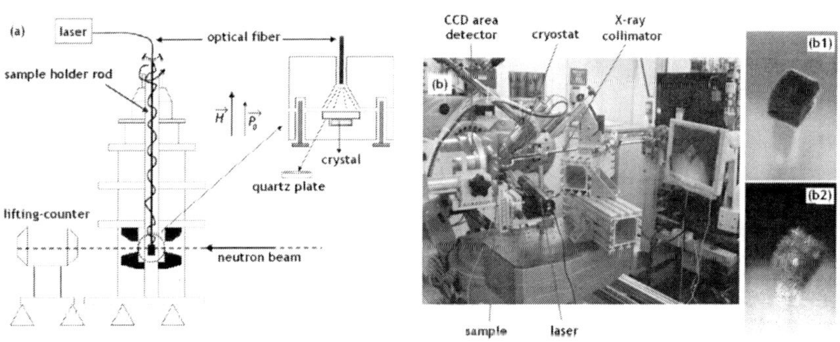

FIGURE 4. a) Schematic experimental setup of the polarized neutron diffractometer; in the insert, the sample holder allowing light irradiation (Reprinted figure with permission from [Goujon et al., *Phys. Rev. B*67, 220401(R) 2003]. Copyright 2012 by the American Physical Society). (b) The photocrystallographic experiment used by Legrand et al. (2007) showing the 6-circle diffractometer (BM01 beamline, ESRF - France) equipped with a large CCD area detector, an open flow He cryosystem and a He-Ne laser; inserts show the measured single-crystal (b1) without and (b2) with laser irradiation.

Real time investigation of temporally varying molecular structures during chemical reactions and phase transitions is a great challenge due to their ultrashort time scales. Collet et al. shown that it is possible to track complex reactions by time-resolved X-ray diffraction to temporal and spatial resolutions of 100 ps and 0.01 Å. Kind measurements give access to *molecular movies* during the transformation of matter induced by light irradiation. The authors succeed to describe for the first time the different steps of the dynamical process of several molecular materials and give their thought for the future: "these results pave the way for structural studies away from equilibrium and represent a first step toward femtosecond crystallography".

Other examples using the advantages of synchrotron radiations are numerous in structural biology, geology and materials sciences. The atomic structure of the bluetongue virus core (700 Å in diameter), containing about 1000 protein components self-assembled, was determined at a resolution of 3.5 Å (Grimes et al., 1998). The atomic structure of large ribosomal subunits were refined (Ban et al., 2000; Wimberly et al., 2000) providing a wealth of information about RNA and protein structure, protein-RNA interactions and ribosome assembly. The high brilliance of synchrotron sources in combination with fast detectors was used to perform time-resolved radiography of metal foam formation (Banhardt et al., 2001) (figure 5a). The possibility to have a micro-focus beam was applied to the study of the structure of spider silk (Riekel et al., 1999; Riekel & Vollrath, 2001) (figure 5b). The beam coherence properties enable new imaging techniques to do quantitative phase imaging and holotomography to probe materials with a micrometer resolution (Cloetens et al., 1999).

Nowadays, highlight articles are still numerous and new instrumental and device developments allow unthinkable experiments even ten years ago. Using highly brilliant X-ray beams at the APS (Advanced Photon Source, USA), ESRF (European Synchrotron Radiation Facility, France) and SPRING-8 (Super Photon ring-8 GeV, Japan), teams uncovered subtle details of how cyan fluorescent proteins store incoming energy and retransmit it as fluorescent light (Goedhart et al., 2012), or investigated structure-properties relationships of biological mesocrystal in the adult sea urchin spin (Seto et al., 2012). Combination of in situ synchrotron X-ray absorption techniques and molecular dynamics simulations were used to determine the density range of primitive lunar melts at pressures equivalent to those in the lunar interior (4.5 GPa and 1800 K) (Van Kan Parker et al., 2012): these extreme conditions were generated with tiny samples heated thanks to high electric current while squashing them in a press (figure 6); knowing the attenuation of the synchrotron X-ray beam through both the solid and molten parts of the sample, the density at high pressure and high temperature could be measured. High pressure in situ X-ray diffraction and specific volume measurements on isotactic poly(4-methyl-1-pentene) melt have uncovered abrupt changes in the pressure dependence of microscopic structure as well as that of macroscopic density (Chiba et al., 2012). It was proved another time that pressure has an essential role in the production and control of superconductivity (Sun et al., 2012): it is reported that in the superconducting iron chalcogenides, a second superconducting phase suddenly re-emerges at a critical temperature Tc reaching 48.0 K and above 11.5 GPa, after the Tc drops from the first maximum of 32 K at 1 GPa. Metallic liquid silicon at 1787 K was investigated using X-ray scattering (Okada et al., 2012): the results show persistence of covalent bonding in liquid silicon and provide support for the occurrence of theoretically predicted liquid-liquid phase transition in supercooled states. To finish, it can be pointed out the crystallographic structure refinement of a protein, PthXo1 (which is a transcription activator–like, TAL, effector), encoded by an important group of harmful plant pathogens (Mak et al., 2012): understanding DNA recognition by TAL effectors may facilitate rational design of DNA-binding proteins with biotechnological applications.

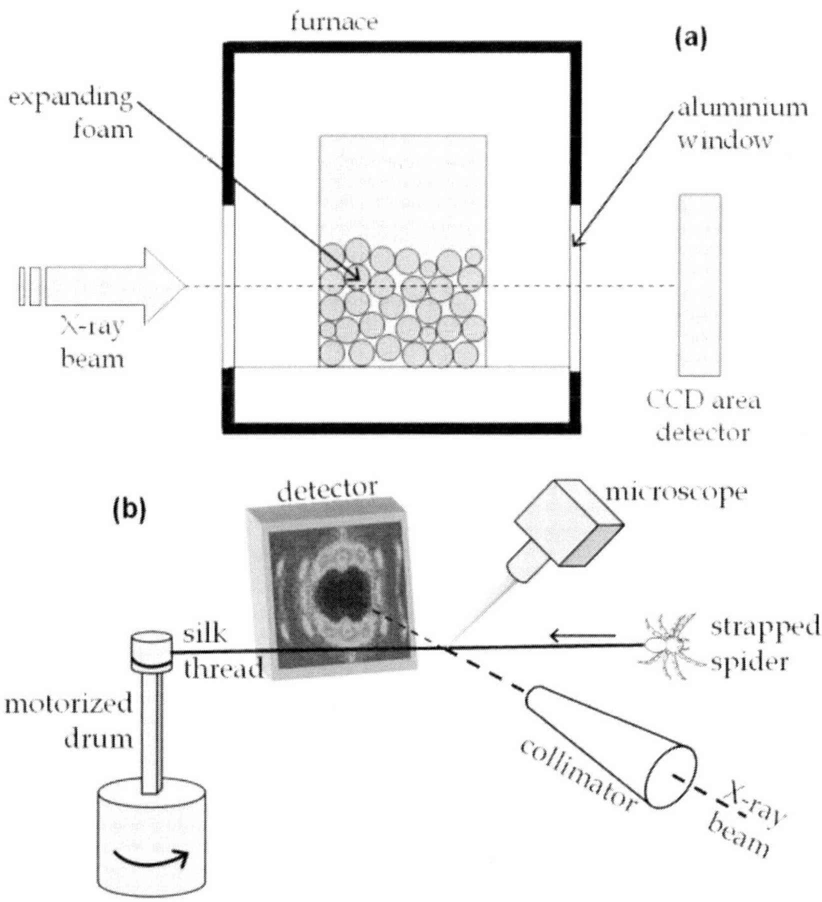

FIGURE 5. Schematic views of the experimental setup for real-time studies (a) of metal foam formation (Banhardt et al., 2001) and (b) of structural properties of spider silk (Riekel & Vollrath, 2001).

Using neutron diffraction on reactors and spallation sources, researchers also push back the limits of measurements as for examples in the observation of a roton-like excitation in a monolayer of liquid ^3He, a Fermi liquid (Godfrin et al., 2012). A new class of magnetic ionic liquid surfactants showing remarkable effects on surface and interfacial tension and allowing access to magneto-responsive emulsions and new methods of separation, recovery, catalysis, and potential magnetophoretic applications were analyses by small-angle neutron scattering (SANS) (Brown et al., 2012). Using neutron diffraction studies, along with computational chemistry, the mystery of lead oxide, the battery anode material of lead-acid batteries, was solved (Scanlon et al., 2011): oxygen vacancies give metallic behaviour by freeing up electrons to carry electrical current, neutron experiment put in evidence that commercial lead oxide powder is oxygen deficient by showing that oxygen sites were 1.6% vacant. Using high resolution neutron diffraction under high temperature and high magnetic field, it was shown that the thermal evolution of the helimagnetic state in CoMnSi is accompanied by a change in interatomic distances of up to 2%, the largest ever found in a metallic magnet (Barcza et al., 2010); this important result could lead to more efficient fridges and cooling systems.

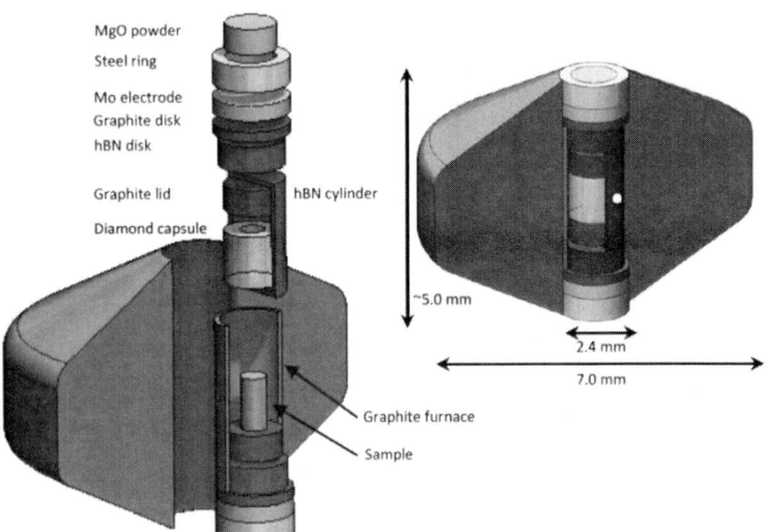

FIGURE 6. Drawing of the high-pressure cell assembly for the synchrotron X-ray experiments used to study the density range of primitive lunar melts (**Van Kan Parker et al., 2012**). The artificial moon rock samples (orange) were placed inside the ring-shaped, natural diamond sample holder (grey) which in turn was surrounded by a large, disk-shaped container (red).

(Reprinted by permission from Macmillan Publishers Ltd: *Nature Geoscience*, copyright 2012)

The above overview has to remain incomplete with a number of fields left out. However, one can realize the wealth of extreme condition experiments which can be performed nowadays at laboratories and, particularly, using Large scale facilities which benefit of the last innovations in terms of sample environments.

2.3. Crystallographic Experiments: Advances In Sample Environments

Sample environment equipments like pressure cells, cryostats, etc, are essential adjuncts to perform with success the analysis of materials into a state or a phase with special behaviours and properties. This section is a brief overview of the most employed stately sample environments available mainly on Large scale facilities. The technical points of the discussed devices are not detailed here and the reader could find further information in the references given through section 2.3 and in specific reviews as written by Bailey (2003) or Mignot (2008). Specific Information is also related in reviews from the Workshop "sample environments in neutron and X-ray experiments" held at the Institut Laue-Langevin (France) in 1984 (Vettier et al., 1984). There are various ways of tuning materials properties. In this section, the most employed sample environments are described, but the possibilities are miscellaneous and it is not possible to make an exhaustive list. *Exotic* sample environments like, among others, electric fields (Guillot et al., 2002; Hansen et al., 2004) or in situ reactions using gas handling (Walker et al., 2009; Price et al., 2010), as well as experiments performed with time resolved methods (Gembicky & Coppens, 2006; Nozawa et al., 2007) are not discussed. The investigations using these increasingly sophisticated experimental methods contribute enormously to the understanding of structural properties and progress in solid state physics and chemistry, biology and in geosciences. Advances in sample environments help to perform crystallographic experiments in extreme conditions and extend current areas of investigations.

2.3.1. Low Temperature

Cryogenic conditions are by far the most used sample environments principally to reduce vibrations and thermal motions of atoms, thus the structural refinement is improved. Moreover, phenomena like phase transitions are also generally induced at low temperature. Different devices could be used for low temperature experiments depending the desired temperature range: liquid helium bath cryostat (Orange Cryostat (Brochier,

1977)) for 300 K to 1.5 K; ^3He sorption systems for 3 K to 300 mK; $^3He/^4He$ dilution refrigerator systems below 1 K. Of course, there exist many others systems allowing to reach low and very low temperatures as N_2 or He gas-flow systems, helium closed cycle refrigerators (Displex), etc.

The Orange cryostat is now regarded as the international standard for low temperature experiments performed above 1.5 K. Some variations exist differing in the sample stick geometry (diameters and length). Cryofurnace and Cryomagnet versions were also developed. Nowadays, more and more experiments are based on combining sample environments. Thus, special versions of the Orange cryostat or Displex were built to meet the demand of users. Among the list of special purposes, it can be noted versions allowing the use of a high pressure cell (even Paris-Edinburgh cell) (figure 7), a gas handling for in situ chemical reactions analysis, an omega rotating for single crystals measurements, a sample changer. Because these cryostats are very common at Large scale facilities and are user friendly, recently improvements were made to have the possibility to do ultra-low temperature measurements using new sample stick inserts including high performance dilution refrigerator with a base temperature of less than 15 mK (Neumaier et al., 1984).

2.3.2. High Temperature

Contrary to cryogenic devices, there is no standard high temperature apparatus. The designs depend of the working temperature range, of the type of experiment, of the diffraction beam (X-ray or neutron). However, a typical furnace configuration places a sample into the centre of a thin cylindrical metal foil element, of about 50 µm thick, surrounded by heat shields. According to the chosen metal, the achieved temperature could be typically up to 1400-2000 K. Closed-shell furnaces (Kuhs et al., 1993) and mirror furnaces (Lorenz et al., 1993) were developed for powder and single-crystal analysis at high temperature. This last furnace type has the advantage to allow air-operated experiments with no heat shield, thus removing apparatus diffraction. As for cryogenic, variations are developed to combine for example high temperature and high pressure (Falconi et al., 2005). Finally, the last developments concern the design and built of furnace using aerodynamic levitation for reaching ultra-high temperatures up to 3000 K. This technique is well described by Hennet et al. (2007): “A spherical sample (3mm in diameter) is placed on a levitator that contains a convergent-divergent nozzle enabling the diffusion of a regulated gas flow onto the sample from below. This enables the sphere to remain in a stable position without any contact with the nozzle. When the sample levitates, it is heated by two CO_2laser from the bottom through the hole in the nozzle. The temperature is measured with one or two optical pyrometers”. Aerodynamic levitation apparatus were optimized for neutron (Hennet et al., 2006) and X-ray (Hennet et al., 2002; Sakai et al., 2005) scattering measurements. They were applied with success for examples to study the liquid states and the solidification of $CaAl_2O_4$ (Hennet et al., 2008) and Y_2O_3 (Hennet et al., 2003; Cristiglio et al., 2007).

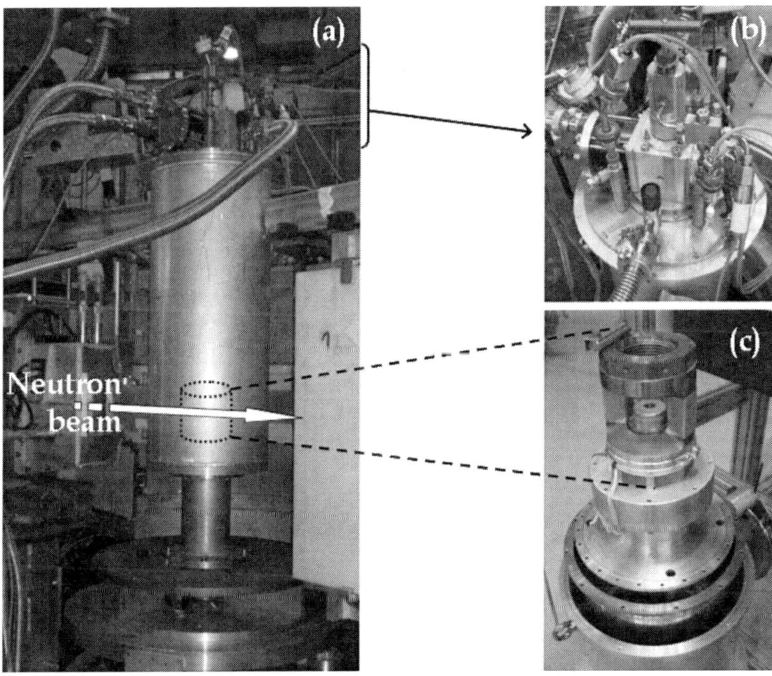

FIGURE 7. a) Details of the high-intensity 2-axis diffractometer D20 (**Hansen et al., 2008**) at the Institut Laue-Langevin (France) with a special version of the helium closed cycle refrigerator (Displex) allowing the use of (c) Paris-Edinburgh high pressure cell. Cryostats are now fully self-governed (b) with automated cold valve, pressure transducer which monitors the exhausting helium gas pressure and temperature controller which adjust the liquid helium flow to the set temperature value.

2.3.3. High Magnetic Field

Studying materials under high magnetic field could be done ether by X-ray diffraction (Katsumata, 2005) or by neutron scattering (Brown et al., 2002; Givord et al., 2004). However, one of the advantages of neutrons is their possible interaction with the unpaired electron magnetic field. Thus, magnetic studies deal with a wide proportion of the total neutron diffraction experiments. As magnetic properties variations are generally observed at low temperature, superconductor cryomagnets are used for kind studies, with applied vertical or horizontal fields. The available magnetic field range is wide and nowadays fields around 10 T are commonly obtained, some devices also produce fields between 14.5 T and 17 T (Nietz, 2003). It can finally be noticed that punctual apparatus are developed through the world allowing 1 s pulsed magnetic fields up to 25 T (Nojiri et al., 1998) or 40 T (Grössinger, et al., 2004), in order to achieve fields up to 70 T in the future. Thermal neutrons are thus largely used to refine magnetic structures in addition to the nuclear structure. Hot polarized neutron diffraction is also a very interesting technique used to obtain crucial information about the magnetic form factors (reflecting the state of the magnetic ions) and magnetisation distributions of para-, ferro- and some antiferro-magnetic single-crystal materials if measurements are performed under high magnetic field configuration. On the other hand, one can performed zero-field neutron polarimetry measurements using Cryopad (CRYOgenic Polarization Analysis Device) (Tasset et al., 1999; Lelièvre-Berna et al., 2005; Takeda et al., 2005) to refine complex antiferromagnetic structures or to perform accurate determination of antiferromagnetic distributions.

2.3.4. High Pressure

High pressure condition is certainly the more used sample environment after low/high temperature. This field of research is also one of the most well-informed facing the number of published papers and reviews concerning experimental results or technical information (see for examples Paszkowicz, 2002; Boldyreva, 2008; Katrusiak, 2008; and references therein). Material studies under pressure are nowadays numerous due to the increase capabilities of modern X-ray diffractometer and the making of dedicated instruments on synchrotrons and neutron facilities. At the present time, about 30 synchrotron beamlines in the world (at 11 synchrotrons) allow high-pressure measurements to be performed, with possibility of high temperature too. It could be pointed out that 2012 marks the 35th anniversary of high pressure diffraction at synchrotron (Buras et al., 1977). The application of pressure on a material may induce major structural variations, particularly a modification in the weak intermolecular contacts and reorganization of the electronic structure, but also phase transitions or chemical reactions. Anyway, the application of pressure on a system induces more effects than the application of cryogenic temperatures, which is of interest to physicists, chemists, biologists and geologists. Pressure can give substantial information about structure-properties relationships and new thermodynamic phenomena (phase diagrams, polymorphism, cohesion forces, transformations...). Moreover, temperature variations can now be applied in combination to high pressure due to the high brilliance of modern sources at facilities which can bypass the unavoidable absorption of the beam by the cryostat or furnace and the pressure cell containing the tiny studied sample. Now, it is possible to perform experiments at 100 GPa and further thousands of Celsius degrees (Fiquet et al., 1999; Akahama et al., 2002; Guo et al., 2002; Hu et al., 2002; Orosel et al., 2012) using synchrotron radiations, this is achieved with the laser heating technique (Bassett, 2001). Due to the relatively large sample volume required for neutron diffraction, high pressure measurement are always a challenge and reached maximum pressures are not so high: it was recently reported a high-pressure single-crystal neutron diffraction to 10 GPa at ambient temperature taking advantage of the neutron diffractometer D9 at the Institut Laue-Langevin (France) (Bull et al., 2011). To succeed those measurements, teams generally use DAC (Diamond Anvil Cell) (Jamieson et al., 1959; Weir et al., 1959), MAC (Moissanite Anvil Cell) (Xu et al., 2002) or LAC (Large Anvil Cell, as Paris-Edinburgh cell) (figure 7c) (Ohtani et al., 1977; Besson et al., 1992; Mezouar et al., 1999; Le Godec et al., 2003) pressure cells. For low and moderate pressures up to about 3 GPa at ambient temperature, there are also piston-in cylinder clamped cells (McWhan et al., 1974) and continuously loaded cells (Paureau, 1975). From the firsts diffraction measurements under pressure (Barnett et al., 1963; Bassett et al., 1966) to the last ones (Basu et al., 2012; Orosel et al., 2012; Tulk et al., 2012) efforts still continue to increase the available pressure range in both static and dynamic conditions, and also to improve instruments (Utsumi et al., 2002) or allow new techniques (Bromiley et al., 2009).

3. NEW INSTRUMENTAL DEVELOPMENTS AT DAWN OF THE THIRD MILLENNIUM

Using extreme conditions, researchers can consider not only to extend the accessible parameters scale to measure some physical constants or to verify existing theories, but also to point out new behaviours and states which increase our knowledge of materials properties. With progresses in the domains of available sources as 3rd generation synchrotrons or powerful neutron sources and of new sample environment devices as explained in part 2., extreme conditions measurements have known since 10 years a meteoric rise in all scientific fields for fundamental studies, for examples, of complex fluids, high correlated electrons systems, geological and biological materials at the origin of earth or life. However, will this keen interest continue for a long time? Are technological innovations always possible in that scientific area to build more efficient instruments and devices? Could it be possible to go further the actual limits of high pressures, high and low temperatures, applied magnetic fields? Is it really necessary to cross these limits as numerous material properties are still unexplored? All these questions have one simple answer: yes. Going further in the field of crystallography under extreme conditions is also expanding our knowledge of the world and life by probing the matter at the border of the unknown. Scientist needs everyday new challenges and the one to push back those limits is a necessity for him, but mostly essential for mankind.

In this way, *extreme conditions beamline* projects have emerged throughout the world at Large scale facilities. It will be pointed out thereafter some examples of instrumental projects which were, or will be, realized to explore materials properties giving access to new non-ambient conditions. The proposed examples are a non-exhaustive list of the actual developments in crystallography under extreme conditions. These examples are chosen on the most powerful Large scale facilities around the world in terms of the most important parameters (energy, current...): SPring-8 (Japan), APS (USA) and ESRF (France) are the chosen synchrotron sources (table 1); ISIS (neutron spallation source, UK) and the ILL (neutron reactor, France) are the chosen neutron facilities.

TABLE 1.. The most important parameters of the 10 world-wide most powerful 3rd generation light sources available in 2012.

Name	Location	Energy (GeV)	Perimeter (m)	Current (mA)	Emittance (mrad)	Number of straights
SPRING-8	Japan	8	1436	100	3	48
APS	USA	7	1060	100	3	40
ESRF	France	6	844	200	3.8	32
SSRF	China	3.5	396	300	4.8	20
ALBA	Spain	3	268.8	250	3.7	24
Australian Synchrotron	Australia	3	216	200	8.6	14
DIAMOND	UK	3	560	300	2.7	24
SPEARS3	USA	3	240	500	18	18
CLS	Canada	2.9	171	500	18	15
SOLEIL	France	2.85	354	500	3.1	24

During the last 10 years, numerous Large scale facilities begun an upgrade of their instruments and installations. This is partly due to the increase demand for high-brilliance X-ray and high-flux neutron beams by user communities requiring always an increase of the instrument performances and of the allocated beamtime. Users want to perform measurements on smaller samples mostly in *exotic*environments. One of the engineering goals for the upgrades is making available advanced sample environments with extreme conditions on the beamlines whilst fitting in the tight space around the sample also detectors and monitoring equipment. The challenge for the facilities will be to commonly supply extreme environments like pressures up to 100 GPa, low temperature below 1 K, high temperature above 3000 K, high magnetic field up to 50 T in pulsed mode and 30 T in continuous mode.

To achieve these goals, facilities must also invest in the engineering of new dedicated instruments and particularly in the development of new detectors which have to be more sensitive, more efficient (especially for high-energy X-rays) and faster (notably in the sub-millisecond range to perform time resolved measurements). Upgraded a Large scale facility is difficult, long and very expensive (table 2).

TABLE 2. Estimated time and total cost of the upgrade or construction of some synchrotron and neutron facilities around the world.

Name	Type of facilities		Period	Total cost
SPRING-8	synchrotron	upgrade	2012-2019	375 M€
APS	synchrotron	upgrade	2010-2018	275 M€
ESRF	synchrotron	upgrade	2008-2017	290 M€
DIAMOND	synchrotron	construction	2002-2007	460 M€
SOLEIL	synchrotron	construction	2000-2006	403 M€
ILL	neutron reactor	upgrade	2001-2011	85 M€
ISIS- TS2	neutron spallation source	construction	2003-2009	200 M€

SPring-8 light source will be upgraded in order to advance promising science and to support industrial innovations that will improve our life and contribute to a more sustainable society (SPring-8, 2012). This upgrade programme is principally focused on the construction of a new ring in place of the actual one, retaining the existing insertion device beamlines. In 2019, the new ring (SPring-8 II) would produce 1000 times higher brilliance than the present SPing-8 (new stored current of 300 mA instead of 100 mA presently) with a short X-ray pulse option of around 1 ps. The frontiers of science will be definitely opened. Instruments like BL02B1 (designed for analyses of single crystal structures and for investigations of phase transitions under external fields), BL04B1 (dedicated to high pressures and high temperatures researches in geoscience), BL10XU (designed to perform X-ray diffraction structure analyses under high pressure - 300 GPa - and low temperature - 10 K - or high temperature - 3000 K -) or BL40XU (high flux beamline for time-resolved x-ray diffraction) will benefit of new opportunities.

On April 2010, the US Department of Energy approved the Advanced Photon Source (APS) upgrade project (APS, 2011). The upgrade will provide high energy, high brilliance, short pulse, new and upgraded instruments. The objective is to push the stored current to 150 mA, to upgrade 6 beamlines and to build at least 6 new ones between 2010 and 2018 (Mills, 2011). The new beamlines will especially deal with the overarching theme *Real materials under real conditions in real time.* It means an easier access for users to high pressures, low/high temperatures and in situ chemical reactions. One frontier for X-ray science in the

21st century is to combine atomic-level spatial and temporal information, a frontier which would be crossed after the upgrade. For extreme conditions experiments, the APS had plane to upgrade 4 instruments and to build 2 new ones. For examples, the outboard branch of 16-ID will be dedicated to high pressure spectroscopy using sub-μm probes (beam size of 100-500 nm with a flux $\geq 10^{12}$ photons.s^{-1} at sample position).

In this context, the ESRF (ESRF, 2007) has planned an upgrade programme for its instruments, like ID20 (beamline for magnetic and resonant X-ray scattering investigations under extreme conditions; Paolasini et al., 2007), ID22 (micro-fluorescence, imaging and diffraction beamline, Martinez-Criado et al., 2007), ID24 (Energy dispersive X-ray absorption spectroscopy beamline) or ID27 (high pressure beamline; Mezouar et al., 2005). The ID27 source was upgraded to the most recently developed cryogenic or superconducting undulator to obtain the maximum flux from the ESRF machine after its upgrade. This very high flux beamline will provide new approaches to very challenging problems such as the search for a new metallic superfluid state of matter predicted at pressures above 400 GPa, or to perform ultra-high pressure and temperature X-ray emission spectroscopy in combination with X-ray diffraction in the laser heated diamond anvil cells. In addition to this upgrade, the ESRF will build two new dedicated high pressure beamlines for diamond anvil cells and Paris-Edinburgh research. Time-resolved diffraction is also a priority for the ESRF which propose to build a dedicated beamline covering the timescale from 10 picoseconds to seconds in studies of physical, chemical and biological processes. The 4-circle diffractometer of this beamline will include Laue diffraction and classical diffraction, a very intense focused beam, a new 3 kHz chopper which isolates single pulses of X-rays from the timing modes and a fast readout (0.1 s) Frelon CCD based camera. An other upgrade example concerns the ID20 magnetic scattering beamline with an extend of its experimental conditions in order to provide new research opportunities. It is also planned to build a dedicated station for very low temperatures below 1 K, high magnetic fields up to 17 T and high pressure up to 5 GPa at 1.5 K.

The world's leading spallation neutron source, ISIS, has contributed significantly to many of the major breakthroughs in materials science, physics and chemistry since it was commissioned in 1985. To keep the UK at the forefront of neutron research, it was decided in 2003 to build a second spallation neutron source, ISIS Target Station 2, where three key areas of science are planned to be investigated in priority: soft matter, bio-sciences and advanced materials. ISIS TS-2 project was completed in 2009 on time and to budget. There are currently 7 available instruments (and 4 instruments under building) at ISIS TS-2 which has a capacity for a total of 18 instruments in the future, adding to the 20 instruments already available at ISIS Target Station 1. Concerning the new instruments dedicated to extreme condition measurements, the available WISH instrument (Chapon et al., 2011) is a long-wavelength diffractometer primarily designed for powder diffraction at long d-spacing in magnetic and large unit cell systems, with the option of enabling single-crystal and polarised beam experiments. The WISH sample environments include a dedicated low background cryostat (with ultra-low temperature inserts), a dedicated 13.6 T vertical magnet, but also the standard ISIS furnaces, pressure cells and gas rigs. Futhermore, there is the IMAT instrument (Kockelmann et al., 2007), in design phase and operating in 2015, which will be a neutron imaging and diffraction instrument for materials science, materials processing and engineering. IMAT will offer a unique combination of imaging and spatially resolved diffraction modes for *tomography-driven diffraction*: residual stresses inside engineering-sized samples can be more effectively analysed if the diffraction scans are guided by radiographic data. Finally, the future EXEED instrument (McMillan and Tucker, 2007) will be a neutron time-of-flight diffractometer optimised for extreme environment studies of materials which will complement the capabilities of WISH on Target Station 2 and PEARL on Target Station 1. The experiments will be performed at high pressure, above 50 GPa using diamond anvil cells, under combined very low temperatures (mK), very high temperatures (2000 K, including laser heating in cells) or intense magnetic field, up to 10 T.

Finally, the neutron reactor installations and instruments at the ILL are undergoing a modernisation phase called the *Millennium programme 2001 2014*, 30 years after the first experiments in 1972. During the first phase (2001-2008) of the ILL upgrade, significant advances had been provided. The efficiency of the instruments has been boosted by a factor of 19. Moreover, 6 new instruments were built and 8 others were upgraded. In the second phase (2008-2014), 7 new instruments are planned to be built and 4 actual instruments upgraded. For extreme conditions experiments, it was planned a new neutron diffractometer XtremeD (Rodriguez-Velamazan et al., 2011). This instrument will be optimized for high pressure, up to 30 GPa, and high magnetic field, up to 15 T in continuous mode, studies for both single crystals and powders. The diffractometer, mainly dedicated to study molecular chemistry and magnetism under extreme conditions, will include a large 2D position-sensitive detector and a radial oscillating collimator to suppress background.

It is clear, with all the above mentioned new instrumental developments at almost worldwide Large scale facilities, that the challenge for crystallography in the next years is to perform non-equilibrium and extreme condition measurements. All the domains of sciences are concerned in order to increase the knowledge and the understanding of materials behaviours and properties under external perturbations. It can be asserted without ambiguity that the present century is the one of *crystallography under extreme conditions.*

4. CONCLUSION

Since the first crystallographic experiment at the beginning of the 20th century, scientists have always whished to push back frontiers of measurements. Obtaining the structure-properties relationships in static and dynamic modes is one fundamental goal to increase our knowledge of materials and verify existing theories. This was largely made possible by the arrival of modern Large scale facilities allowing material investigations under extreme conditions. State-of-the-art crystallographic measurements discussed in this review and recent upgrades achieved at most facilities augur that limits will be increasingly crossed in the next years, giving rise to unsuspected highlight researches in all domains of sciences. The considered perspectives will offer new opportunities in crystallography.

REFERENCES

1. Y. Akahama, H. Kawamura, T. Le Bihan, 2002A new distorded body-centred cubic phase of titanium (δ-Ti) at pressure up to 220 GPa. J. Phys.: Condens. Matter, 141058310588
2. APS2011Conceptual design report. Advanced Photon Source upgrade project. http://aps.anl.gov/Upgrade/Documents/
3. I. F. Bailey, 2003A review of sample environments in neutron scatteringZ. Kristallogr., 2188495
4. N. Ban, P. Nissen, J. Hansen, P. B. Moore, T. A. Steitz, 2000The complete atomic structure of the large ribosome subunit at 2.4 Å resolution. Science, 289905
5. J. Banhart, H. Stanzick, L. Helfen, T. Baumbach, (2001, 2001Metal foam evolution studied by synchrotron radioscopyAppl. Phys. Letters, 7811521154
6. A. Barcza, Z. Gercsi, K. S. Knight, K. G. Sandeman, 2010Giant Magnetoelastic Coupling in a metallic helical metamagnetPhys. Rev. Letters, 104247202
7. J. D. Barnet, H. T. Hall, 1964High pressure- high temperature, X-ray diffraction apparatus. Rev. Sci. Instrum., 35175182
8. W. A. Bassett, T. Takahashi, P. W. Stook, 1967X-ray diffraction and optical observations on crystalline solids up to 300 kbar. Rev. Sci. Instrum., 383742
9. W. A. Bassett, 2001The birth and development of laser heating in diamond anvil cellsRev. Sci. Instrum., 7212701272
10. A. Basu, A. Chandra, A. K. Tyagi, G. D. Mukherjee, 2012Reppearance of ferroelectric soft modes in the paraelectric phase of Pb1-xCaxTiO3 at high pressures : Raman and X-ray diffraction studies. J. Phys.: Condens. Matter, 24115404
11. J. M. Beson, R. J. Nelmes, G. Hamel, J. S. Loveday, G. Weill, S. Hull, 1992Neutron powder diffraction above 10 GPa. Physica B, 180-181907910
12. D. H. Bilderback, P. Elleaume, E. Weckert, 2005Review of third and next generation synchrotron light sourcesJ. Phys. B: At. Mol. Opt. Phys., 38S773S797

13. E. V. Boldyreva, 2008High-pressure diffraction studies of molecular organic solids. A personal view.Acta Cryst. A, 64218231
14. M. Born, 1933Modified field equations with a finite radius of the electronNature132282
15. M. Born, 1935Quantised field theory and the mass of the protonNature136952
16. M. Born, E. Schrödinger, 1935The absolute field constant in the new field theoryNature135342
17. A. Bousseksou, G. Molnar, L. Salmon, W. Nicolazzi, 2011Molecular spin crossover phenomenon:recent achievements and pospects.Chem. Soc. Rev., 403313
18. W. H. Bragg, 1912X-rays and crystals. Nature,90219
19. W. H. Bragg, 1913aX-rays and crystals. Nature,90572
20. W. H. Bragg, 1913bThe reflection of X-rays by crystalsNature91477
21. W. H. Bragg, 1915aThe distribution of the electrons in atomsNature95344
22. W. H. Bragg, 1915bThe structure of magnetite and the spinelsNature95561
23. W. H. Bragg, W. L. Bragg, 1913The structure of diamond. Nature91557
24. W. L. Bragg, 1920Crystals structure. Nature, 105646648
25. D. Brochier, 1977Cryostat à température variable pour mesures neutroniques ou optiques. ILL Tech. Report 77/74
26. G. D. Bromiley, S. A. T. Redfern, Y. Le Godec, G. Hamel, S. Klotz, 2009A portable high-pressure stress cell based on the V7 Paris-Edinburgh apparatusHigh pressure Researchan International Journal, 29306316
27. P. J. Brown, J. B. Forsyth, E. Lelièvre-Berna, F. Tasset, 2002Determination of the magnetization distribution in Cr2O3 using spherical neutron polarimetryJ. Phys.: Condens. Matter, 1419571966
28. P. Brown, A. Bushmelev, C. P. Butts, J. Cheng, J. Eastoe, I. Grillo, R. K. Heenan, A. M. Schmidt, 2012Magnetic control over liquid surface properties with responsive surfactantsAngewandte ChemieInt. Ed., 5124112416
29. C. L. Bull, M. Guthrie, J. Archer, M. Fernandez-Diaz-T, J. S. Loveday, K. Komatsu, H. Hamidov, R. J. Nelmes, 2011High-pressure single-crystal neutron diffraction to 10 GPa by angle-dispersive techniquesJ. Appl. Cryst., 44831
30. B. Buras, J. S. Olsen, L. Gerward, G. Will, E. Hinze, 1997X-ray energy-dispersive diffractometry using synchrotron radiationJ. Appl. Cryst., 10431438
31. L. C. Chapon, P. Manuel, P. G. Radaelli, C. Benson, L. Perrott, S. Ansell, N. J. Rhodes, D. Raspino, D. Duxbury, E. Spill, J. Norris, 2011Wish: the new powder and single crystal magnetic diffractometer on the second target stationNeutron News, 222225
32. D. Chasseau, P. Guionneau, J. Gaultier, Y. Barrans, L. Ducasse, C. J. Kepert, P. Day, M. Kurmoo, 1997Crystal structures of (BEDT-TTF)3CuBr4 at 10 K and 10 kbar. Synthetic Metals, 8620452046
33. A. Chiba, N. Funamori, K. Nakayama, Y. Ohishi, S. M. Bennington, S. Rastogi, A. Shukla, K. Tsuji, M. Takenaka, 2012Pressure-induced structural change of intermediate-range order in poly(4-methyl-1-pentene) melt. Phys. Rev. E, 85021807
34. E. Clementi, C. Roetti, 1974Roothaan-Hartree-Fock atomic wavefunctions. Atomic data and nuclear data tables, 14177478
35. P. Cloetens, W. Ludwig, J. Baruchel, D. Van Dyck, J. Van Landuyt, J. P. Guigay, M. Schlenker, 1999Holotomography: quantitative phase tomography with mocrometer resolution using hard synchrotron radiation X-rays. Appl. Phys. Letters, 7529122914
36. E. Collet, M. Lorenc, M. Cammarata, L. Guérin, M. Servol, A. Tissot, M. Boillot-L, H. Cailleau, Cointe. M. Buron Le, 2012a100 picosecond diffraction catches structural transients of laser-pulse triggered switching in a spin crossover crystalChem. Eur. J., 1820512055
37. E. Collet, N. Moisan, C. Baldé, R. Bertoni, E. Trzop, C. Laulhé, M. Lorenc, M. Servol, H. Cailleau, A. Tissot, M. Boilot-L, T. Graber, R. Henning, P. Coppens, Cointe. M. Buron Le, 2012bUltrafast spin-state photoswitching in a crystal and slower consecutive process investigated by femtosecond optical spectroscopy and picosecond X-ray diffraction. Phys. Chem. Chem. Phys., 1461926199

38. P. Coppens, D. V. Formitchev, M. C. Carducci, K. Culp, 1998Crystallographiy of molecular excited states. Transition metal nitrosyl complexes and the study of transient species. J. Chem. Soc., Dalton Trans., 865872
39. P. Coppens, I. I. Vorontsov, T. Graber, M. Gembicky, A. Y. Kovalensky, 2005The structure of short-lived excited states of molecular complexes by time-resolved X-ray diffractionActa CrystA, 61162172
40. V. Cristiglio, L. Hennet, G. J. Cuello, I. Pozdnyakova, A. Bytchkov, P. Palleau, H. E. Fisher, D. Zanghi, P. Saboungi-L, D. L. Price, 2007Structural study of levitated liquid Y2O3 using neutron scatteringJ. Non-Crystalline Solids, 353993995
41. Q. Guo, H. K. Mao, J. Hu, J. Shu, R. J. Hemley, 2002The phase transitions of CoO under static pressure to 104 GpaJ. Phys.: Condens. Matter, 141136911374
42. R. Destro, L. Loconte, Presti. L. Lo, P. Roversi, R. Soave, 2004On the role of data quality in experimental charge density studiesActa Cryst. A, 60365370
43. F. R. Elder, A. M. Gurewitsch, R. V. Langmuir, H. C. Pollock, 1947A 70 MeV Synchrotron. J. Appl. Phys., 18810818
44. ESRF2007Science and technology programme 2008-2017.http://www.esrf.eu/AboutUs/Upgrade/documentation/purple-book/,1-2
45. S. Falconi, L. F. Lundegaard, C. Hejny, M. I. Mc Mahon, 2005X-ray diffraction study of liquid Cs up to 9.8 GPa.Phys. Rev. Letters, 94125507
46. E. Fermi, 1930Magnetic moment of atomic Nuclei.Nature12516
47. E. Fermi, 1934Radioactivity induced by neutron bombardmentNature133757
48. E. Fermi, 1940Reactions produced by neutrons in heavy elementsNature146640642
49. E. Fermi, 1947Elementary theory of the chain-reacting pileScience10527
50. P. Fertey, R. Argoud, P. Bordet, J. Reymann, C. Palin, C. Bouchard, R. Bruyère, E. Wenger, C. Lecomte, 2007A mini-goniometer for X-ray diffraction studies down to 4 K on four-circle diffrctometers equipped with two-dimensional detectors. J. Appl. Cryst., 40526531
51. G. Fiquet, D. Andrault, 1999Powder X-ray diffraction under extreme conditions of pressure and temperatureJ. Synchrotron Rad., 68186
52. D. V. Formitchev, I. Novozhilova, P. Coppens, 2000Photo-induced linkage isomerism of transition metal nitrosyl and dinitrogen complexes studied by photocrystallographic techniques. Tetrahedron, 5668136820
53. P. Garcia, S. Dahaoui, P. Fertey, E. Wenger, C. Lecomte, 2005Crystallographic investigation of temperature-induced phase transition of the tetrathiafulvalene-p-bromanil, TTF-BA charge transfer complex. Phys. RevB, 72104115
54. P. Garcia, S. Dahaoui, C. Katan, M. Souhassou, C. Lecomte, 2007On the accuracy of intermolecular interactions and charge transfer: the case of TTF-CA. Faraday Discussions, 135217235
55. M. Gembicky, P. Coppens, 2007On the desing of ultrafast shutters for time-resolved synchrotron experiments. J. Synchrotron Rad., 14133137
56. F. Givord, J. Boucherle-X, E. Lelièvre-Berna, P. Lejay, 2004The cerium magnetic form facto rand diffuse polarization in CeRh3B2 as functions of temperature. J. Phys.: Condens. Matter, 1612111230
57. H. Godfrin, M. Meschke, H. Lauter-J, A. Sultan, H. M. Böhm, E. Krotscheck, M. Panholzer, 2012Observation of a roton collective mode in a two-dimensional fermi liquid.Nature483576579
58. J. Goedhart, D. Stetten, M. Noirclerc-Savoye, M. Lelimousin, L. Joosen, M. A. Hink, L. Weeren, T. W. J. Gadella Jr, A. Royant, 2012Structure-guided evolution of cyan fluorescent proteins towards a quantum yield of 93%. Nature Communications, 3751
59. A. Goujon, B. Gillon, A. Gukasov, J. Jeftic, Q. Nau, E. Codjovi, F. Varret, 2003Photoinduced molecular switching studied by polarized neutron diffractionPhysical ReviewB, 67220401R). DOI:PhysRevB.67.220401http: //prb.aps.org/abstract/PRB/v67/i22/e220401
60. A. Goujon, B. Gillon, A. Debede, A. Cousson, A. Gukasov, J. Jeftic, G. Mc Intyre, F. Varret, 2006Neutron Laue diffraction on the spin crossover crystal [Fe(1-n-

propyltetrazole)6](BF4)2 showing continuous photoinduced transformation. Physical Review B, 73104413

61. F. K. Goward, D. E. Barnes, 1946Experimental 8 MeV synchrotron for electron acceleration.Nature158413
62. H. Graafsma, S. O. Svensson, A. Kvick, 1997An X-ray charge density feasibility study at 56 keV of magnesium formate dihydrate using a CCD area detectorJ. Appl. Cryst., 30957962
63. J. M. Grimes, J. N. Burroughs, P. Gouet, J. M. Diprose, R. Malby, S. Zientara, Merten, P. P. C. , D. I. Stuart, 1998The atomic structure of the bluetongue virus core.Nature395470478
64. R. Grössinger, H. Sassik, O. Mayerhofer, E. Wagner, M. Schrenk, 2004Austromag: pulsed magnetic fields beyond 40 T. Physica B, 346-347609613
65. R. Guillot, P. Allé, P. Fertey, N. K. Hansen, E. Elkaïm, 2002Diffraction measurements from crystals under electric fields: instrumentationJ. Appl. Cryst., 35360
66. P. Guionneau, C. Brigouleix, Y. Barrans, A. E. Goeta, J. Létard-F, J. A. K. Howard, J. Gaultier, D. Chasseau, 2001High pressure and very low temperature effects on the crystal structures of some iron(II) complexes. C. R. Acad. Sci. Paris, Chimie, 4161171
67. P. Guionneau, D. Le Pévelen, M. Marchivie, S. Pechev, J. Gaultier, Y. Barrans, D. Chasseau, 2004Laboratory high pressure single-crystal X-ray diffraction- recent improvements and examples of studies. J. Phys.: Condens. Matter, 16S1151S1159DOI:http://iopscience.iop.org/0953-8984/16/14/025
68. P. Gütlich, H. A. Goodwin, 2004Spin crossover in transition metal compounds. Eds. Topics in Current Chemistry, Springer-Verlag: Berlin, 233-234
69. N. K. Hansen, P. Coppens, 1978Testing aspherical atom refinements on small-molecule data setsActa Cryst. A, 34909921
70. N. K. Hansen, P. Fertey, R. Guillot, (2004, 2004Studies of electric field induced structural and electron-density modifications by X-ray diffractionActa Cryst. A, 60465
71. T. C. Hansen, P. F. Henry, H. E. Fischer, J. Torregrossa, P. Convert, 2008The D20 instrument at the ILL: a versatile high-intensity two-axis neutron diffractometerMeas. Sci. Technol., 19034001
72. L. Hennet, D. Thiaudière, M. Gailhanou, C. Landron, J. Coutures-P, D. L. Prince, 2002Fast X-ray scattering measurements on molten alumina using a 120° curved position sensitive detector. Rev.Scientific Instruments, 73124129
73. L. Hennet, D. Thiaudière, C. Landron, P. Melin, D. L. Price, J. Coutures-P, J. Bérar-F, M. Saboungi-L, 2003Melting behaviour of levitated Y2O3. Appl. Phys. Letters, 8333053307
74. L. Hennet, I. Pozdnyakova, A. Bytchkov, V. Cristiglio, P. Paleau, H. E. Fischer, G. J. Cuello, M. Johnson, P. Melin, D. Zanghi, S. Brassamin, M. Saboungi-L, 2006Levitation apparatus for neutron diffraction investigations on high temperature liquidsRev. Scientific Instruments, 77053903
75. L. Hennet, I. Pozdnyakova, V. Cristiglio, S. Krisnan, A. Bytchkov, F. Albergamo, G. J. Cuello, J. Brun-F, H. E. Fischer, D. Zanghi, S. Brassamin, M. Saboungi-L, D. L. Price, 2007Structure and dynamics of levitated liquid aluminatesJ. Non-Crystalline Solids, 35317051712
76. L. Hennet, I. Pozdnyakova, A. Bytchkov, V. Cristiglio, D. Zanghi, S. Brassamin, J. F. Brun, M. Leydier, D. L. Price, 2008Fast X-ray scattering measurements on high temperature levitated liquidsJ. Non-Crystalline Solids, 35451045017
77. J. Hu, J. Xu, M. Somayazulu, Q. Guo, R. Hemley, H. K. Mao, 2002X-ray diffraction and laser heating: application of a moissanite anvil cellJ. Phys.: Condens. Matter, 141047910481
78. J. C. Jamieson, A. W. Lawson, N. D. Nachtrieb, 1959New device for obtaining X-ray diffraction patterns from substances exposed to high pressure. Rev. Sci. Instrum., 3010161019
79. A. Katrusiak, 2008High-pressure crystallographyActa Cryst. A, 64135148

80. K. Katsumata, 2005Synchrotron X-ray diffraction studies of magnetic materials under extreme conditionsPhysica Scripta71CC7C13
81. W. Kockelmann, E. C. Oliver, P. G. Radaelli, 2007IMAT- an imaging and materials science & engineering facility for TS-II. Draft proposal for discussion. http://www.isis.stfc.ac.uk/Instruments/Imat/
82. W. F. Kuhs, J. Archer, D. Doran, 1993A closed-shell furnace for neutron single-crystal diffractionJ. Appl. Cryst., 26730733
83. Y. Le Godec, M. T. Dove, S. A. T. Redfern, M. G. Tucker, W. G. Marshall, G. Syfosse, S. Klotz, 2003Recent developments using the Paris-Edinburgh cell for neutron diffraction at high pressure and high temperature and some applicationsHigh Pressure Research23281287
84. V. Legrand, 2005Cristallographie et photo-cristallographie haute résolution de composés moléculaires à transition de spin : propriétés structurales, électroniques et mécanismes de conversion. thesis, Univ. Henri Poincaré, Nancy 1 (France), n°1132
85. V. Legrand, C. Carbonera, S. Pillet, M. Souhassou, J. Létard-F, P. Guionneau, C. Lecomte, 2005Photo-crystallography: from the structure towards the electron density of metastable statesJournal of Physics: Conf. Series, 217380
86. V. Legrand, S. Pillet, M. Souhassou, N. Lugan, C. Lecomte, 2006Extension of the experimental electron density analysis to metastable states : a case example of the spin crossover complex Fe(btr)2(NCS)2.H2O. J. Am. Chem. Soc., 12613921
87. V. Legrand, S. Pillet, H. Weber-P, M. Souhassou, J. Létard-F, P. Guionneau, C. Lecomte, 2007aOn the precision and accuracy of structural analysis of light-induced metastable statesJ. Appl. Cryst., 4010761088
88. V. Legrand, S. Pillet, C. Carbonera, M. Souhassou, J. Létard-F, P. Guionneau, C. Lecomte, 2007bOptical, magnetic and structural properties of the spin crossover complex [Fe(btr)2(NCS)2].H2O in the light-induced and thermally quenched metastable states. European Journal of Inorganic Chemistry, 56935706
89. V. Legrand, F. Le Gac, P. Guionneau, Létard, J. -F, 2008Neutron powder diffraction studies of two spin transition Fe(II)-complexes under pressure. Journal of Applied Crystallography, 41637640
90. E. Lelièvre-Berna, E. Bourgeat-Lami, P. Fouilloux, B. Geffray, Y. Gibert, K. Kakurai, N. Kernavanois, B. Longuet, F. Mantegezza, M. Nakamura, S. Pujol, L. Renault-P, F. Tasset, M. Takeda, M. Thomas, X. Tonon, 2005Advances in spherical neutron polarimetry with CryopadPhysica B356131135
91. G. Lorenz, R. B. Neder, J. Marxreiter, F. Frey, J. Schneider, 1993A mirror furnace for neutron diffraction up to 2300 KJ. Appl. Cryst., 26632635
92. A. N. Mak-S, P. Bradley, R. A. Cernadas, A. J. Bogdanove, B. L. Stoddard, 2012The crystal structure of TAL effector PthXo1 bound to its DNA targetScience335716719
93. G. Martinez-Criado, R. Steinmann, B. Alén, A. Labrador, D. Fuster, J. M. Ripalda, A. Homs, S. Labouré, J. Susini, 2007New cryogenic environment for beamline ID22 at the European Synchrotron Radiation FacilityRev. Sci. Instrum., 78025106
94. P. Mc Millan, M. Tucker, 2007EXEED- an extreme sample environment diffractometer. Draft proposal for discussion. http://www.isis.stfc.ac.uk/instruments/Exeed/
95. D. B. Mc Whan, D. Bloch, G. Parisot, 1974Apparatus for neutron diffrction at high pressure. Rev. Sci. Instrum., 45643646
96. M. Mezouar, T. Le Bihan, H. Libotte, Y. Le Godec, D. Häusermann, 1999Paris-Edinburgh large volume cell coupled with a fast imaging-plate system for structural investigation at high pressure and high temperatureJ. Synchrotron Rad., 611151119
97. M. Mezouar, W. A. Crichton, S. Bauchau, F. Thurel, H. Witsch, F. Torrecillas, G. Blattmann, P. Marion, Y. Dabin, J. Chavanne, O. Hignette, C. Morawe, C. Borel, 2005development of a new state-of-the-art beamline optimized for monochromatic single-crystal and powder X-ray diffraction under extreme conditions at the ESRFJ. Synchrotron Rad., 12659664

98. J. Mignot-M, 2008Diffusion neutronique sous conditions extrêmes. Collection SFN 9. EDP Sciences, Les Ulis. DOI:10.1051/sfn:2008012
99. D. M. Mills, 2011The Advanced Photon Source- where we are and where we are going. Nuclear Instruments and Methods in Physics Research A, 6492224
100. S. W. Muchmore, 1999Experiences with CCD detectors on a home X-ray source.Acta Cryst. D, 5516691671
101. K. Neumaier, A. Heidemann, A. Mageri, 1984A dilution refrigerator insert for standard ILL cryostat. Rev. Phys. Appl. (Paris), 19773774
102. V. Nietz, 2003Prospects for the use of the pulsed fields in neutron research of condensed matterJ. Magnetism and Magnetic Materials, 26084104
103. H. Nojiri, K. Takahashi, T. Fukuda, M. Jujita, M. Arai, M. Motokawa, 199825 T repeating pulsed magnetic fields system for neutron diffraction experimentsPhysicaB, 241-243210212
104. S. Nozawa, S. Adachi, J. Takahashi, R. Tazaki, L. Guérin, M. Daimon, A. Tomita, T. Sato, M. Chollet, E. Collet, H. Cailleau, S. Yamamoto, K. Tsuchiya, T. Shioya, H. Sasaki, T. Mori, K. Ichiyanagi, H. Sawa, S. Koshihara, 2007Developing 100 ps-resolved X-ray structural analysis capabilities on beamline NW14A at the Photon Factory Advanced Ring.J. Synchrotron Rad., 14313319
105. A. Ohtani, M. Mizukami, M. Katayama, A. Onodera, N. Kawai, 1977Multi-anvil apparatus for high pressure X-ray diffractionJpn. J. Appl. Phys., 161843
106. J. T. Okada, P. H. Sit-L, Y. Watanabe, Y. J. Wang, B. Barbiellini, T. Ishikawa, M. Itou, Y. Sakurai, A. Bansil, R. Ishikawa, M. Hamaishi, T. Masaki, P. Paradis-F, K. Kimura, T. Ishikawa, S. Nanao, 2012Persistence of covalent bonding in liquid silicon probed by inelastic x-ray scattering.Phys. Rev. Lett., 108067402
107. D. Orosel, R. E. Dinnebier, V. A. Blatov, M. Jansen, 2012Structure of a new high-pressure- high temperature modification of antimony(III)oxyde, from high-resolution synchrotron powder diffraction data. Acta Cryst. B, 6817
108. L. Paolasini, C. Detlefs, C. Mazzoli, S. Wilkin, P. P. Deen, A. Bombardi, N. Kernavanois, F. de Bergevin, F. Yakhou, J. P. Valade, I. Brslavetz, A. Fondacaro, G. Pepellin, P. Bernard, 2007ID20: a beamline for magnetic and resonant X-ray scattering investigations under extreme conditions.J. Synchrotron Rad., 14301312
109. W. Paszkowicz, 2002High pressure powder X-ray diffraction at the turn of the century. Nucl. Instr. And Meth. In Phys. Res. B, 198142182
110. J. Paureau, 1975Nouveau dispositif d'étanchéité pour hautes pressions. Revue de Physique Appliquée, 10475478
111. S. Pillet, V. Legrand, H. Weber-P, M. Souhassou, J. Létard-F, P. Guionneau, C. Lecomte, 2008Out-of-equilibrium charge density distribution of spin crossover complexes from steady-state photocrystallographic measurements : experimental methodology and results. Z. Kristallogr., 223235249
112. T. C. Price, D. M. Grant, V. Legrand, G. S. Walker, 2010Enhanced kinetics for the LiBH4:MgH2 multi-component hydrogen storage system- the effects of stoichiometry and decomposition environment on cycling behavior. International Journal of Hydrogen Energy, 3541544161
113. C. Riekel, M. Müller, F. Vollrath, 1999In situ X-ray diffraction during forced silking of spider silkMacromolecules3244644466
114. C. Riekel, F. Vollrath, 2001Spider silk fibre extrusion: combined wide- and small-angle X-ray microdiffraction experiments.Inter. J. Biological Macromol., 29203210
115. J. A. Rodriguez-Velamazan, J. Campo, J. Rodriguez-Carvajal, P. Noguera, 2011XtremeD- a new neutron diffractometer for high pressures and magnetic fields at ILL developed by Spain. J. Phys.: Conference Series, 325012010
116. I. Sakai, K. Murai, L. Jiang, N. Umesaki, T. Honma, A. Kitano, 2005Aerodynamic levitation apparatus for structure study of high temperature materials coupled with Debye-Scherrer Camera at BL19B2 of Spring-8J. Electron Spectroscopy and Related Phenomena, 114-14710111013
117. D. O. Scanlon, A. B. Kehoe, G. W. Watson, M. O. Jones, W. I. F. David, D. J. Payne, R. G. Egdell, P. P. Edwards, A. Walsh, 2011Nature of the band gap and origin of the

conductivity of PbO2 revealed by theory and experiment.Phys. Rev. Letters, 107246402

118. J. Seta, J. , S. A. Davis, F. Meldrum, A. Gourrier, Y. Kim-Y, U. Schilde, M. Sztucki, M. Burghammer, S. Maltsev, C. Jäger, H. Cölfen, 2012Structure-preoperty relationships of a biological mesocrystal in the adult sea urchin spine. PNAS, 10936993704
119. SPring-82012SPring-8 upgrade plan preliminary report. http://www.spring8.or.jp/en/about_us/whats_sp8/spring-8_II/.
120. L. Sun, W. Chen-J, J. Guo, P. Gao, Q. Huang-Z, H. Wang, M. Fang, X. Chen, G. Chen, C. Zhang, D. Gu, X. Dong, L. Wang, K. Yang, A. Li, X. Dai, H. Mao-K, Z. Zhao, 2012Re-emerging superconductivity at 48 kelvin in iron chalcogenidesNature4836769
121. H. Suzuki, S. Naher, T. Shimogushi, M. Mizuno, A. Ryu, H. Fujishita, 2002X-ray diffraction measurement below 1 KJ. of Low Temperature Physics, 12817
122. L. Szilárd, 1935Absorption of residual neutrons. Nature, 136950951
123. M. Takeda, M. Nakamura, K. Kakurai, E. Lelièvre-Berna, F. Tasset, L. Renault-P, 2005Cryopad on the triple-axis spectrometer TAS-1 at JAERIPhysica B356136
124. F. Tasset, 1989Zero-field neutron polarimetryPhysica B156-157627630
125. F. Tasset, P. J. Brown, E. Lelièvre-Berna, T. Roberts, S. Pujol, J. Allibon, Lami. E. Bourgeat, 1999Spherical neutron polarimetry with Cryopad-II. Physica B,267-2686974
126. C. A. Tulk, D. D. Klug, Santos. A. M. dos, G. Karotis, M. Guthrie, J. J. Molaison, N. Pradhan, 2012Cage occupancies in the high pressure structure H methane hydrate: a neutron diffraction study.J. Chem. Phys., 136054502
127. W. Utsumi, K. I. Funakoshi, Y. Katayama, M. Yamakata, T. Okada, O. Shimomura, 2002High-pressure science with a multi-anvil apparatus at Spring-8J. Phys. Condens. Matter, 141049710504
128. Parker. M. Van Kan, C. Sanloup, N. Sator, B. Guillot, A. J. Tranche, J. Perrillat-P, M. Mezouar, N. Rai, W. Westrenen, 2012Neutral buoyancy of titanium-rich melts in the deep lunar interiorNature Geoscience5186189DOI:ngeo1402http://www.nature.com/ngeo/journal/vaop/ncurrent/full/ngeo1402.html
129. C. Vettier, A. Wright, 1984Workshop "sample environments in neutron and X-ray experiments"Revue de Physique Appliquée, 19643836
130. G. S. Walker, D. M. Grant, T. C. Price, X. Yu, V. Legrand, 2009High capacity multicomponent hydrogen storage materials: investigation of the effect of stoichiometry and decomposition conditions on the cycling behaviour of LiBH4-MgH2. Journal of Power Sources19411281134
131. C. E. Weir, E. R. Lippincott, A. Van Valkenburg, E. N. Bunting, 1959Infrared studies in the 1- to 15-micron region to 30,000 atmospheresJ. Res. Nat. Bur. Stand. A, 635562
132. M. A. White, M. P. Pressprich, P. Coppens, 1994Apparatus for the measurement of the electronic excited-state structure of single crystals using X-ray diffractionJ. Appl. Cryst., 27727732
133. B. T. Wimberly, D. E. Brodersen, W. M. Clemons, R. J. Morgan-Warren, A. P. Carter, C. Vonrhein, T. Hartsch, V. Ramakrishnan, 2000Structure of the 30S ribosomal subunit.Nature407327339
134. J. Xu, H. K. Mao, R. J. Hemley, E. Hines, 2002The moissanite anvil cell: a new tool for high-pressure researchJ. Phys.: Condens. Matter, 141154311548
135. V. V. Zhurov, E. A. Zhurova, A. A. Pinkerton, 2008Optimization and evaluation of the data quality for charge density studies. J. Appl. Cryst., 41340349

CHAPTER 9

Energy and Wavelength Scaling of Shock-Ignited Inertial Fusion Targets

S Atzeni[1,3], A Marocchino[1], A Schiavi[1] and G Schurtz[2]

[1] *Dipartimento SBAI, Universita di Roma 'La Sapienza' and CNISM, ` Via A Scarpa, 14–16, I-00161 Roma, Italy*

[2] *Centre Lasers Intenses et Applications, Universite Bordeaux 1-CNRS-CEA, ´ 351 Cours de la Liberation, Talence F-33405, France*

[3] *Dipartimento SBAI, Universita di Roma 'La Sapienza' and CNISM, ` Via A Scarpa, 14–16, I-00161 Roma, Italy*

ABSTRACT

In inertial fusion shock ignition, separation of the stages of fuel compression and hot spot creation introduces some degree of design flexibility. A lower implosion velocity can be compensated for by a more intense ignition pulse. Flexibility increases with target (and driver) size and allows for a compromise between energy gain and risk reduction. Having designed a reference ignition target, we have developed an analytical model for (up)-scaling targets as a function of laser energy, while keeping under control parameters related to hydro- and plasma instabilities. Detailed one-dimensional simulations confirm the model and generate gain curves. Options for increasing target robustness are also discussed. The previous results apply to UV laser light (with wavelength $\lambda = 0.35$ μm). We also show that our scaling model can be used in the design of targets driven by green laser light ($\lambda = 0.53$ μm).

Introduction and background. This paper concerns shock-ignition laser-driven inertial confinement fusion (ICF). In conventional laser fusion powerful laser beams implode to high velocity a hollow spherical target, containing a mixture of deuterium and tritium fusion fuel. At the end of the implosion the fuel is compressed to high density and a central hot spot is generated where nuclear fusion reactions initiate. In shock-ignition, the stages of compression and ignition are separated. The fuel is first imploded to lower velocity and then the hot spot is created by a strong converging shock driven by an intense laser 'spike'. The lower implosion velocity reduces risks associated with Rayleigh–Taylor instabilities (RTIs), which are a major obstacle to the achievement of thermonuclear ignition. In addition, shock ignition has potential for higher energy gain than conventional ICF. On the other hand, spike intensity has to be limited to avoid laser-plasma instabilities (LPIs).

Main results. We have developed, and tested versus simulations, a model to scale a reference target to different size (and laser wavelength). We find that the possibility of adjusting two parameters, namely implosion velocity and laser spike intensity, introduces significant design flexibility. Such flexibility increases with target (and laser) size (see figure).

Wider implications. The model indicates a possible path for the development of shock ignition. Smaller targets can be used to demonstrate the feasibility of the scheme and to evaluate the major risks. Such targets can then be scaled at larger size, to achieve higher energy gain and, simultaneously, reduce risks (see figure), i.e. achieve the reliability required for energy production applications.

1. INTRODUCTION

Shock ignition [1] is a recently proposed approach to laser-driven inertial confinement fusion (ICF), in which an intense laser spike drives a converging shock wave which helps the formation of a central hot spot in an already compressed fuel. Hot spot ignition and substantial fuel compression are indeed essential ingredients of any inertial fusion scheme [2, 3] aiming at the high energy gain (ratio of released fusion energy to driver energy) required for commercial energy production. In the standard laser fusion scheme [2, 3], both fuel compression and creation of the ignition hot spot are achieved by irradiating the fuel-containing capsule with a properly time-shaped laser pulse, driving a high-velocity implosion. Velocities exceeding 300–400 km s^{-1} are required (the actual value depending on the fuel mass and the entropy of the compressed fuel; see also section 3).

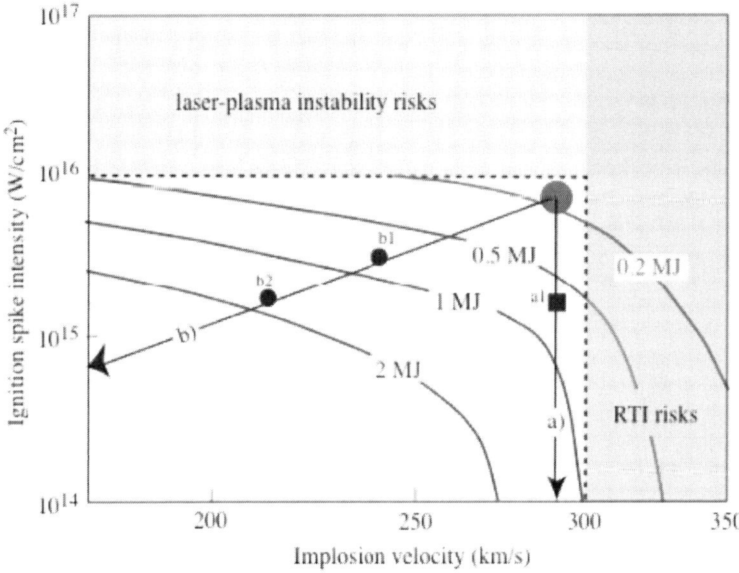

Figure. Scaling a reference shock ignition target (red circle) to larger size (and laser energy). Ignition spike intensity versus implosion velocity for different values of the laser compression incident energy. The labelled filled points represent cases discussed in the paper. Laser spike intensity should be limited to reduce LPIs, while the implosion velocity should be limited to reduce RTI. The HiPER baseline target working point is indicated by the red circle. By increasing the target size (and then compression energy) risks can be reduced. The arrows in the figure indicate two scaling strategies considered in the paper: (a) scaling at constant implosion velocity; (b) scaling at fixed value of the ratio of the implosion velocity to the self-ignition energy.

However, the higher the implosion velocity, the higher the risks associated with the development of the hydrodynamic Rayleigh–Taylor instability (RTI), threatening the integrity of the imploding shell or causing deleterious material mixing. In addition, higher velocity also implies higher driving pressure and hence higher laser intensity, which can result both in degradation of the absorption of laser light and in fuel preheating (making compression less efficient). The detrimental effects of laser plasma instabilities (LPIs), mainly backscattering and generation of hot electrons, may grow with intensity. (In fact, a figure of merit often used to estimate risks related to LPIs is the irradiance $I\lambda^2$, with I the intensity and λ the laser wavelength, which enters the gain coefficient of linear LPIs; see e.g. [4].)

Fuel preheat may result from hot electrons, hard x-rays and mistimed shocks, all more likely at higher intensities.

ICF schemes requiring lower implosion velocity are therefore highly desirable [1, 5]. The required compression can still be achieved [6], provided the average entropy of the imploding fuel is kept low. Generation of the igniting hot spot, instead, requires a separate mechanism. In shock ignition [1], a hollow spherical target is first imploded at lower velocity by a standard time-shaped laser pulse, with intensity in the range (0.3–1) $\times 10^{15}$ W cm^{-2}. Towards the end of the implosion, the target is irradiated by a more powerful, shorter, laser pulse (a *spike*, see figure 2), with intensity up to ten times larger than that of the compression pulse. (The quoted intensities refer to a laser wavelength of 0.35 μm, the third harmonic of Nd:glass laser.) Absorption of this spike about the critical density generates an ablation pressure of 200–300 Mbar, driving a strong converging shock wave. Collision of this inward-directed shock, intensified by convergence, with the shock bouncing from the centre of the imploded fuel results in compression and heating of the hot spot.

Shock ignition was proposed by Betti *et al* [1] a few years ago; a first proof-of-principle experiment was performed soon afterwards [7]. Subsequently, several groups designed different targets, studied aspects of their robustness and investigated the physics of shock pressure multiplication [8–15]. Most studies referred to a laser wavelength of 0.35 μm, but also wavelengths of 0.25 and 0.53 μm were considered in [11] and [16], respectively. Gain curves (gain versus driver energy) were also generated by considering families of scaled targets [8, 11, 12]. Different authors scaled targets in somewhat different ways and with different assumptions. We consider a more systematic analysis of target scaling, enlightening the opportunities offered by shock ignition.

In this paper, we derive general relations for scaling targets with size (hence with energy) and laser wavelength and generate gain curves based on different scaling options, and also for different laser wavelengths. We reckon, and show with a few examples, that the above-quoted flexibility can be exploited to increase target robustness to asymmetries and instabilities. The scaling procedure is applied to the so-called HiPER baseline target [17], which was studied in detail in earlier papers on shock ignition [9, 13]. For completeness, the main target and pulse parameters are briefly recalled in section 2.

The main issue for the feasibility of shock ignition concerns the efficient generation of the high-pressure shock, which occurs in a laser–plasma interaction regime not yet studied experimentally. At the relevant intensity (more precisely, the values of the product $I\lambda^2$ of the intensity and wavelength squared), LPIs might degrade absorption [4]. Also, the generation of a significant amount of hot electrons is expected. We do not study these problems in this paper. However, we study the scaling of I and $I\lambda^2$, and show that in any considered option they decrease with target size.

A comment is probably in place concerning reliability of studies based on analytical models and one-dimensional (1D) numerical simulations, in light of the recent results [18, 19] of the National Ignition Campaign (NIC) [20]. NIC experiments so far have shown implosion velocity smaller than predicted (at a given laser power) [18]. Also, inferred hot spot pressures have been found to be smaller than expected (at a given implosion velocity) [19]. The results on the smaller implosion velocity are not necessarily relevant to the direct-drive implosions discussed here, since laser absorption and coupling mechanisms are different in indirect- (where x-ray conversion, x-ray transport and x-ray coupling to the capsule play a key role) and direct-drive laser fusion. The results on the hot spot pressure, instead, should not depend on the specific drive mechanism. According to the Administrator of the US National Nuclear Security Administration [19] (page 9) 'the reason for the (pressure) deficit is not clear but is due, most likely, to low-mode fuel asymmetries and to the hot spot experiencing more mixing than expected'. Experiments and improved analyses are required to solve these issues. One may wonder whether and how these effects might alter the conclusions of the present study. If all effects are due to asymmetries and mixing, the model developed here will be unaffected, being 1D, and being designed to scale a given igniting target to larger size. However, the reduction of asymmetries and mixing may turn out to be more difficult than previously assumed. This means that ignition of the reference target to be scaled may require either somewhat higher implosion velocity and/or a more powerful spike than computed here for the baseline target. Provision for these effects can be included in our model by increasing the value of the safety factor S_m, introduced in section 2.2.

2. HIPER BASELINE TARGET

The scaling relations studied in this paper can, in principle, be applied to any target and laser pulse. In this paper, we apply them to design targets up-scaled from the so-called HiPER baseline target.

2.1. Baseline case

This target, a simple cryogenic deuterium-tritium (DT) shell (see figure 1), was originally proposed for fast ignition [17] and later shown to be suitable for shock ignition [9, 13].

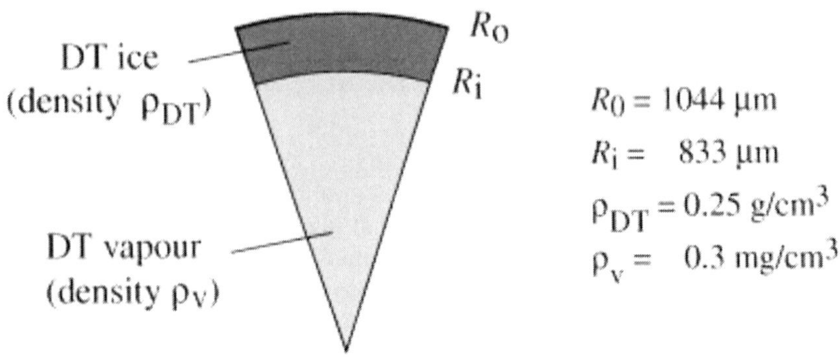

FIGURE 1. HiPER baseline target. All targets considered in this paper are scaled from this target as discussed in the main text.

The target was designed to satisfy the following constraints:

- compression intensity at half radius below 5×10^{14} W cm^{-2} (with laser wavelength $\lambda = 0.35$ μm), to control laser-plasma instabilities;
- in-flight-aspect-ratio (at 2/3 of the radius, $\mathrm{IFAR}_{2/3} < 35$ and growth factor of the fastest mode of the ablation front RTI smaller than seven, to make RTI growth tolerable;
- in-flight isentrope parameter $\alpha_{\mathrm{if}} \leq 1.2$ to reduce compression work. (The isentrope parameter is the ratio of the actual material pressure to the pressure of a cold material, with Fermi degenerate electrons, at the same mass density.)

Target design was based on 1D simulations performed with the DUED code [21, 22], summarily described in appendix A. Target structure and laser power temporal shape are shown in figures 1and 2, respectively. Also listed are the numerical values of the main quantities defining the baseline laser pulse used in the present study. Timing and power levels depend (weakly) on the material equation-of-state and on the flux limiter coefficient *f*. The laser pulse parameters refer to simulations with flux-limited thermal conduction, with flux limitation factor $f = 0.07$ (sharp cut-off model). The dependence of simulation results on the flux-limiter *f* is discussed in appendix A.

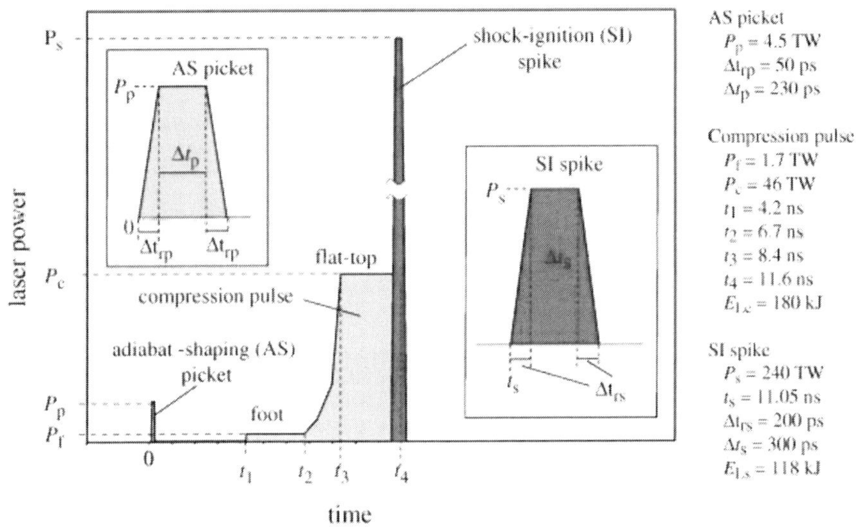

FIGURE 2. Temporal shape of the laser power for shock ignition of the targets considered in this paper. The parameters listed on the right-hand side refer to the HiPER baseline target shown in figure 1. The baseline laser light wavelength is 0.351 μm. The final spike adds to the compression spike.

In the baseline design, the target is irradiated by 48 beams [13, 23, 24], generating a nearly perfectly symmetric irradiation pattern. Each compression beam has a Gaussian profile $I \propto \exp -(r/w)^2$ with width $w = w_c = 640$ μm for the compression pulse and $w = w_s = 400$ μm for the final ignition spike.

Without the final spike, the shell implodes at velocity $u_{imp} = 291$ km s^{-1} and achieves a confinement parameter $\langle \rho R \rangle = 1.58$ g cm^{-2}, but does not ignite. Ignition and gain $G = 50$–70 are achieved with the addition of a spike with power $P_s \geq 100$ TW. In the reference case, $P_s = 240$ TW and $G = 65$. Other performance parameters will be listed later in table 3, column H.

2.2. Performance versus implosion velocity and spike power

In shock ignition, the implosion velocity is below the ignition threshold. The igniting hot spot is created with the help of the additional powerful spike. The smaller the implosion velocity, the higher the laser power required for ignition. This is clearly shown by figure 3, showing gain contours in the plane of implosion velocity and additional spike power for the HiPER baseline target. The thick red curve marks the spike power threshold for gain $G > 50$. The figure is the result of parametric numerical simulations. The implosion velocity has been varied by varying the compression power. The duration of the compression plateau and the launch time of the spike have been optimized to maximize the gain. The spike duration has been kept constant. In all cases, $\alpha_{if} = 1.2$.

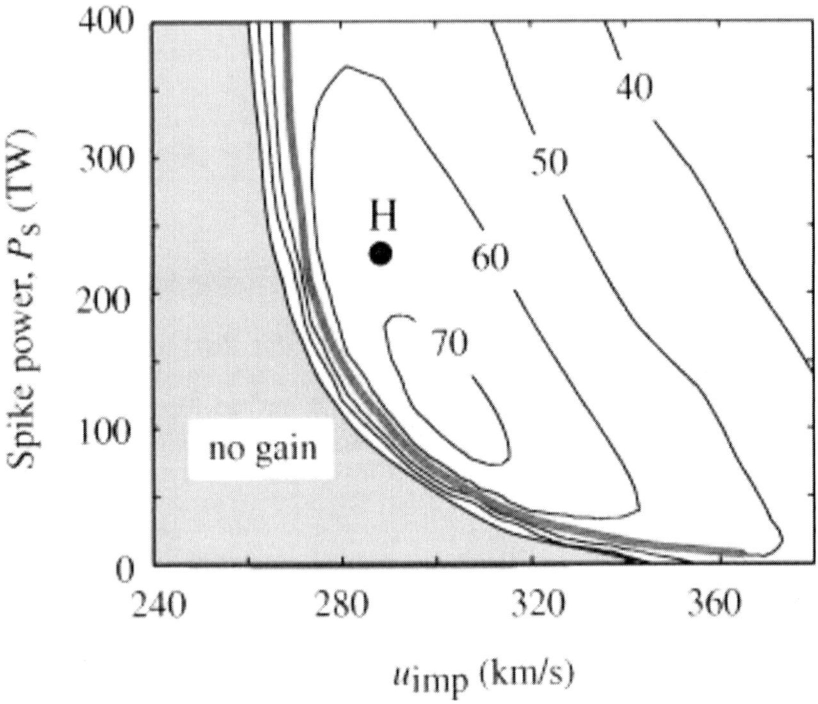

FIGURE 3. Gain contours in the plane of implosion velocity and spike power for the HiPER baseline target, obtained from 1D numerical simulations. The red curve indicates the spike power required to achieve high gain ($G = 50$) versus implosion velocity; this line is very close to the gain threshold. The point labelled H indicates the HiPER baseline operating point, chosen with a rather large safety margin.

Analysis of two cuts of the gain contours of figure 3 shows interesting features. Figure 4(a) shows gain versus implosion velocity for $P_s = 0$, i.e. conventional implosions, without a spike. We see that ignition occurs for $u_{imp} > u_{ig*} = 332$ km s^{-1}, while peak gain is achieved at $u_{imp} = u_{pg} \simeq$ 370 km s^{-1}. For $u_{imp} < u_{ig*}$ ignition (or for $u_{ig*} < u_{imp} < u_{pg}$ efficient burn) requires the spike. The HiPER target operates at $u_{imp} = 291$ km s^{-1}. As shown by figure 4(b), a spike of 100 TW is already sufficient for large gain.

However, we have chosen a baseline spike of 240 TW so as to have some safety margin. As a measure of this margin, we take the ratio $S_m = p_h/p_{ig}$ of the pressure of the hot spot (computed in a run with alpha-particle heating switched off) to the minimum pressure required for substantial gain ($G = 50$). For the baseline case, $S_m \simeq 1.25$. The decrease of the gain as spike power exceeds some optimal value (clearly shown by both figure 3 and 4(b)) is easily explained. At some point excess shock pulse energy is not compensated by an increase of the fusion yield (but it increases ignition safety).

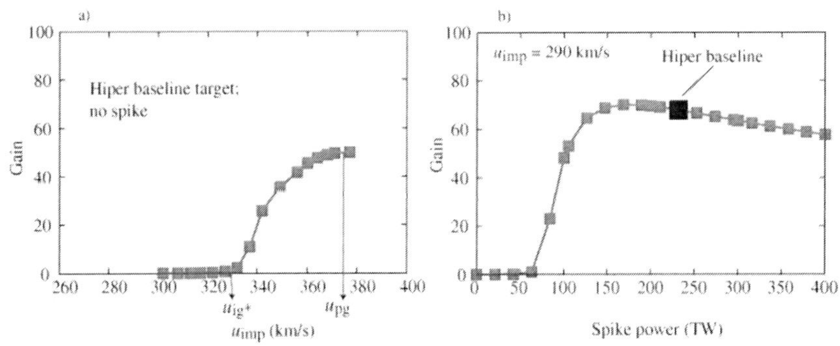

FIGURE 4. For the HiPER baseline target, (a) energy gain versus implosion velocity for a conventional implosion, without the final spike; (b) energy gain versus laser spike power, at implosion velocity u_{imp} = 290 km s^{-1}.

3. IGNITION REQUIREMENTS

3.1. Implosion velocity for self-ignition

The previous discussion highlights the crucial role played by implosion velocity in both conventional central ignition and shock ignition. For conventional central ignition, Herrmann *et al* [25] studied the relation between shell energy for ignition and implosion velocity u_{imp}. They performed systematic parametric simulations and found that the minimum shell kinetic energy for ignition is fitted by the expression

$$E_{k-ig} \propto u_{imp}^{-5.89\pm0.12} \alpha_{if}^{1.88\pm0.05} p_a^{-0.77\pm0.12}, \tag{1}$$

where p_a is the pressure driving the implosion of the shell. Assuming (to a good approximation) that $p_a \propto u^2_{imp}$ and writing the kinetic energy as $E_k = (1/2)m_{imp}u^2_{imp}$, we find that the minimum implosion velocity u_{ig*} for ignition of a mass m_{imp} of fuel scales as

$$u_{ig*} \propto m_{imp}^{-0.106} \alpha_{if}^{0.2}, \tag{2}$$

where we have omitted the error bars in the exponents. Herrmann *et al* [25] also found that the kinetic energy required to achieve peak gain with a given fuel mass is instead fitted by

$$E_{k-pg} \propto u_{imp}^{-3.3\pm0.4} \alpha_{if}^{1.60\pm0.2} p_a^{-0.9\pm0.1}, \tag{3}$$

from which we obtain the implosion velocity for peak gain as

$$u_{pg} \propto m_{imp}^{-0.141} \alpha_{if}^{0.225}. \tag{4}$$

As shown in figure 5, the above dependences of the ignition and peak gain velocities on the mass are nicely reproduced by a series of simulations of ours referring to targets scaled from the HiPER baseline targets. All targets have the same payload aspect ratio $A_p \simeq 10$ and isentrope α_{if} in the range 1.2–2.3. (Here the payload is the imploded mass m_{imp}, initially contained in a layer of thickness ΔR_p, and we define the payload aspect ratio as the ratio $A_p = (R_i + \Delta R_p)/\Delta R_p$ of its initial outer radius to its initial thickness.) The imploded mass, instead, varies over the large range 0.28–5.1 mg. Figure 5clearly shows that velocities required for threshold ignition and peak gain get closer to each other as target size increases. This is further illustrated by figure 6, showing target gain as a function of implosion velocity for targets with imploded masses of 0.28 and 1.54 mg, respectively.

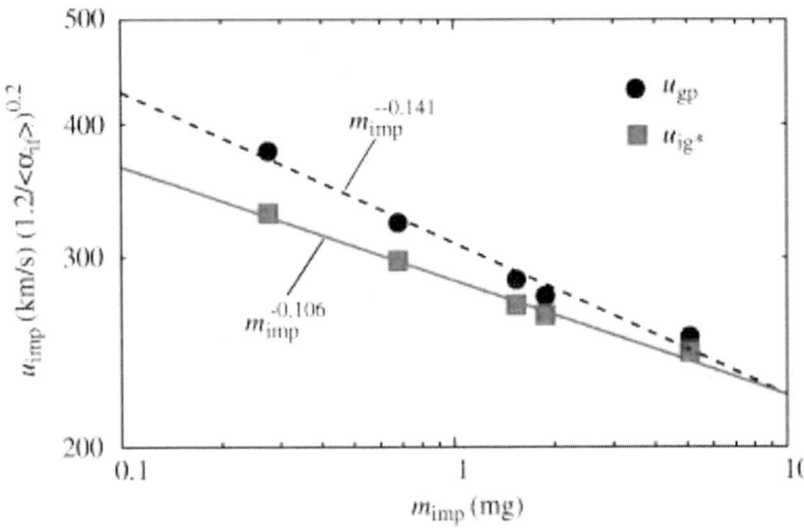

FIGURE 5. Implosion velocity for ignition versus imploded fuel mass. Points: simulations; lines: scaling laws.

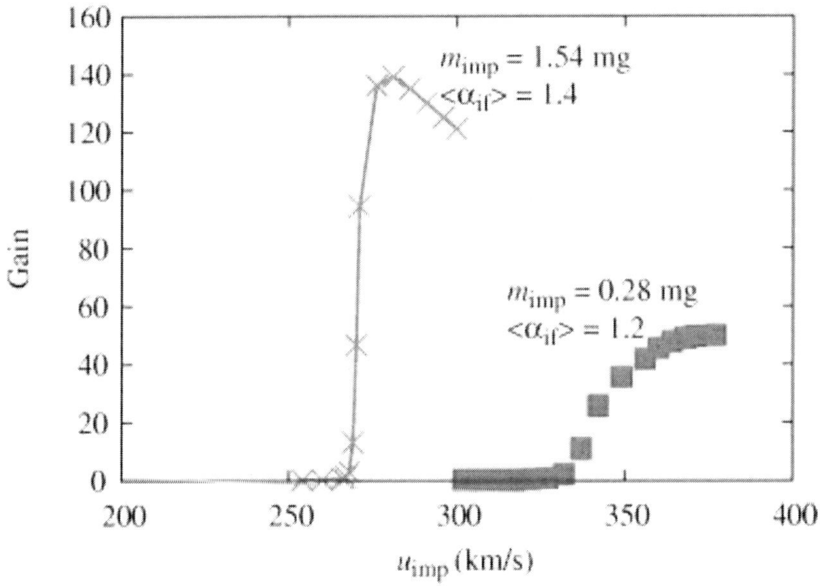

FIGURE 6. Gain versus implosion velocity for two targets with different mass, driven by a conventional laser pulse (no spike). Velocity for peak gain and threshold ignition velocity become closer as the mass increases.

3.2. Hot spot pressure for ignition

Both in conventional ICF and in shock ignition, ignition occurs in a central hot spot, surrounded by denser fuel, and approximately at rest. In both cases, ignition occurs when the hot spot parameters satisfy a certain ignition condition, which is essentially a condition on the hot spot pressure. For simplicity, we assume that the compressed stagnating fuel is a DT spherical assembly, initially at rest, consisting of a spherical hot spot with radius R_h, temperature T_h and density ρ_h, surrounded by cold fuel at density ρ_c. The relevant, approximate ignition criterion is [2, 26]

$$\rho_h R_h T_h > 5\sqrt{\rho_h/\rho_c}\ \text{g keV cm}^{-2}, \tag{5}$$

which, considering the hot spot as an ideal gas, can also be written as

$$p_h > p_{ig} = 470\left(\frac{R_0}{1\ \text{mm}}\right)^{-1}\frac{C_r}{30}\sqrt{\frac{6}{\rho_c/\rho_h}}\ \text{Gbar}, \tag{6}$$

where R_0 is the initial shell outer radius and $C_r = R_0/R_h$ is the hot spot convergence ratio.

3.3. Hot spot stagnation pressure

In conventional inertial fusion, the hot spot is generated at the centre of the fuel at the end of the implosion, when the kinetic energy of the fuel shell is converted into internal energy. The hot spot pressure at stagnation is a strong function of the implosion velocity u_{ig}. According to a self-similar model [27, 28],

$$p_{\mathrm{stagn}} \propto u_{\mathrm{imp}}^{3} \alpha_{\mathrm{if}}^{-9/10} p_{\mathrm{a}}^{-2/5}, \tag{7}$$

where pressure p_{a} is the pressure applied to the shell surface and α_{if} is the in-flight isentrope parameter of the shell. Fuel self-ignition (conventional ignition) occurs when the above stagnation pressure exceeds the ignition threshold illustrated in the previous subsection. In shock ignition, where the hot spot stagnation pressure is insufficient for ignition, the pressure deficit is compensated for by the strong shock driven by the laser spike, as previously discussed in section 2.2.

It is interesting to show that using the simple ignition condition (6) and the self-similar result for stagnation pressure, we find an expression for the scaling of the (self-)ignition implosion velocity with fuel mass and adiabat analogous to equation (2). Assuming $p_{\mathrm{a}} \propto u^2$ and, as appropriate for a thin shell, $m_{\mathrm{imp}} \approx 4\pi R^3{}_0 A_{\mathrm{p}}{}^{-1}$, we obtain

$$u_{\mathrm{ig}} \propto m_{\mathrm{imp}}^{-0.09} \alpha^{0.24} \left[C_{\mathrm{r}}^{0.26} \left(\frac{\rho_{\mathrm{h}}}{\rho_{\mathrm{c}}} \right)^{0.13} A_{\mathrm{p}}^{-0.09} \right]. \tag{8}$$

4. TARGET SCALING

In this section, we describe the scaling of a given target (and related laser pulse) to different size and laser wavelength. Before deriving the scaling model, we discuss general aspects of scaling strategies for shock-ignited targets.

4.1. Scaling strategies and risk reduction

As we have seen in section 2.2, two parameters, namely the implosion velocity and the spike pulse power, can be adjusted to achieve ignition. This allows some flexibility in target scaling. Such a flexibility increases with target size, and, in principle, allows for significant risk reductions, as illustrated in figure 7 for targets scaled from the HiPER baseline target. The figure shows curves of the spike intensity required to ignite a shell versus the implosion velocity for different values of the energy E_{Lc} of the compression laser pulse. In order to reduce risks associated with laser–plasma instabilities and with RTI, spike intensity and implosion velocity should be kept below some threshold values, e.g. 10^{16} W cm^{-2} and 300 km s^{-1}, respectively, which constrains the *safe* operating space. We see that at low E_{Lc}, e.g. 200 kJ, both the intensity and the velocity are not far from the instability boundaries. However, at larger energies, the window enlarges considerably, and different scaling options are available. For instance, targets can be scaled at constant implosion velocity, option (a) in the figure. In this case the required spike intensity decreases with energy and LPI risks are greatly reduced. Another option, (b) in the figure, consists in scaling targets by keeping constant the ratio u_{imp}/u_{ig^*} of the implosion velocity to the self-ignition velocity. In this case both the implosion velocity and the spike intensity decrease with increasing target size, leading to reduced risks for both LPIs and hydrodynamic instabilities. The curves shown in figure 7 refer to targets imploded at the same α_{if} = 1.2. The curves are estimated, while the filled points have been obtained by the detailed 1D simulations illustrated in section 5.

4.2. Characteristic scales: dimensional scaling

In this paper, we refer to the simple target shown in figure 1. We scale the target inner radius by a factor s and keep fixed the densities ρ_{DT} of the cryogenic fuel and ρ_v of the DT vapour, as well as the payload aspect ratio A_p. Therefore, the imploding mass m_{imp}, equal to the mass of the effective fuel, scales as s^3. In addition to the length scale s, we introduce the dimensionless implosion velocity w, time τ, compression laser energy ε, compression laser power Π, laser wavelength Λ and laser irradiance β. This last quantity is the dimensionless value of the product $I\lambda^2$, and can be written as $\beta = (\Pi/s^2)\Lambda^2$. Of course, $\tau = s/w$ and $\Pi = \varepsilon/\tau$ (see table 1). For the reference target, e.g. the HiPER H target, $s = w = \tau = \varepsilon = \Pi = \beta = 1$. The imploding mass is a fraction y of the total mass m_t.

4.2.1. Compression pulse

On dimensional grounds compression pulse energy E, intensity I, target radius R and pulse time t are related by $E \approx IR^2t$. On the other hand, the kinetic energy of the imploding fuel is a fraction η_h of the absorbed laser energy $\eta_{ac}E$:

$$E\eta_{ac}\eta_h = \frac{1}{2}m_{imp}u_{imp}^2, \tag{9}$$

where η_{ac} is the absorption efficiency of the compression laser pulse. Our simulations (see e.g. table 3) show that the absorption efficiency η_{ac} is nearly independent of target size and laser wavelength for the considered targets. We therefore neglect η_{ac} in the scaling procedure.

For the hydrodynamic efficiency η_h, we use

$$\eta_h \propto u_{imp}^{3/4}(I_a\lambda^2)^{-1/4} \quad (10)$$

which was proposed by Betti and Zhou [6] as a fit to detailed numerical simulations. Here I_a is the absorbed light intensity and λ the laser wavelength. In fact, our simulations are better reproduced

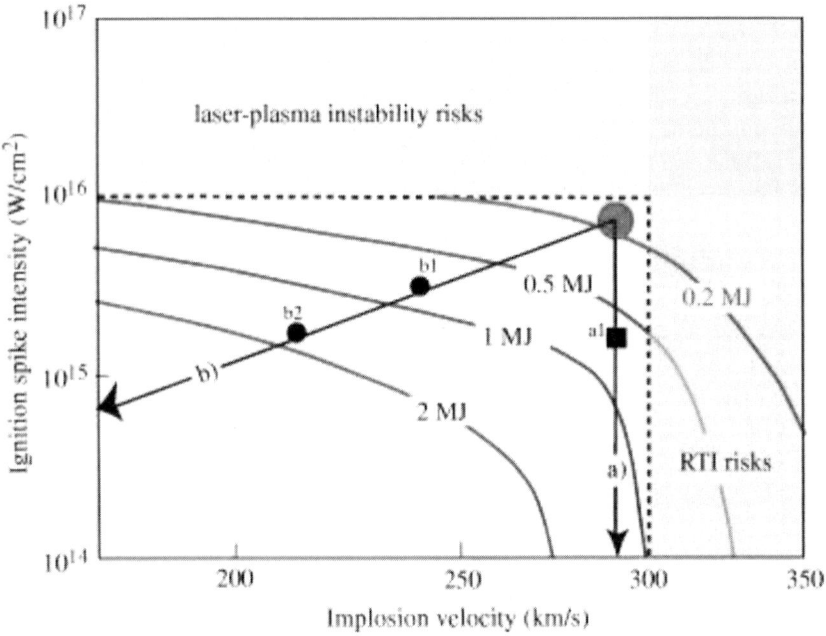

FIGURE 7. Ignition spike intensity versus implosion velocity for different values of the laser compression incident energy. The labelled filled points represent cases discussed in section 5. Laser spike intensity should be limited to reduce laser–plasma instabilities (LPI), while the implosion velocity should be limited to reduce hydrodynamic instabilities (in particular, RTI). The HiPER baseline target operating point (u_{imp} = 291 km s^{-1}; E_{Lc} = 180 kJ) is indicated by the red circle. On increasing the target size (and then compression energy), risks can be reduced. The arrows in the figure indicate two scaling strategies considered in this paper: (a) scaling at constant implosion velocity; (b) scaling at a fixed value of the ratio of the implosion velocity to the self-ignition energy.

Table 1. Dimensionless variables characterizing target and laser compression pulse.

Quantity	Scaling variable	Our model
Laser wavelength	Λ	
Length	s	
Implosion velocity	w	$w = \beta^{1/3}\Lambda^{-8/9}$
Time	$\tau = s/w$	
Compression laser energy	ε	$\varepsilon = s^3 w^{5/4}\beta^{1/4}$
Compression laser power	$\Pi = \varepsilon/\tau$	
Compression laser irradiance ($I\lambda^2$)	$\beta = \Pi\Lambda^2/s^2$	
Payload mass	s^3	
Ablator mass		$\beta^{0.27}\Lambda^{-0.71}s^{2.78}$

by $\eta_h \propto u_{imp}^{3/4}(I_a\lambda^2)^{-0.2}R^{-0.08}$. However, as we show in appendix B, the scaling relations are nearly insensitive to the differences between these two expressions for η_h.

Introducing the above dimensionless variables and using equation (10) for the hydrodynamic efficiency, we obtain

$$w = \beta^{1/3}\Lambda^{-8/9} \tag{11}$$

and the expression of ε reported in the right-hand-side column of table 1. According to the model, all power levels of the compression laser pulse (e.g. P_p, P_f, P_c; see figure 2) are scaled by a factor Π, and all times by τ; simulations show that minor adjustments are required to picket duration and foot onset only.

4.2.2. Ablator mass

Since the fuel aspect ratio is kept fixed, the imploded mass scales as s^3; a different law instead applies to the ablator mass, which can be estimated as

$$m_a \simeq 4\pi\dot{\mu}R^2 t_p, \tag{12}$$

with $\dot{\mu}$ the mass ablation rate (ablated mass per unit time and unit surface) and t_p a characteristic pulse duration, proportional to τ. According to simulations of ours, referring to targets analogous to those considered in this paper, the mass ablation rate can be written as

$$\mu \propto I_a^{0.6} \lambda^{-0.4} R^{-0.22} \propto (I_a \lambda^2)^{0.6} \lambda^{-1.6} R^{-0.22}, \tag{13}$$

where R is the radius of the spherical target. Note that this agrees with the expressions obtained by early theoretical models [29, 30]. Inserting equation (13) into (12), we obtain

$$m_a \propto \beta^{0.6} \Lambda^{-1.6} s^{-0.22} s^2 \tau = \beta^{0.27} \Lambda^{-0.71} s^{2.78}, \tag{14}$$

where the relation $\tau = s/w = s\beta^{-1/3}\Lambda^{8/9}$ has been used. The mass of the scaled target will then be obtained from that of the HiPER baseline target according to

$$m_t = m_{t0} s^3 \left[y_0 + (1 - y_0) \beta^{0.27} \Lambda^{-0.71} s^{-0.22} \right], \tag{15}$$

where $m_{t0} = 0.576$ mg and $y_0 = 0.48$ are, respectively, the total mass and the payload fraction of the target with $s = \beta = \Lambda = 1$.

4.3. Shock ignition spike

Accurate determination of the power of the shock ignition laser spike relies on numerical simulations. However, approximate scaling expressions can be obtained in the two important cases discussed in section 4.1, namely, (a) targets scaled at constant implosion velocity and (b) targets scaled at constant ratio u_{imp}/u_{ig*}.

4.3.1. Spike power. Targets scaled at constant implosion velocity

We first consider targets scaled in size, but imploding at the same velocity and driven by light with the same wavelength. We proceed as follows [31].

(i) We require hot spot pressure $p_h = S_m p_{ig}$ or, for equation (6), $p_h \propto S_m C_r / R_0$, where we have neglected the dependence on the density ratio.

(ii) We assume that, for scaled targets, the final pressure p_h is proportional to the product of the ablation pressure and the convergence ratio [12]: $p_h \propto C_r p_a$, and then using the result of item (i) above we also have $p_a \propto S_m / R_0$.

(iii) The ablation pressure, in turn, is related to the absorbed intensity, and then to the absorbed laser power P_a by $p_a \propto I^{2/3}{}_a \propto (P_a R^{-2} a)^{2/3} \propto (P_a R^{-2}{}_0)^{2/3}$. From this last equation and the previous expression for p_h, we obtain $P_a \propto S^{3/2}{}_m R^{1/2}{}_0$ and $P_a = k S^{3/2}{}_m s^{1/2}$, where k is a constant to be determined as shown below.

(iv) The absorbed laser power is related to the incident power of the spike pulse and compression pulse by $P_a = P_s \eta_{as} + P_c \eta_{ac}$, where η_{as} and η_{ac} are the absorption efficiencies for the spike and compression pulses, respectively. At constant implosion velocity, $P_c \propto R^2$ (see table 1), or $P_c = P_{c0} s^2$, where $P_{c0} = P_c(s = 1)$ is the power of the laser compression pulse at scale $s = 1$, and then

$$P_a = P_s\eta_{as} + P_{c0}\eta_{ac}s^2 = kS_m^{3/2}s^{1/2}.$$

The constant k is then found by imposing $P_s(s = 1, S_m) = P_{s0}$, i.e. the power of the laser spike at scale $s = 1$ and safety margin S_m. The dimensionless spike power $\pi_{spike} = P_s/P_{s0}$ then scales with target size as

$$\pi_{spike} = \left(1 + \frac{P_{c0}\,\eta_{ac}}{P_{s0}\,\eta_{as}}\right)s^{1/2} - \frac{P_{c0}\,\eta_{ac}}{P_{s0}\,\eta_{as}}s^2. \tag{16}$$

In the above derivation, we have assumed absorption efficiencies independent of target scale.

4.3.2. Spike power. Targets scaled at constant ratio u_{imp}/u_{ig^*}

Next, we consider targets scaled in size (and wavelength, too), but imploded at constant ratio u_{imp}/u_{ig^*}, and assume that the power of the compression laser pulse is much smaller than the spike power. We estimate the spike power scaling as follows.

1. Given the constant value of the velocity ratio u_{imp}/u_{ig^*} (and then of the stagnation pressure to the minimum ignition pressure), the shock has to multiply the stagnation pressure by approximately the same factor F_h.

2. We assume, as in the previous subsection, that the shock pressure is proportional to the product of the spike ablation pressure p_{as} and the convergence ratio (constant for scaled targets, as confirmed by the simulations discussed in section 5). This also means that the shock pressure at the ablation front is proportional to the pressure driving the implosion and therefore, for energy conservation, $p_{as} \propto u^2_{imp}$ for targets with the same aspect ratio.

3. The implosion velocity, in turn, scales with target size as $u_{imp} \propto R^{-0.32}$, since $u_{imp} \propto u_{ig^*}$ and, for equation (2), $u_{ig^*} \propto m^{-0.106}_{imp}$.

4. In the relevant high intensity regime, to a good approximation, $p_{as} \propto (I_{as}/\lambda)^{2/3}$, and then $p_{as} \propto (P_{as}/R_{as}^2\lambda)^{2/3}$, where I_{as} and P_{as} are the absorbed laser power and intensity, respectively.

In conclusion, we find $P_{as} \propto R^{1.04}\lambda$ or, in dimensionless form,

$$\pi_{spike} = s^{1.04}\Lambda. \tag{17}$$

To test the above scaling, we have performed parametric simulations of scaled targets irradiated with spikes of different power P_s. We have considered the baseline HiPER target (scale $s = 1$) and two larger targets, with scales $s = 1.75$ and 2.43, respectively. (Details on the scaling are given in the following sections.) Figure 8 shows target energy gain as a function of the spike power for each of the three targets. Points of marginal ignition ($G = 1$), high gain threshold (80% of peak gain) and peak gain are evidenced. Figure 8(b) shows that the spike power for marginal ignition scales approximately proportionally to s, just as predicted by equation (17). The spike power for high gain, instead, scales approximately proportionally to $\sqrt{s}$. Based on these simulation results, we have chosen to scale spike power according to

$$\pi_{\mathrm{spike}} = \sqrt{s}\Lambda. \tag{18}$$

In future studies, we plan to evaluate how different choices of the spike power scaling affect ignition margins.

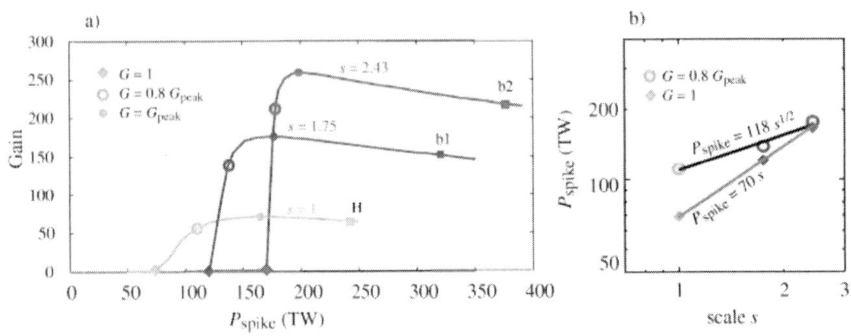

FIGURE 8. (a) Gain versus spike power for three scaled targets, driven at the same value of the ratio $u_{\mathrm{imp}}/u_{\mathrm{ig}^*}$ (1D numerical simulations). The labels H, b1 and b2 refer, respectively, to the HiPER baseline case, and to the scaled targets b1 and b2 discussed in section 5. (b) Spike power for marginal ignition and for high gain versus target linear scale s.

5. SCALED TARGETS—FIXED LASER WAVELENGTH

The general scaling relations derived in the previous section and summarized in table 1 can be used to up- or down-scale a specific target (as discussed in the present section) or design targets driven by laser light of different wavelength (as shown in section 6).

Here we consider the two specific scaling strategies discussed above, namely (a) scaling at constant implosion velocity and (b) scaling at constant ratio $u_{\mathrm{imp}}/u_{\mathrm{ig}^*}$. The expressions for the dimensionless target and laser parameters for the two cases are listed, respectively, in columns (a) and (b) of table 2. We now consider the two cases separately. In both cases, we also perform parametric simulations to verify the scaling laws and compute target gain.

Table 2. Scaling relations—fixed wavelength.

	(a)	(b)
Fixed	u_{imp}; λ	u_{imp}/u_{ig*}; λ
w	1	$s^{-0.32}$
β	1	$s^{-0.96}$
τ	s	$s^{1.32}$
ε	s^3	$s^{2.36}$
Π	s^2	$s^{1.04}$
Payload mass	s^3	s^3
Ablator mass	$s^{2.78}$	$s^{2.52}$
π_{spike}	Equation (16)	$\sqrt{s}$*
ε_{spike}	$\pi_{spike}\tau$	$s^{1.82}$
β_{spike}	$\pi_{spike}s^{-2}$	$s^{-3/2}$

* See the discussion in section 4.3.2.

5.1. (a) Scaling at fixed implosion velocity

Scaling at fixed laser wavelength and implosion velocity is achieved by setting $w = \beta = \Lambda = 1$. The relevant scaling relations for the other dimensionless quantities concerning target size and compression pulse are immediately obtained using the relations in table 1 and are listed in column (a) of table 2. The spike pulse power scales according to equation (16).

The variation of laser power versus target scale is illustrated in figure 9 for targets (up-)scaled from the HiPER baseline target. For the spike power the figure shows curves for $S_m = 1$ and 1.25, i.e. for targets scaled from the baseline target and laser pulses with spikes of 100 and 240 TW, respectively. We see that the spike power decreases with target scale and no spike is necessary for scale $s > s^*(S_m)$. The total power has a shallow minimum at $s = s^*$, while the total laser energy increases monotonically with s.

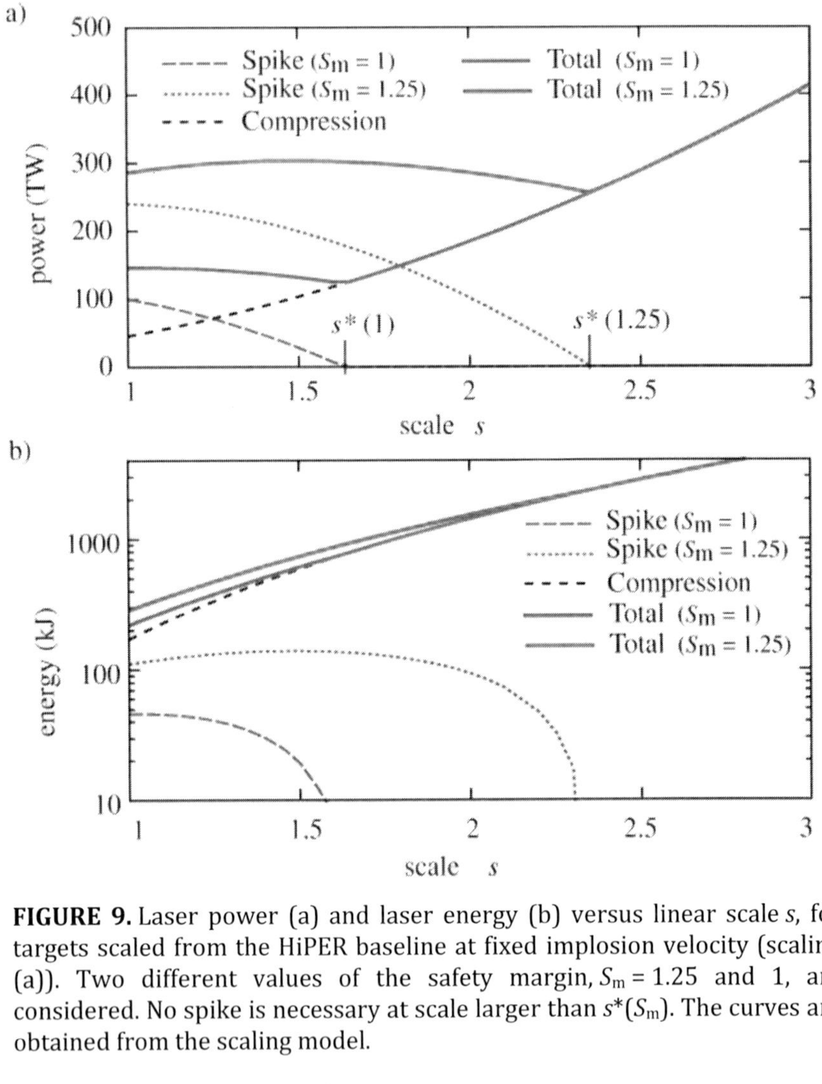

FIGURE 9. Laser power (a) and laser energy (b) versus linear scale s, for targets scaled from the HiPER baseline at fixed implosion velocity (scaling (a)). Two different values of the safety margin, $S_m = 1.25$ and 1, are considered. No spike is necessary at scale larger than $s^*(S_m)$. The curves are obtained from the scaling model.

Detailed 1D simulations verify the accuracy of the scaling laws. Scaled targets driven by scaled pulses attain nearly exactly the same velocity. No adjustment was required for target parameters, pulse power levels, timing of the main pulse and of the ignition spike. Only the duration t_p of the initial picket and the onset of the foot (time t_1) required adjustments (usually obtained with no more than two iterations). Scaled targets have nearly the same in-flight-aspect-ratio; the hot spot convergence ratio decreases slightly with target size.

Figure 10 shows the gain curve (gain versus total laser energy) for targets scaled from the HiPER target H, with margin $S_m = 1.25$ (red squares). The figure also shows the results of simulations of the same targets driven by pulses without the final spike (brown triangles). In agreement with the model, a large gain ($G = 120$) is attained at a laser energy of about 800 kJ, or target scale $s = 1.71$, in agreement with the model for $S_m = 1$. The addition of the spike, however, introduces a safety margin that eventually might compensate for possible material mixing or implosion asymmetries.

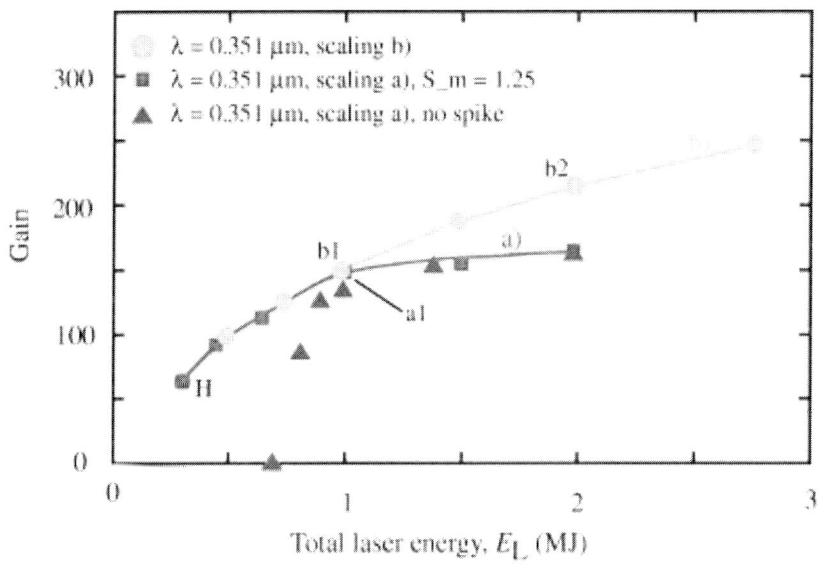

FIGURE 10. Gain curves for shock-ignited targets upscaled from the HiPER target H (from 1D numerical simulations). Red curve and squares: targets scaled at constant implosion velocity (according to scaling (a) of table 2); blue curve and circles: targets scaled at constant u_{imp}/u_{ig*}, (scaling (b) of table 2); brown triangles: targets scaled from H at constant implosion velocity, without the ignition spike. Labels H, a1, b1 and b2 refer to cases with parameters listed in table 3.

A possible further bonus provided by the use of the spike is shown in figure 11. The figure compares the implosion flowcharts for two identical targets (with scale $s = 1.71$), driven by pulses without a spike (a), and with a 164 TW spike (b). The two cases achieve nearly the same energy gain, but in the case with the spike ignition is faster, and the distance ΔR_h *travelled* by the unstable hot spot surface is shorter. In case (a), the hot spot boundary starts decelerating when its radius is 190 μm; it shrinks to a minimum radius of 60 μm and burn propagates when the radius is about 90 μm, and then the total travelled distance is $\Delta R_h = 160$ μm. In case (b), the corresponding radii are 130, 35, and 45 μm, respectively, and $\Delta R_h = 105$ μm. The reduction of ΔR_h is expected to have a beneficial effect on the growth of RTI at stagnation. For a rough estimate, we use linear growth and neglect any stabilizing mechanism. Perturbations with wavenumber k then grow by a factor $\exp(\sqrt{2k\Delta R_h})$ (see e.g. [2], p 242).

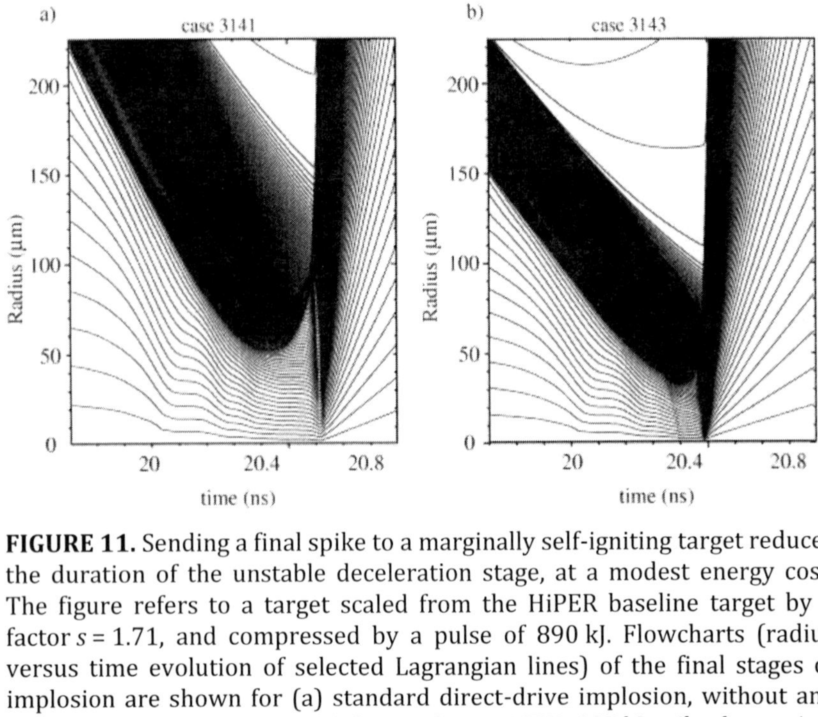

FIGURE 11. Sending a final spike to a marginally self-igniting target reduces the duration of the unstable deceleration stage, at a modest energy cost. The figure refers to a target scaled from the HiPER baseline target by a factor $s = 1.71$, and compressed by a pulse of 890 kJ. Flowcharts (radius versus time evolution of selected Lagrangian lines) of the final stages of implosion are shown for (a) standard direct-drive implosion, without any final spike, (b) the same as (a), but with a 164 TW, 137 kJ spike (case a1 of table 3 and figure 10).

5.2. (b) Scaling at fixed ratio u_{imp}/u_{ig^*}

For targets scaled at fixed values of the ratio of the implosion velocity to the ignition velocity $u_{imp}/u_{ig^*} < 1$, the spike is always needed. The relevant scaling relations listed in column (b) of table 2are obtained from the general relations by observing that $u_{imp}/u_{ig^*} \propto u_{imp}/m^{-0.106}{}_{imp} \propto ws^{0.32}$, and then

$$w = s^{-0.32}. \tag{19}$$

The spike power is computed using equation (18). Notice that spike energy and power increase with size, but the intensity decreases rapidly ($I \propto s^{-3/2}$).

The main advantage of this option is in the higher achievable gain. Indeed, the simulations show that the 1D gain exceeds 200 for laser energy $E_L \geq 1.5$ MJ.

The main target parameters and simulation results for the reference case (HiPER baseline, case H, scale $s = 1$), and several scaled targets are listed in table 3. The laser parameters include the intensity at peak power and radius reduced to one half of the initial radius, I^*_p, and the laser irradiance $I_{p15}\lambda\mu$, with the intensity measured in units of 10^{15} W cm^{-2} and the laser wavelength in units of μm. The quantities IFAR$_{2/3}$ and IFAR$_{1/2}$ are the in-flight-aspect-ratio when the radius of the shell is 2/3 and 1/2, respectively, of the initial inner radius. (Here the in-flight-aspect-ratio is defined as the ratio of the shell radius to the thickness of the portion of the shell with density exceeding a fraction 1/e of the instantaneous peak density.)

Table 3. Selected targets: parameters and simulation results.

Case	H	a1	b1	b2	$2\omega S$	$2\omega L$
Scaling procedure	–	a	b	b	H-S$^{(+)}$	a$^{(*)}$
Scale s	1	1.71	1.75	2.43	1.96	2.67
$\lambda (\mu m)$	0.351	0.351	0.351	0.351	0.527	0.527
m_t (mg)	0.574	2.76	2.76	6.89	3.50	8.60
E_{Lc} (kJ)	180	890	670	1450	970	2440
E_{Ls} (kJ)	118	137	327	590	630	430
E_L (kJ)	298	1027	997	2040	1600	2870
P_c (TW)	46	175	82	116	94	174
P_s (TW)	240	164	317	374	490	260
P_p (TW)	286	339	399	490	584	434
I^*_p (PW cm^{-2})	8.36	3.45	3.97	2.58	4.77	1.92
$I^*_{p15}\lambda^2_\mu$	1.03	0.43	0.49	0.32	1.32	0.53
η_{ac}	74%	75%	77%	78%	73%	75%
η_{as}	38%	35%	45%	51%	56%	50%
$u_{imp-NoSpike}$ (km s^{-1})	291	292	240	213	215	216
Gain G	65	140	152	217	118	205
IFAR$_{2/3}$	32.6	31.2	27.4	27.0	26.9	24.0
IFAR$_{1/2}$	16.7	17.3	16.0	16.4	13.7	13.8
$C_r = R_0/R_h$	38.7	29.0	38.9	39.0	32.1	32.4

$^{(+)}$ To scale target $2\omega S$ from the HiPER baseline target H: see table 4.
$^{(*)}$ To generate the gain curve for $\lambda = 0.527\ \mu$m using $2\omega S$ as a reference, and with spike power optimized to maximize gain.

6. WAVELENGTH SCALING

An important feature of the scaling model is the capability of scaling targets with laser light wavelength. To be specific, we consider scaling of the HiPER baseline target from wavelength $\lambda_0 = 0.351\ \mu$m to $\lambda = 0.527\ \mu$m (i.e. green light). We scale the target by keeping the ratio u_{imp}/u_{pg}fixed, and setting $\Lambda = \lambda/\lambda_0 = 1.5$. (Alternatively, we could have scaled the target at fixed u_{imp}/u_{ig^*}; this would have led to a bigger target, requiring larger laser energy, with a larger safety margin.) The relevant scaling relations are listed in column (a) of table 4. Numerical values for a specific case ($\Lambda = 1.5, \beta = 1.2$), leading to the design of target $2\omega S$, are listed in column (b). It is apparent that such a 50% increase in wavelength implies an increase of the laser energy by a factor about 5.5.

Table 4. Scaling relations—wavelength scaling at constant ratio u_{imp}/u_{pg}.

Fixed	u_{imp}/u_{pg}	u_{imp}/u_{pg} $\Lambda = 1.5, \beta = 1.2$
Free	$\Lambda;\ \beta$	
s	$\Lambda^{1.98}\beta^{-0.74}$	1.95
w	$s^{-0.45} = \Lambda^{-8/9}\beta^{1/3}$	0.74
w	$\Lambda^{-8/9}\beta^{1/3}$	0.74
β		1.20
τ	$\Lambda^{2.87}\beta^{-1.07}$	2.63
ε	$\Lambda^{4.83}\beta^{-1.53}$	5.36
Π	$\Lambda^{1.96}\beta^{-0.46}$	2.04
Payload mass	$\Lambda^{5.94}\beta^{-2.22}$	7.42
Ablator mass	$\Lambda^{4.79}\beta^{-1.78}$	5.04
$\pi_{spike}{}^{(+)}$	$s\sqrt{\Lambda} = \Lambda^{1.99}\beta^{-0.37}$	2.09
$\varepsilon_{spike}{}^{(+)}$	$\pi_{spike}\tau = \Lambda^{4.86}\beta^{-1.44}$	5.52
$\beta_{spike}{}^{(+)}$	$\pi_{spike}s^2 = \Lambda^{-1.97}\beta^{1.11}$	0.55

$^{(+)}$ Approximate values; they apply at constant u_{imp}/u_{ig*}.

Next, we have generated a gain curve for targets driven by green light, by scaling the $2\omega S$ target at constant implosion velocity (see figure 12). We see that a large gain (about 200) is achieved with a laser energy of about 3 MJ and peak power below 500 TW. The main parameters of two targets driven by green light are listed in table 3: see columns labelled $2\omega S$ and $2\omega L$. We note, in particular, that the $2\omega L$ target appears quite robust: IFAR, C_r and irradiance $I\lambda^2$ take smaller values than those characterizing the HiPER baseline target. More details can be found in [16].

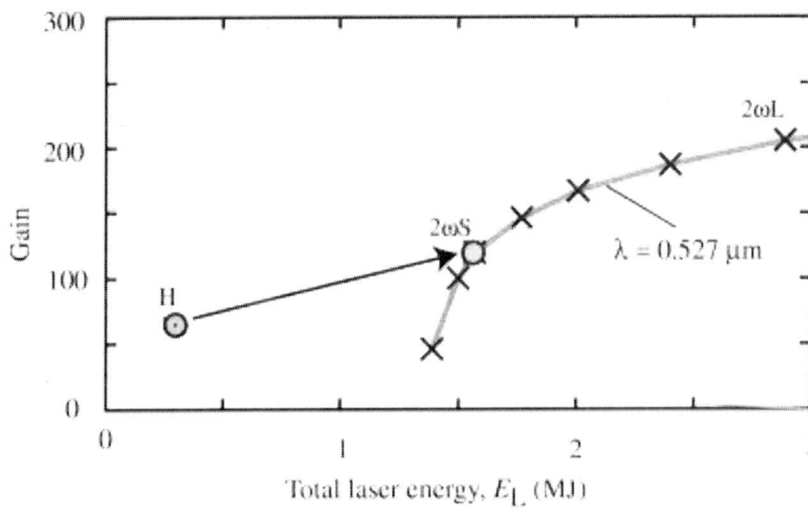

FIGURE 12. Scaling the HiPER baseline target to laser wavelength $\lambda = 0.527\ \mu$m and gain curve for scaled targets driven by laser light with $\lambda = 0.527\ \mu$m. From 1D numerical simulations.

7. CONCLUSIONS

In this paper, we have studied the scaling of shock-ignited targets with energy and wavelength. We have derived simple general scaling laws, and shown that shock-assisted ignition offers some degree of freedom in the choice of the scaling strategy. This may allow one to either increase the gain or ignition margins, and/or to decrease the risks associated with plasma instabilities and/or with hydrodynamic instabilities. In particular, we have proposed scaling laws for targets imploded at fixed implosion velocity and at a fixed ratio of the implosion velocity to the self-ignition velocity u_{ig*}, respectively. In the first case, the power of the shock driving spike decreases with increasing target scale. For sufficiently large targets no spike is necessary. However, addition of the spike increases ignition robustness. The second option leads to larger gain and reduces the risks associated with hydrodynamic instabilities, at the expense of increasing spike power with target size.

The relation between spike power and ignition margins has been briefly discussed. However, a detailed study is necessary for a quantitative analysis. It will be the objective of a future paper. Of course, multidimensional studies will be necessary for an appraisal of symmetry and stability issues.

Finally, as already stated in the introduction, we remark that a proper assessment of shock ignition requires in-depth investigation of laser-plasma instabilities in the interaction of the laser spike with the plasma corona. These processes have been neglected so far in shock-ignition target simulation. Recent studies [34], however, suggest that their detrimental effects may be limited to acceptable values.

ACKNOWLEDGMENTS

This work was partially supported by the Italian MIUR project PRIN 2009FCC9MS and by HiPER project and Preparatory Phase Funding Agencies (EC, MSMT and STFC). Simulations were performed on the STD11-480 CASPUR grant.

APPENDIX A.: DUED CODE MODEL

Simulations were performed with the 1D version of the code DUED [21, 22]. DUED is a two-dimensional, quadrilateral-zone, Lagrangian hydrocode. It includes a standard two-temperature hydrodynamic model, with classical transport coefficients and flux-limited Spitzer thermal conductivity, the real-matter equation of state, (flux-limited) multi-group radiation diffusion, fusion reactions and fuel burn-up. Charged fusion products (in this case the DT fusion 3.5 MeV alpha particles) deposit their energy according to a (flux-limited) multigroup diffusion model. The code neutron transport model is not used; neutron energy is assumed to be completely lost. Laser–target interaction is modelled using a three-dimensional laser ray-tracing [32], taking into account plasma refractivity and inverse bremsstrahlung absorption.

In all the simulations reported in this paper, we have used coefficient $f = 0.07$ and sharp cut-off flux limitation, i.e. the absolute value of the electron thermal flux is the minimum between the classical flux with

Spitzer conductivity and a fraction f of the free streaming flux. Parametric simulations show that absorption efficiency and hydrodynamic efficiency decrease with decreasing f. The same target performance and yield as for $f = 0.07$ are recovered by increasing the power of the compression pulse and reducing the size of the focal spot of the ignition spike. For example, for the HiPER baseline target and $f = 0.06$, we had to increase the compression pulse power (and energy) by 10% and reduce the focal spot of the spike by 15%. The gain turned out to be reduced by 12%.

APPENDIX B.: SCALING RELATIONS FROM THE ROCKET MODEL

To obtain ablation parameters for laser irradiation conditions relevant to the compression stage of shock ignition targets, we performed 1D parametric simulations of hollow shell targets with the same structure and aspect ratio as the HiPER baseline targets. According to the simulations, ablation pressure, areal mass ablation rate and hydrodynamic efficiency scale as

$$p_a \propto I_a^{0.75} \lambda^{-0.4}, \tag{B.1}$$

$$\dot{\mu} \propto I_a^{0.6} \lambda^{-0.4} R^{-0.22}, \tag{B.2}$$

$$\eta_h \propto u_{imp}^{3/4} (I_a \lambda^2)^{-0.2} R^{-0.08}, \tag{B.3}$$

respectively. The scaling of the ablation pressure with intensity agrees with earlier models [29, 30] and with a few recent experiments [33]. The mass ablation rate agrees with the predictions of [29, 30]. Equation (B.3) for the hydrodynamic efficiency differs somewhat from equation (10), used in this paper. Notice, however, that this difference only marginally affects the scaling law for the implosion velocity. Indeed, using (B.3) in our scaling model, we would have obtained $w = \beta^{0.356} \Lambda^{-0.92} s^{0.04}$, very close to equation (11).

It is also interesting to compare the above two expressions for the hydrodynamic efficiency with the one obtained from the rocket model, using the above fits for the ablation pressure and the mass ablation rate. From the rocket model, with the assumptions detailed in [17], one obtains

$$\eta_h = \frac{1}{2} \frac{\dot{\mu} u_{ex}^2}{I_a} \left(\frac{u_{imp}}{u_{ex}} \right)^{3/4} \tag{B.4}$$

with $u_{ex} = p_a / \dot{\mu}$. Substituting equations (B.1) and (B.2),

$$\eta_h \propto u_{imp}^{3/4}(I_a\lambda^2)^{-0.21}\lambda^{-0.025}R^{0.06}. \tag{B.5}$$

This last expression shows practically the same dependences on velocity, intensity and wavelength as equation (B.3), and a somewhat different dependence on the radius.

REFERENCES

1. Betti R et al 2007 Phys. Rev. Lett. **98** 155001
2. Atzeni S and Meyer-ter-Vehn J 2004 The Physics of Inertial Fusion (Oxford: Clarendon)
3. Lindl J D 1995 Phys. Plasmas **2** 3933
4. Kruer W L 1988 The Physics of Laser Plasma Interactions (Redwood City, CA: Addison Wesley)
5. Tabak M et al 1994 Phys. Plasmas **1** 1626
6. Betti R and Zhou C 2005 Phys. Plasmas **12** 110702
7. Theobald W et al 2008 Phys. Plasmas **15** 056306
8. Perkins L J, Betti R, LaFortune K N and Williams W H 2009 Phys. Rev. Lett. **103** 045004
9. Ribeyre X et al 2009 Plasma Phys. Control. Fusion **51** 015013
10. Ribeyre X et al 2009 Plasma Phys. Control. Fusion **51** 124030
11. Schmitt A J, Bates J W, Obenschain S P, Zalesak S T and Fyfe D E 2010 Phys. Plasmas **17** 042701
12. Lafon M, Ribeyre X and Schurtz G 2010 Phys. Plasmas **17** 052704
13. Atzeni S, Schiavi A and Marocchino A 2011 Plasma Phys. Control. Fusion **53** 035010
14. Canaud B and Temporal M 2010 New J. Phys. **12** 043037
15. Atzeni S and Schurtz G 2001 Proc. SPIE **8080** 808022
16. Atzeni S, Marocchino A and Schiavi A 2012 Phys. Plasmas **19** 090702
17. Atzeni S, Schiavi A and Bellei C 2007 Phys. Plasmas **14** 052702
18. Hicks D G et al 2012 Phys. Plasmas **19** 122702
19. US Department of Energy 2012 National Nuclear Security Administration's path forward to achieving ignition in the inertial confinement fusion program Report to Congress (Washington, DC: US Department of Energy) (http://fire.pppl.gov/NIF_Path_Forward_Rpt_120712.pdf)
20. Landen O L et al 2011 Phys. Plasmas **18** 051002
21. Atzeni S 1986 Comput. Phys. Commun. **43** 107
22. Atzeni S et al 2005 Comput. Phys. Commun. **169** 153
23. Hallo L et al 2009 Plasma Phys. Control. Fusion **51** 014001
24. Atzeni S et al 2009 Nucl. Fusion **49** 055008
25. Herrmann M C, Tabak M and Lindl J D 2001 Nucl. Fusion **41** 99
26. Atzeni S 1995 Japan. J. Appl. Phys. **34** 1980
27. Meyer-ter-Vehn J and Schalk C 1982 Z. Naturf. a **37** 955
28. Kemp A, Meyer-ter-Vehn J and Atzeni S 2001 Phys. Rev. Lett. **86** 3336
29. Caruso A and Gratton R 1968 Plasma Phys. **10** 867
30. Mora P 1982 Phys. Fluids **25** 1051
31. Schurtz G et al in preparation
32. Schiavi A, Atzeni S and Marocchino A in preparation
33. Fratanduono D E et al 2011 J. Appl. Phys. **110** 073110
34. Klimo O, Weber S, Tikhonchuk V T and Limpouch J 2010 Plasma Phys. Control. Fusion **52** 055013

CITATIONS

1. *Target heating due to the shock produced hot electrons in the shock ignition scheme Somayeh Rezaei et al 2016 The European Physical Journal D* ***70***
2. *Influence of laser induced hot electrons on the threshold for shock ignition of fusion reactions A. Colaïtis et al 2016 Physics of Plasmas* ***23*** *072703*
3. *Inertial confinement fusion with lasers R. Betti and O. A. Hurricane 2016 Nature Physics 435*
4. *Improved robustness study of a shock ignited target, with DUED code including non-local electron transport and 3D laser ray-tracing Stefano Atzeni et al 2016 Journal of Physics: Conference Series* ***688*** *012005*
5. *Radiation Hydrodynamic Simulations in the Planar Scheme for the Fundamental Studies of Shock Ignition Dong Yunsong et al 2016 Plasma Science and Technology* ***18*** *376*
6. *Optimization of x-ray emission from under-critical CH foam coated gold targets by laser irradiation Wanli Shang et al 2016 Nuclear Fusion* ***56*** *086002*
7. *Shock ignition: a brief overview and progress in the design of robust targets S Atzeni et al 2015 Plasma Physics and Controlled Fusion* ***57*** *014022*
8. *Light for controlled fusion energy: A perspective on laser-driven inertial fusion Stefano Atzeni 2015 EPL (Europhysics Letters)* ***109*** *45001*
9. *Improvement of non-isobaric model for shock ignition Amir Hossein Farahbod et al 2014 The European Physical Journal D* ***68***
10. *Effects of non-local electron transport in one-dimensional and two-dimensional simulations of shock-ignited inertial confinement fusion targets A. Marocchino et al 2014 Physics of Plasmas* ***21*** *012701*
11. *Focus on high energy density physics R Paul Drake and Peter Norreys 2014 New Journal of Physics* ***16*** *065007*
12. *Low initial aspect-ratio direct-drive target designs for shock- or self-ignition in the context of the laser Megajoule V. Brandon et al 2014 Nuclear Fusion* ***54*** *083016*
13. *Shock ignition of thermonuclear fuel: principles and modeling S. Atzeni et al 2014 Nuclear Fusion* ***54*** *054008*
14. *Instantaneous x-ray radiation energy from laser produced polystyrene plasmas for shock ignition conditions Wanli Shang et al 2013 Physics of Plasmas* ***20*** *102702*
15. *Lattice BGK kinetic model for high-speed compressible flows: Hydrodynamic and nonequilibrium behaviors Yanbiao Gan et al 2013 EPL (Europhysics Letters)* ***103*** *24003*

CHAPTER 10

Astrophysics and Dark Matter Theory

Auguste Meessen

UCL, Louvain-la-Neuve, Belgium

ABSTRACT

Space-Time Quantization implies that the cosmic dark matter gas is subjected to pressure effects. We prove this by accounting for the mass-density distribution of dark matter in galactic halos. It can be directly deduced from observed rotation curves and coincides with theoretical predictions for dark matter atmospheres in hydrostatic equilibrium. Through embedding, the pressure of the cosmic dark matter gas prevents also the gravitational collapse of the Oort cloud, globular star clusters and cosmic filaments. The Sun has only a very small dark matter atmosphere, but observations confirm that dark matter is orbiting around the Sun. Other facts are explained by planetary dark matter disks. Space-Time quantization accounts also for dark matter-electron interaction, which allowed already for direct detection of galactic dark matter particles.

Keywords: Dark Matter, Space-Time Quantization, Dark Matter Density Distribution, Dark Matter Rings, Pioneer Anomaly, Flyby Anomaly

1. INTRODUCTION

The idea that our universe could contain "dark matter" was unexpected. It was formulated by Zwicky in 1933, since his measurements of the velocities of nebula in the Coma cluster revealed a very high average value and a great dispersion. He analyzed this data by means of the virial theorem of statistical mechanics, which applies to any spherical distribution of equal masses, subjected to gravitational interactions. He stated [1] that "if these optical observations were confirmed, they would yield the surprising result that dark matter (dunkle Materie) is present at much greater densities than luminous matter."

This seemed to be unbelievable, but the breakthrough came in the 1970s, because of the tenacity of Vera Rubin. She measured the redshift of many stars in the Andromeda galaxy [2] [3] [4] . The resulting dependence of orbital velocities on the distance to the galactic center required the existence of Dark Matter (DM). It had even to be present far beyond the visible boundaries of galaxies. Albert Bosma confirmed this conclusion by means of the 21-cm hydrogen line [5] [6] . He determined the mass M, which causes the observed orbital velocities by gravity. He measured also the optically detectable light intensity L and concluded that "there is a large amount of high M/L material in the outer parts." There is not only more mass than the visible one, but it "may not be distributed in the disk at all". Actually, it constitutes a very great spherical halo.

Nevertheless, the existence of DM in our universe is still subject to controversies. The main reason is that DM particles have not yet been identified and that measured rotation curves can also be fitted by assuming a modification of Newton's law of gravity [7] . A recent analysis of excellent empirical data for 153 galaxies of different morphologies, masses, sizes and gas content led even to a very puzzling result [8] . The observed gravitational acceleration, determined by means of the rotation curves, was always correlated in a simple way with the baryonic acceleration, calculated by means of the distribution of luminous matter. This can suggest that DM may not be needed [9] and requires an explanation. There are also other ambiguities that have to be clarified.

Could the theory of Space-Time Quantization (STQ) be helpful in this regard? It generalizes Relativistic Quantum Mechanics by considering a finite limit for the smallest measurable distance. It is sufficient that its value $a \neq 0$, to prove that STQ accounts for elementary particle physics [10] . It yields also insight into the nature and properties of DM particles. This solves basic cosmological problems [11] , but we want also to test the proposed theory by means of a variety of astrophysical observations. This is the essential aim of the present article.

It may thus be useful to recall some basic consequences of the proposed theory. All elementary particles are characterized by four new quantum numbers (u_x, u_y, u_z, u_{ct}). They specify possible variations of ψ functions at the smallest possible scale along the chosen reference axes. These u-quantum numbers are subjected to a very strict conservation law when particles are transformed by annihilation and creation processes. This law determines all possible types of interactions and the constitution of compound particles [10] . We can thus distinguish three levels. 1) The basic level concerns only elementary particles. There are quarks, but also elementary DM particles, which interact with one another by exchanging gluons. Since they are electrically neutral, we called them narks. 2) On the second level, quarks get bound to one another inside nucleons, while narks can constitute various types of neutralons. Nucleons interact with one another by exchanging π mesons and neutralons by exchanging N_2 bosons, composed of a nark and an antinark. This allows in both cases for scattering and the formation of compound particles. 3) The third level concerns the resulting particles. Baryonic matter is constituted of single nucleons or nuclei, where nucleons are bound together. The cosmic DM gas contains neutralons and compound DM particles, where neutralons are bound to one another.

DM particles of any type do not interact with baryonic matter [10] , but they are subjected to gravitational interactions, because of their mass. Moreover, compound DM particles allow for fusion and fission processes everywhere in the cosmic DM gas. This accounts for "dark energy" and is important for cosmology [11] . However, DM particles are usually only scattered by one another. Because of energy quantization, elastic scattering is predominant. DM particles behave thus like molecules in ordinary gasses and the comic DM gas has some pressure. This concept will be essential for this article.

It is organized in the following way. In chapter 2, we show that the density distribution of DM inside spherically symmetric galactic halos can be directly deduced from rotation curves. This empirical result is then explained by combining gravity with pressure effects. We discuss also related problems, especially for the Solar system. In chapter 3, we consider cases where gravitational collapse of ordinary matter is prevented, since it is imbedded in the cosmic DM gas. This mechanism is especially important for the gigantic cosmic filaments, where galaxies remain separated, although they attract one another. Chapter 4 is devoted to the study of DM rings. Chapter 5 completes the previous discussion [10] of detection or production of DM particles. Instead of DM-nucleon interactions, we consider now DM-electron interactions. They explain why galactic DM particles could already be detected by means of scintillators.

2. SPHERICALLY SYMMETRIC DARK MATTER ATMOSPHERES

2.1. Empirical DM Density Distribution in Galactic Halos

To determine the radial distribution $\rho(r)$ of the DM mass density in spiral galaxies, we adopt the model of Figure 1. There are many visible stars and baryonic particles in the galactic disk, but we consider only one of them. The orbital velocity v of this "test body" can be determined by measuring the Doppler shift of its radiation. Assuming a circular orbit of radius r, this implies a radial acceleration $g = v^2/r$. It is due to gravitational attraction towards the center of the galaxy, but we can consider that all masses in the galactic disk are negligible with respect to the mass M of the central supermassive black hole. It follows then from Newton's law of gravity

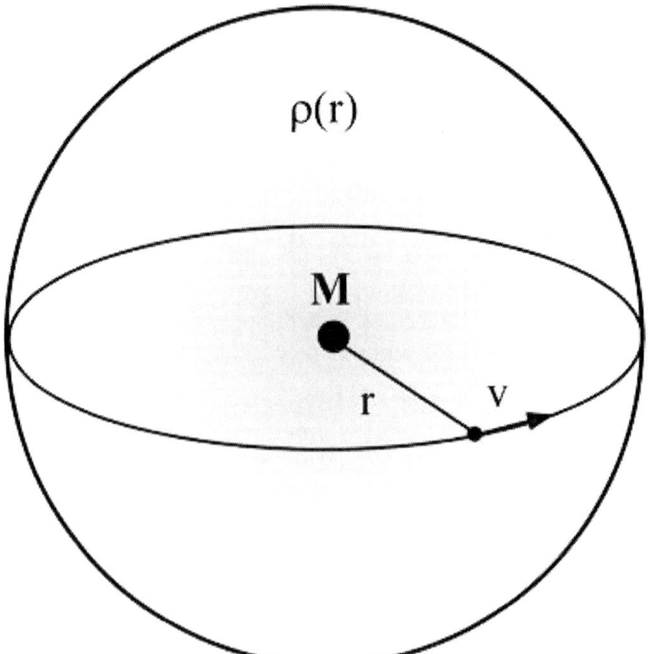

Figure 1. The DM atmosphere in the sphere that has the same radius r as an orbit in the galactic disk.

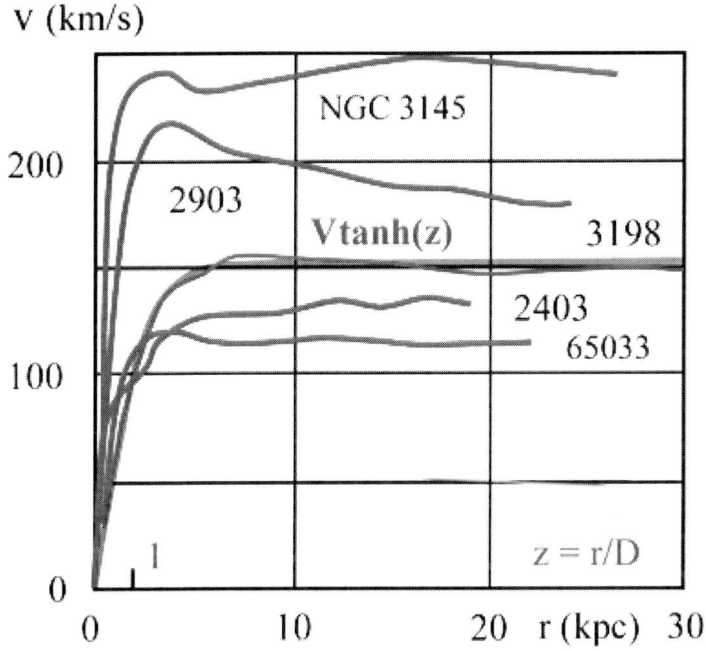

Figure 2. Observed rotation curves (in blue) and proposed model function (in red).

that $g = GM/r^2$. This would yield Kepler's law $v^2 = GM/r$, but the observed rotation curves $v(r)$ are totally different (Figure 2). When $r \to 0$, we get a linear variation: $v(r) = \omega r$, while $v(r)$ tends towards a constant value V when $r \to \infty$. Galactic rotation curves are said to be "flat". Sometimes the linear increase of $v(r)$ near the center of the galaxy is followed by a "bump" and the spiral arms can lead to oscillations of $v(r)$. Some rotation curves display only the rising part [7] , but this can be interpreted as resulting from a low value of ω for these galaxies. We propose therefore that all observed rotation curves are (in first approximation) of the same type as the red one in Figure 2. It represents empirical results by means of the simple universal function:

$$v(r) = V \tanh(r/D) \qquad \text{or} \qquad y(z) = \tanh(z) \tag{1}$$

There are only two parameters: V and r, while $y = v/V$ and $z = r/D$. The intermediate curvature is adequate and the distance D corresponds to the value of r where the extrapolated straight line $v(r) = \omega r$ meets the horizontal line $v(r) = V$. The parameters D and V define natural units for the horizontal and vertical axes in Figure 2. Different galaxies are characterized by particular values of D and V.

Relation (1) summarizes empirical data in a unified and flexible way, by separating essential features from secondary ones. We can express this result in another way, since the rotation curve $v(r)$ is determined by the condition of dynamical equilibrium. The orbiting test body is attracted towards the center of the galaxy, but not only by the mass M of the central black hole. It is also necessary to take into account the total mass Mr of the spherically symmetric distributed DM up to the radius r (Figure 1). The term "halo" does simply result from an analogy with light of decreasing intensity for greater and greater distances. Actually, we should consider the radial mass-density distribution ρ(r) of DM. This function is related to the measured rotation curves, since the radial acceleration is

$$g(r) = \frac{G}{r^2}(M + Mr) = \frac{v^2}{r} \text{ where } Mr = 4\pi\int_0^r \rho(r) r^2 dr \qquad (2)$$

We expect that the mass-density $\rho(r)$ reaches its highest value ρ_o at the center of the galaxy and that it is there nearly constant. It follows that

$$GM + \frac{4\pi G\rho_o}{3} r^3 = rv^2 = \left(\frac{V}{D}\right)^2 r^3 \text{ for } r \to 0$$

Since the mass M is independent of r, the density ρ_o is so enormous that Mr □ M, even for small values of r. Moreover, it is possible to deduce the central mass density ρ_o from the measured values of D and V, but it will be more useful to define D in terms of ρ_o and V. We can thus express $\rho(r)$ in natural units, so that:

$$\rho(r) = \rho_o f(z) \text{ where } z = \frac{r}{D} \text{ and } D^2 = \frac{3V^2}{4\pi G\rho_o} \tag{3}$$

Because of (2), we get then an integral equation:

$$3\int_0^z f(z) z^2 dz = zy^2 \text{ or } f(z) = \frac{y^2 + 2zyy'}{3z^2} \tag{4}$$

The shape of the galactic DM density profile $\rho(r)$ can thus be directly deduced from the observed rotations curves $v(r)$. The function $f(z)$ follows from $y(z)$ and vice-versa. When $z \to 0$, we get $y(z) = z$ and thus $f(z) = 1$, as expected. We get even $y(z) = z - z^3/3$ and $f(z) = 1 - (10/9)z^2$, as long as z □ 1. When z □ 1, it follow from (4) and $y(z) = 1$ that the asymptotic variation of $f(z) = 1/3z^2$. This can also be deduced from (2).

The associated functions $y(z)$ and $f(z)$ are represented in Figure 3. In the following section, we will compare this empirical mass density distribution for DM in galactic halos to theoretical predictions (Figure 4). It would have been useful to compare the function $\rho(r)$, which can be directly derived from observations, to results of N-body computer simulations. When $x = r/R$, where R is a natural unit of length, they can be expressed in the following way

$$\rho_N(x) = \frac{\rho_1}{x(1+x)^2} \qquad \rho_B(x) = \frac{\rho_o}{(1+x)(1+x^2)}$$

and $\rho_E(x) = \rho_o \exp(-Ax^\alpha)$

The first expression was obtained by Navaro, Frenk and White [12] in 1996 and the second one in 1999 by Burkert and Silk [13] . These functions decrease, but the asymptotic variation is $1/x^3$ instead of $1/x^2$. The more general expression $\rho_E(x)$ was already proposed in 1993 by Einasto [14] . An evaluation by Merritt et al. concluded in 2006 that the Einasto profile is the best [15] , but for this parametrization, A and α depend on the total mass of DM halos [16] .

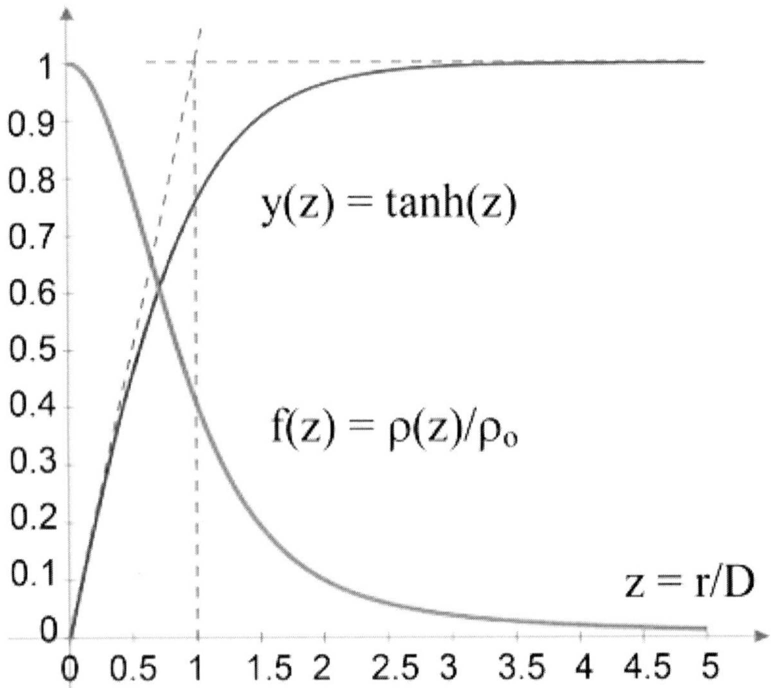

Figure 3. The empirical functions y(z) and f(z).

2.2. Theoretical Mass Density Profile in DM Atmospheres

We will now compare the empirical profile $f(z)$, determined by (4), with a theoretical one. It is deduced from the definition of DM pressure by $p = nkT$, where n is the local density of DM particles and $3kT/2$ their average kinetic energy. Since DM particles cannot interact with photons and with particles of ordinary matter, local heating is excluded. The cosmic DM gas is isothermal in the whole universe and tends to be distributed as evenly as possible. However, it is also gravitationally attracted by the masses M and Mr. The relevant force depends on the local mass-density $\rho = nm$ of the DM gas. It is proportional to the average mass m of DM particles, but also to their density n. We get thus $p = \tau\rho$, where $\tau = kT/m$ refers to thermal agitation.

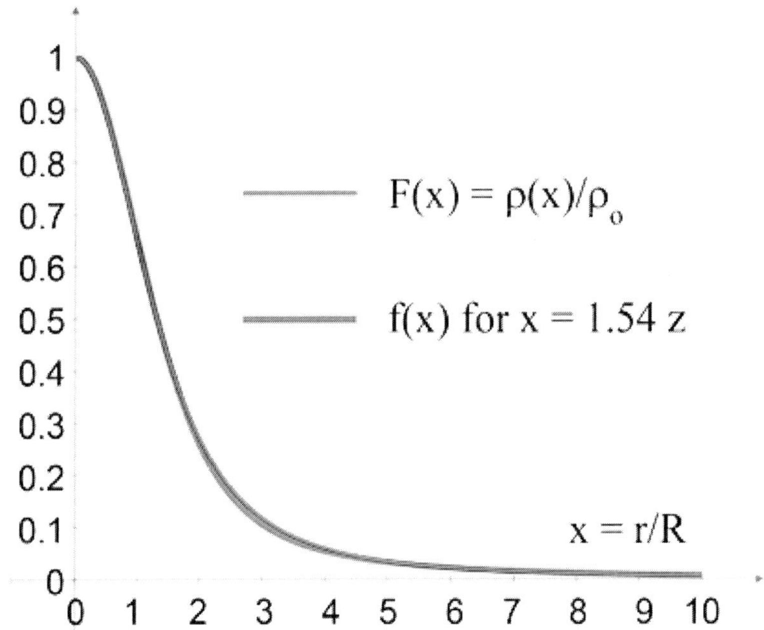

Figure 4. Theoretical and empirical density profiles.

Since gravitational attraction towards the center of the galaxy increases the particle density n, this increases also the local pressure p. The cosmic DM gas will thus behave like the air atmosphere of the Earth. Greater pressure at lower altitude, allows for hydrostatic equilibrium. To express this condition for galactic DM atmospheres, we consider again the model of Figure 1. The mass Δm of DM contained in a portion of a very thin spherical layer of radius r will be attracted towards the center of the galaxy. For a layer of thickness dr and a portion that is delimited by a central solid angle dΩ this mass is $\Delta m = \rho(r) r^2 \mathrm{d}r \mathrm{d}\Omega$. The radial gravitational force is $g(r)\Delta m$, when $g(r)$ is the local gravitational acceleration, but this force is equilibrated by the difference of pressure forces $\Omega r^2 p(r)$ exerted on the lower and upper surfaces. Since the pressure decreases when r increases,

$$g(r)\rho(r)\mathrm{d}r = [p(r) - p(r+\mathrm{d}r)] = -\frac{\mathrm{d}p}{\mathrm{d}r}\mathrm{d}r$$

Because of $p(r) = \tau\rho(r)$, we get a differential equation for $\rho(r)$:

$$g(r)\rho(r) = -\tau\frac{d\rho}{dr} \quad \text{or} \quad \rho(r) = \rho_o \exp[u(r)] \tag{5}$$

Galactic DM mass density profiles are always of the Einasto type, but

$$\frac{du}{dr} = -\frac{g(r)}{\tau} \quad \text{or} \quad x^2\frac{du}{dx} = -3\int_0^x e^u x^2 dx \tag{6}$$

This follows from (2), where the mass M is negligible when

$$r = xR \quad \text{where} \quad R^2 = \frac{3\tau}{4\pi G\rho_o} \tag{7}$$

Derivation of the integral Equation (6) yields

$$u'' + \frac{2u'}{x} + 3e^{u(x)} = 0 \quad \text{when} \quad \frac{\rho(x)}{\rho_o} = F(x) = \exp[u(r)] \tag{8}$$

The differential equation for $u(r)$ can be solved by numerical integration, with the boundary conditions $u(0)=0$ and $u'(0)=0$, since $\rho(r)=\rho_o$ when $r \to 0$. The resulting function $F(x)$ is represented by the red curve of Figure 4. It is practically identical to the green curve for $f(z)$ when z is replaced by $x = \beta z$, where $\beta = 1.54$. The agreement between theoretical and empirical curves validates the concept of DM pressure and STQ. We can even see why D and V depend on specific properties of DM particles in different galaxies. Since $x = \beta z$ implies that $r/R = r/D$, it follows from (3) and (7) that $(D/R)^2 = V^2/\tau = \beta^2 = 2.37$. The flat rotation curves imply that the value of V determines the value of $w = kT/mc^2$. Because of (3), D is greater and w is smaller for low values of ρ_o.

Actually, $V^2/3\tau = (V/c)^2/3w \approx 1$. According to Figure 2, V ≈ 150 km/s or $V/c = 0.5\times10^{-3}$. This means that $w \approx 10^{-7}$. The cosmic DM gas is cold. If we did know kT, we could determine the average mass m of DM particles. Since DM particles cannot interact with photons, they are not in thermal equilibrium with the cosmic microwave background. However, color neutral n_e narks could be converted into ν_e neutrinos [10]. If the cosmic DM gas were in thermal equilibrium with the cosmic neutrino background (T =1.95 K), the average mass of galactic DM particles would be such that $mc^2 \approx 3.2$ keV. This figure could be helpful for detecting DM particles.

The results of numerical simulations led to different density profiles $\rho(x)$, since they did not account for DM pressure. They were based on the Standard model of cosmology, where it is assumed that DM particles do not interact with one another. The concept of self-interacting DM has been proposed to improve the agreement with observed galactic density profiles [17] . Bullock and his team confirmed this by simulations [18] . STQ explains why DM particles interact with one another and why elastic collisions are predominant in DM atmospheres.

2.3. Luminous Matter and Dark Matter

Initially, we thought that the very great mass M of the central black hole of galaxies is responsible for the attraction of huge amounts of DM and the spherical symmetry of halos for spiral galaxies. Because of (2), it appeared that the actual value of M is irrelevant. This suggests that a spherically symmetric DM halo could even be autonomous. It is sufficient that the central mass density ρ_0 is very great and that the DM atmosphere is not rotating. Actually, there are so-called "voids", where the density of luminous matter is quite low, but there is DM that could be attracted towards a small local concentration of baryonic matter. Since cosmic DM gas tends to be distributed as evenly as possible, there is no DM void. More and more DM could thus be accumulated, even when there is only a small amount of baryonic matter (BM).

This explains some recent observations. It is now possible, indeed, to localize massive DM halos by gravitational lensing. This led already in 1999 to the discovery of a "truly massive dark clump" in the cluster Abell 1942 [19] . It was said to represent a new class of objects with unusual properties, since the baryon density is there very low [20] . Even a so-called "ghost galaxy" has been detected in 2016 [21] . This Dragonfly 44 galaxy belongs to the Coma cluster and contains nearly no stars. It is "relatively round" and "nearly dark", although the mass of DM is similar to that of the Milky Way. It seems to be a "failed galaxy", prevented from building a normal stellar population. Its low luminosity and lack of a classical disk and bulge were said to be "anomalous", but there are also other faint galaxies in the Coma galaxy cluster.

Dragonfly 44 was said to be "representative of an entirely new class of objects", indicating that our understanding of the formation of galaxies is not yet sufficiently complete [22] . However, the constitution of spiral galaxies and DM halos will usually result from a synergy, leading to a statistical correlation between the amount of baryonic matter and DM in spiral galaxies. The Tully Fischer relation [23] expresses it in terms of the luminosity L and the orbital velocity V.

We mentioned in the introduction the puzzling claim [8] that rotation curves of galaxies seem to be due to baryonic matter alone. Actually, the gravitational acceleration g_{obs} was determined by means of rotation curves and compared to an acceleration g_{bar} that was calculated by means of the radial variation of the surface brightness of baryonic matter. It appeared that g_{obs} and g_{bar} were strongly correlated. For large values, they were even equal to one another. This may suggest that the orbital velocity V does not depend on DM, but g_{bar} was calculated by considering a spherically symmetric gravitational potential. Its radial gradient would then determine the centripetal acceleration. However, luminous matter is essentially distributed in the galactic disk and we will show in section 4.1 that gravitational effects are then very different from those of spherically symmetric mass distributions. The existence of DM does not have to be questioned.

2.4. The Small Solar DM Atmosphere

The motions of planets in our Solar system were correctly described without considering any DM. The mass M_o of the Sun is so modest, indeed, that it cannot attract much DM from outer space. Its density ρ_s is there very low. Nevertheless, the Sun attracted some DM and constituted its own spherically symmetric DM atmosphere. Since it has to be in hydrostatic equilibrium, the mass density $\rho(r)$ varies according to (5), where $g(r)$ is defined by (2). However, Mr is negligible with respect to the mass M_o of the Sun. Thus,

$$\frac{d\rho}{dr} = -\frac{L}{r^2}\rho(r) \quad \text{and} \quad \rho(r) = \rho_s e^{L/r} \quad \text{where} \quad L = \frac{G}{\tau}M_o \tag{9}$$

The boundary condition is satisfied, since $\rho(r) \to \rho_s$ when $r \to \infty$, but this solution is only valid outside the Sun ($r \geq r_1$). Contrary to the air atmosphere of the Earth, the Solar DM atmosphere penetrates into this star. At a distance r from the center of the Sun, DM is attracted by the mass $M(r)=(r/r_1)^3 M_o$, when we consider only the average mass density of baryonic matter inside the Sun. The reduced attraction yields the equation

$\frac{d\rho}{dr}=-br\rho(r)$ and $\rho(r)=\rho_{so}e^{-br^2/2}$ where $b=L/(r_1^3)$

ρ_{so} is the DM mass-density at the center of the Sun and $\rho(r)$ decreases like a Gaussian curve inside the Sun. At its surface, $\rho(r_1)=\rho_{so}\exp(-L/2r_1)=\rho_s\exp(L/r_1)$.

Thus, $\rho_{so}=\rho_s\exp(3L/2r_1)$. For the Sun, $GM_o/c^2=1.47$ km. When $mc^2/kT\approx 10^7$ as for galaxies, the length $L\approx 1.5\times 10^7$ km $=0.1$ AU. Since $r_1=4.65\times 10^{-3}$ AU, $\rho_{so}\approx\rho_s\exp(32)$.

Evaluations of ρ_s display wide variations [24] , but $\rho_s\approx 10^{-24}$ g/cm³. This would yield $\rho_{so}\approx 8\times 10^{-11}$ g/cm³, while the Sun's average mass density is 1.4 g/cm³. The contribution of the solar DM atmosphere to its mass M_o is negligible.

Pitjec and Pitjeva [25] evaluated the upper limit for the DM density $\rho(r)$, allowed by present-day measurements when r is the average orbital radius for Mercury, Venus, Earth, Mars, Jupiter or Saturn. For Mercury (r_2 = 0.387 AU), they found that $\rho(r_2)<9.3\times 10^{-20}$ g/cm³. We get $\rho(r_2)\approx 1.3\times 10^{-24}$ g/cm³. Planetary motions are thus not affected by the mall Solar DM atmosphere.

2.5. The Stability of the Oort Cloud

The Kuiper belt is composed of small icy objects, orbiting around the Sun in its ecliptic plane between about 30 and 55 AU. It constitutes a reservoir for short-period comets, formed about 4.5 billion years ago. They are "fossils" of the origin of the Solar system and some of them can be ejected from the Kuiper belt, because of gravitational interactions with outer planets or mutual scattering. There are also long-period comets, moving around the Sun on very elongated elliptical orbits with an isotropic distribution. They were assumed to come from a spherically symmetric layer, surrounding the Sun between 50 and 100 thousand AU. The average radius of this "Oort cloud" is immense, since the closest star (Proxima Centauri) is situated at 265 thousand AU. How is it possible that the Sun attracted ordinary matter from far away and stored it in a spherical shell?

A constant mass-density ρ_s of DM would even imply that the total mass Mr of DM inside a sphere of radius r increases like $\rho_s r^3$. Beyond a certain distance, Mr would become greater than the mass M_o of the Sun. More and more DM and BM would be attracted towards the Sun. Actually, the value of ρ_s that we considered in the previously section, would imply that $Mr \approx M_o$ when $r \approx 500$ thousand AU. Inside the Oort cloud, the mass-density of the DM gas remains equal to ρ_s, but why can baryonic matter be stored in the Oort cloud and remain there in equilibrium?

The spherical Oort cloud protects the Solar system from undesirable intrusions, like the membrane of biological cells. However, the Oort cloud is attracted by the Sun and should collapse. This reminds us of soap bubbles. They should collapse because of surface tension, but this is opposed by a somewhat higher air pressure inside these bubbles than outside. The Oort cloud can be stabilized by DM pressure, since that requires only that the mass density ρ_s of the DM gas is somewhat higher inside than outside the Oort cloud.

Actually, the DM mass-density decreases progressively from ρ_s inside the Oort cloud to ρ_{is} in the surrounding interstellar space, while the mass-density of baryonic matter (BM) increases and decreases there like a Gaussian curve. The associated variations can be determined by considering that BM is embedded in the DM gas, but for our present purpose, it is sufficient to treat the Oort cloud like the liquid membrane of a soap bubble. When the Oort is mentally reduced to a spherical membrane of radius $r = X$ and small thickness δ, every part of it is attracted towards the Sun, but this force is equilibrated by the difference of pressure forces. Equation (5) is then reduced to

$$\frac{GM_o}{X^2}\rho_s \approx \frac{kT}{m\delta}(\rho_s - \rho_{is})$$

3. EMBEDDING DARK MATTER ATMOSPHERES

3.1. Globular Star Clusters and Elliptical Galaxies

Globular star clusters are spherically symmetric groups of 10^4 to 10^7 very old stars. It is thus astonishing that these stars could remain separated from one another. The astronomer Jeans assumed that these stars are moving around as if they were subjected to thermal agitation. He recognized, indeed, that gravitational interactions between stars should lead to mutual scattering and that their motions will be randomized [26] . He thought that globular star clusters are in hydrostatic equilibrium because of the resulting pressure. This contributes to its stability, but is not sufficient. Since BM can lead to irreversible aggregation, the equilibrium would be unstable.

Globular star clusters are actually situated in galactic DM halos, outside the visible disk. They did thus attract DM and constituted their own DM atmosphere. Although stars and DM are then subjected to common gravitational attractions towards the center of the globular cluster, they would resist collapse by means of two different pressure effects. Indeed, Jean's thermal agitation of stars can be combined with DM pressure. The mass density $\rho(r)$ of the DM gas and the average mass density $\hat{\rho}(r)$ of ordinary matter vary like $\rho(r) = \rho_o \exp[u(r)]$ and $\hat{\rho}(r) = \hat{\rho}_o \exp[u(r)]$. The central values are different, but the radial distributions are identical. Because of (5),

$$\frac{\mathrm{d}u}{\mathrm{d}r} = \frac{-4\pi G}{r^2} \int_0^r \left(\mathrm{C} + \hat{\mathrm{C}}\right) \mathrm{e}^{u(r)} r^2 \mathrm{d}r \tag{10}$$

The constant $C = \rho_o m/kT$ and $\hat{C} = \hat{\rho}_o \hat{m}/k\hat{T}$. Setting $r = x\lambda$ and $\lambda^2 = 3/4\pi G\left(\mathrm{C} + \hat{\mathrm{C}}\right)$, we get the differential Equation (6). This yields the function $F(x) = \exp[u(r)]$ of Figure 4. The average distribution of visible matter in globular star clusters provides thus an adequate image of the DM profile, but there can be much more DM than BM. Recent observations confirmed that DM is an important component of globular star clusters. ESO's Very Large Telescope in Chile made it possible to analyze 125 compact stellar systems near a giant elliptical galaxy. They suggested also that globular star clusters should have a "significant dark gravitating component" [27] .

Elliptic galaxies are similar to globular star clusters, but they contain much more DM and stars. Many stars were so strongly attracted towards the global center of gravity, that they constituted there a supermassive black hole. Other stars remained in orbit, but not like they do in spiral galaxies. Their motions were constantly perturbed by gravitational interaction. Collapse of this swarm was not only prevented by quasi-thermal agitation. It resulted also from the fact that moving stars were embedded in a DM atmosphere. By participating in the global rotation, it justifies also the spheroidal shape of elliptical galaxies. Recent astrophysical observations confirmed that they are rotating [28] and that there is some kind of "conspiracy" between DM and stars [29] .

3.2. Cosmic Filaments and Collisions of Galactic Clusters

Cosmic filaments are the greatest structures in our Universe. They are composed of galaxies that contain stars and planetary systems, dust and hot gas. These entangled filaments constitute a cosmic web, since they are "knotted" together at places where there are superclusters of galaxies. These filaments are gigantic, but they are also associated with DM. It facilitated their formation and insures their stability. For any rectilinear portion of cosmic filaments and cylindrical coordinates, Equation (10) is replaced by

$$\frac{du}{dr} = -\frac{2\pi G}{r} \int_0^r \left(C + \hat{C}\right) e^{u(r)} r dr$$ where $r = x\Lambda$

Setting $\Lambda^2 = 3/2\pi G\left(C + \hat{C}\right)$, this yields the integral Equation (6). The density distributions of DM and ordinary matter are thus also universal function for cosmic filaments. Galactic clusters can contain up to several thousand galaxies. They are associated with DM, but we recall that the mass Mr of DM atmospheres can be much greater than the mass M of BM. Galaxies and galactic clusters are moving. They can even pass through one another or be affected by tidal interactions. These are slow processes on terrestrial timescales, since huge masses imply not only enormous gravitational forces, but also great inertia. It is possible, today, to localize galaxies by means of the emitted electromagnetic radiation and big clumps of DM by means of gravitational lensing, but we get only "snapshots". It can thus be difficult to distinguish between configurations before or after close encounter. Anyway, DM atmospheres can become autonomous entities.

Galactic collisions seem to provide evidence for self-interacting DM [30] , but they should be described by pressure effects.

4. ROTATING DM RINGS IN THE SOLAR SYSTEM

4.1. Formation of Planetary Systems

The Sun is at least a second generation star. When the initial stars exploded, they dispersed DM as well as BM. Both types of matter were gravitationally attracted towards the center of a new stellar system, but angular momentum had to be separately conserved. Indeed, DM particles do not collide with particles of BM and they tend to be distributed as evenly as possible. This led to the formation of a protoplanetary disk, where DM could remain in dynamical equilibrium by means of adequate orbital velocities. We have thus to expect that there is not only a small Solar DM atmosphere, but also a rather great amount of DM that is still orbiting in the ecliptic plane.

A very instructive picture of a protoplanetary disk was obtained in 2014 for the HL Tau star by means of ESOs ALMA Telescope [31] . This star is only about 1 million years old and visible matter is still distributed in large circular rings. The formation of planets resulted from gravitational accretion of baryonic matter, with a tendency to be radially distributed according to the Titius-Bode law. It is due to resonance effects for orbital motions at different frequencies [32] . However, DM particles will remain dispersed, because of elastic scattering. This means that DM rings are rotating around the Sun like planets, but they constitute a quasi-continuous radial distribution of the surface mass-density $\sigma(r)$. Since this function was determined by the conditions that prevailed during the formation of the Solar system, it may reach its maximum value in the region of the big planets, but DM pressure implies that $\sigma(r)$ becomes only negligible far beyond the Kuiper belt.

Are there signs of the presence of rotating DM rings in our Solar system? They are not obvious, but we know that asteroids are relatively small rocky objects, orbiting around the Sun between Mars and Jupiter. They constitute a flat circular belt, extending from about 2.2 to 3.3 AU. To assume that they result from the explosion of a planet is not plausible, since the debris would have been attracted toward a local center of gravity. It is also unbelievable that they were initially distributed in a homogenous way in vacuum, but failed to merge.

Actually, we know that the inner zone of the Solar system allowed for the formation of compact solid planets, while big gaseous planets were created in the outer zone. The intermediate zone allowed only for the formation of smaller solid objects. Their mutual gravitational interactions were quite modest compared to the stabilizing effect of surrounding DM. Asteroids could thus remain separated from one another, by being embedded in rotating DM rings. They were not only limited to this region, but present everywhere in the ecliptic plane. They stabilized also the Kuiper belt, composed of small icy objects.

4.2. Gravitational Effects of a DM Disk

The existence of rotating DM rings in the ecliptic plane of the Solar system escaped attention, since their gravitational effects are small and differ from those of customary spherical matter distributions. Any material object that is situated at a distance D from the center of the Sun would be subjected to the centripetal acceleration $g(D) = GM_o/D^2$, where M_o is the mass of the Sun. It accounts for dynamic stability of planetary motions, but there is also an additional radial acceleration $a(D)$. It results from all DM rings and its value is

$$a(D) = G\int_0^\infty \sigma(r) f(r,D)\,\mathrm{d}r \quad \text{where} \quad f(r,D) = \int_0^{2\pi} \frac{r\mathrm{d}\varphi}{\delta^2}\cos\theta \tag{11}$$

The radial distribution of the surface mass density $\sigma(r)$ in the ecliptic plane is assumed to be a widely smeared out continuous function. However, every DM ring of radius r produced a gravitational acceleration at a distance D from the Sun. The function $f(r,D)$ specifies its value for one DM ring of radius r, carrying a constant and unitary mass density when $G=1$. The resulting acceleration is positive like $g(r)$, when the probe is attracted towards the center of a smaller circle. It is negative for attraction towards the closest part of any greater circle. Since these effects tend to compensate one another, they are easily overlooked. To calculate $a(D)$ for a given distribution function $\sigma(r)$, we have to know the function $f(r,D)$. The angles φ and θ, as well as the distance δ are defined in Figure 5. The triangle yields $r^2 = \delta^2 + D^2 - 2\delta D\cos\theta$, where $\delta^2 = r^2 + D^2 - 2rD\cos\varphi$. Eliminating $\cos\theta$, we get

$$f(r,D) = 2r\int_0^{\pi} \frac{(D - r\cos\varphi)\,d\varphi}{\left(r^2 + D^2 - 2rD\cos\varphi\right)^{3/2}}$$

The function $f(r,D)$ can be calculated for any value of r and D by means of elliptic functions [33] , but numerical integration is sufficient. A typical curve $f(r,D)$ is represented in Figure 6. The values of r and D are expressed in arbitrary units, which could thus be AU. The singularity for $D \approx r$ is obvious, since the test object would then be strongly attracted by the nearest parts of the ring, as if it were a straight line. Since $1/(D-r)$, DM rings have thus long- range effects, although the surface mass density $\sigma(r)$ may by quite small and can lead to compensation of attractions towards the Sun and away from it. At the center of any DM rings, $f(r,0) = 0$, while far away, it acts as if its total mass were concentrated at its center. Thus, $f(r,D) \to 2\pi/D^2$ when $D \gg r$.

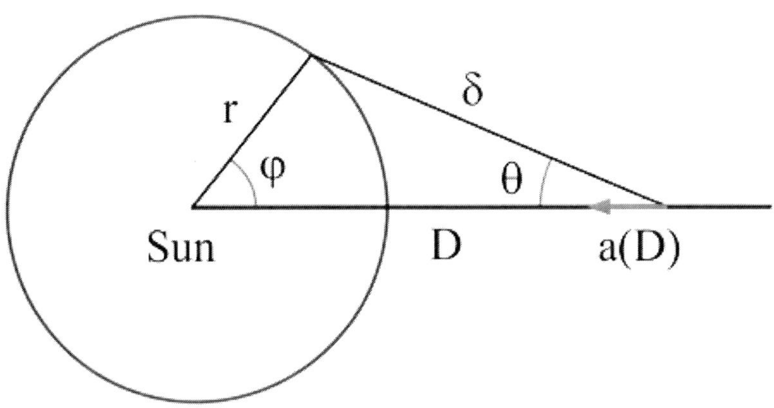

Figure 5. Gravitational attraction by a single ring of DM.

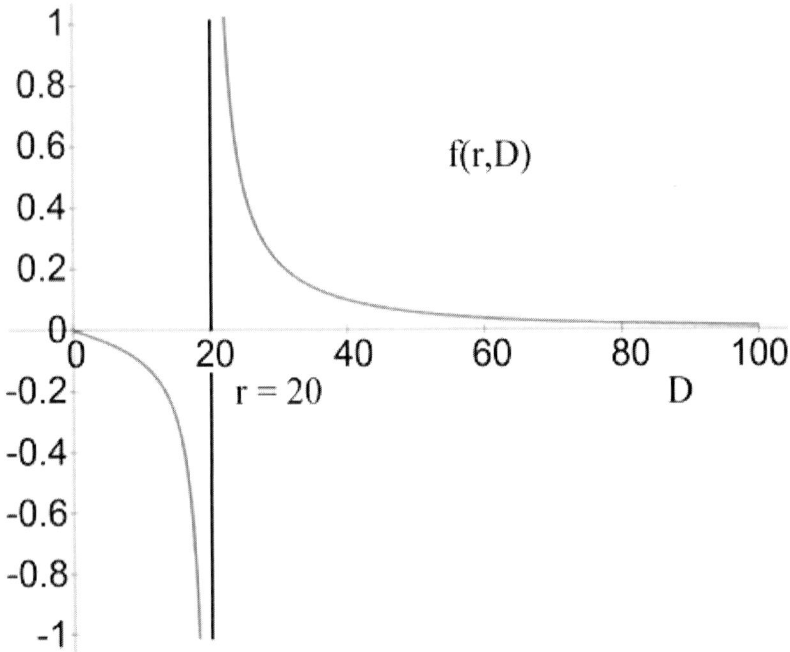

Figure 6. $f(r,D)$ for a DM circle of radius r = 20.

4.3. The Perplexing Pioneer Anomaly

Telemetric measurements of the velocity of the Pioneer 10 and Pioneer 11 spacecraft revealed a small deceleration when they were moving away from the Sun [34] . This is equivalent to an additional positive acceleration, oriented towards the Sun. It was unexpected and thus said to be "anomalous" with respect to Newton's law of gravity. When this result was published in 1998, simple causes and conceivable errors were discarded. Similar effects were even reported for other spacecraft. Figure 7 shows that the anomaly was detected for Pioneer 11, after its flyby of Jupiter (D = 5.2 AU) and Saturn (D = 9.6 AU). It increased up to the orbit of Uranus (D = 19.2 AU) and remained then nearly constant. The same value was also found when measurements were started and pursued for Pioneer 10, but there was then a slight decrease.

Unfortunately, the so-called "onset" of the Pioneer anomaly was only observed for Pioneer 11. Since Pioneer 10 had been launched one year earlier, its anomalous acceleration had not been noticed and was only measured for this spacecraft when D > 25 AU. Although the onset is a prominent feature of Figure 7, attention was only focused on the plateau. It was stated in 1998 that the "canonical" value of the anomalous acceleration is 8.74 (±1.33) × 10^{-10} m/s^2 [35] . This value applies to both spacecraft and the essential problem was to decide between two possibilities.

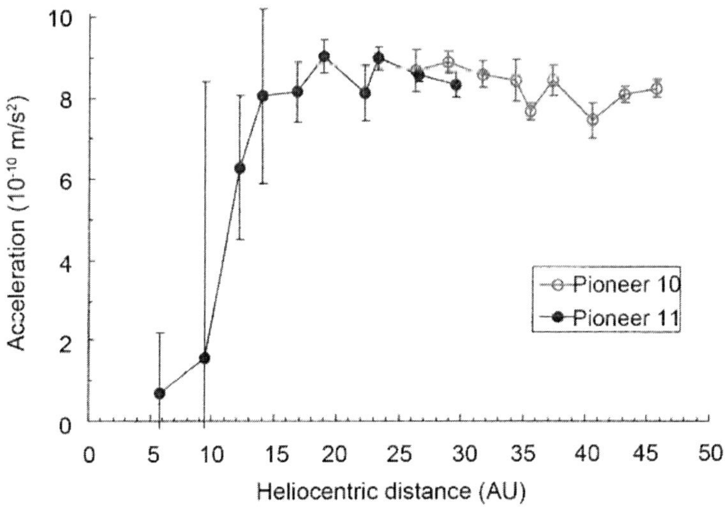

Figure 7. The "onset" of the Pioneer anomaly. The anomaly was not immediately noticed for Pioneer 10.

Does the anomaly result from a force of external origin that is directed towards the Sun, or should it be attributed to a small braking force, produced by the spacecraft themselves. Being aware of the possible existence of rotating DM rings in the ecliptic plane, we wanted to see if this hypothesis could explain the Pioneer anomaly. We developed thus the theory of the preceding section and calculated $a(D)$ by means of (11) for various distributions $\sigma(r)$. It appeared that the "onset" could easily be explained by nearly equal gravitational attractions towards smaller and greater DM rings. They tend to compensate one another until the inner DM rings attract more than the remaining outer rings. Figure 8 shows the result for the indicated distribution $\sigma(r)$ when G = 1. To perform the numerical integration in a simple way, we superposed here only two Gaussian curves, respectively centered at $r=5$ and $r=15$, with a width of 200 and 2000 and a central magnitude of 1.7 and 2.3. Since these distances could be expressed in AU, $\sigma(r)$ displays a broad maximum near the orbits of the great planets and vanishes only far beyond the Kuiper belt.

The Pioneer 10 and 11 spacecraft had been constructed in exactly the same way and they carried two external Pu^{238} radioisotope power sources. The possibility of anisotropic heat radiation had been examined and rejected [34] , but a short comment of Katz [36] stated that this effect might have been "underestimated". The NASA specialists answered [37] that enough power was available, but that the design and position of the power sources would not allow for sufficiently strong heating and anisotropic radiation of the large antenna. These sources "see" it edge on. Nevertheless, the hypothesis of a purely thermal effect came more and more in vogue.

An estimation of thermal effects by means of model calculations was published in 2008. It was asserted that recoil of IR photons could account for 35% to 57% of the canonical value [38] . This result was considered as providing "a clear indication of the possible thermal origin of the so-called Pioneer anomaly". In April 2012, the same authors included also diffuse reflection. This accounted for 26% - 66% of the plateau [39] . The "onset" was disregarded and the slow decrease of the anomaly was attributed to the radioactive decay of the Pu^{238} power sources. Rievers presented in January 2012 his doctoral thesis [40] , where he claimed that thermal effects "can explain the so-called Pioneer anomaly within a modeling accuracy of 11.5%".

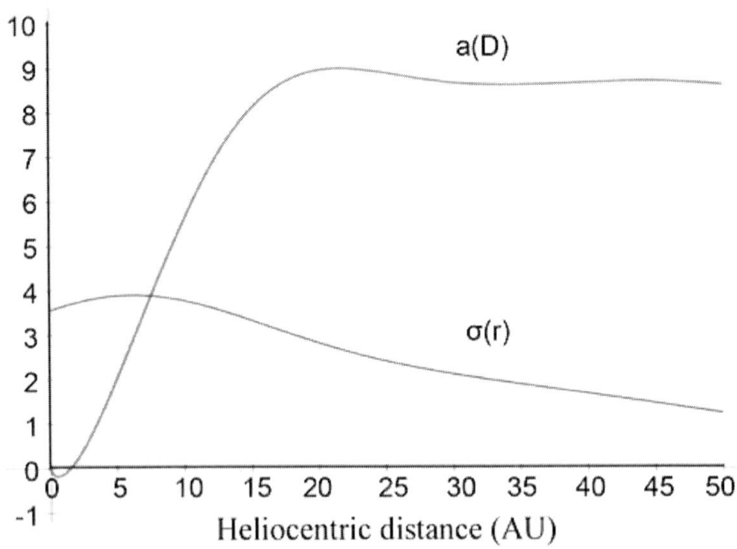

Figure 8. Calculated acceleration $a(D)$ and mass density $\sigma(r)$.

In April 2012, Turyshev et al. published the result of their own computations, based on a comprehensive model of Pioneer 10 and 11 and data available at the JPL. Evaluating thermal effects at various heliocentric distances, they found that the numerical estimate of the recoil force amounts to about 80% of the measured values [41] . However, they stated that "once the thermal recoil force is properly accounted for, no anomalous acceleration remains" and concluded even that the Pioneer anomaly "is consistent with known physics." Another computation [42] , published in July 2013, strengthened the conviction that "solutions may be achieved that do not require the addition of any 'unknown' acceleration other than the one of thermal origin."

The desire to stick to a conventional explanation was noticeable, but it could be misleading, since thermal modeling depends on many factors and no calculation did account at least for 100% of the canonical value. Moreover, the backscattering of thermal radiation could not suddenly start and increase for Pioneer 11 between 5 and 20 AU. Solar radiation pressure varies also in a smooth and predictable way. To prove or disprove the reality of the "onset" of the anomalous acceleration could have been decisive. Turyshev and his colleagues had recuperated initial data in 2005, but did not report that this input eliminated the threshold [43] . There remained uncertainties.

Since the author of this article had developed the theory of DM rings in 2012, he asked (by email on Dec. 3, 2012) if the onset was still there. Having calculated the acceleration $a(D)$ and found results like that of Figure 8, Meessen mentioned that rings of DM particles, rotating around the Sun in its ecliptic plane can explain the onset of the Pioneer anomaly and also the decrease of $a(D)$ between 20 and 45 AU. Slava Turyshev kindly answered that the onset subsists and confirmed that the decrease of the anomaly does not agree with the radioactive lifetime of the Pu^{238} energy source. He attributed this discrepancy to electronic parts of the energy conversion system. The author waited for proofs of a purely thermal effect by means of laboratory experiments, but the problem seemed to be settled, at least for public opinion.

4.4. Why Are Planets Not Affected?

Checking then the literature in regard to DM rings, we found that Moore and Moore [44] had also considered the gravitational effect of a DM distribution $\sigma(r)$ in the ecliptic plane of the Solar system. They had even determined the function $\sigma(r)$, which yields the closest match with Figure 7.

It is very similar to $\sigma(r)$ in Figure 8. These authors solved also another problem, at least partially. The motion of planets has been correctly described without considering any DM. We have shown in section 2.4 that the spherically symmetric Solar DM atmosphere is too small to influence the motion of planets, but why are they not perturbed by rotating DM rings?

If the acceleration $a(D)$ reported in Figure 7 were due to some additional gravitational force, oriented towards the Sun, it should have been noticed. This was proven in 2006 by Iorio and Guidice [45] . They found that the orbital motions of Uranus, Neptune and Pluto would have been perturbed, but this has not observed. Tangen [46] confirmed.

Moore and Moore showed that these planets are protected when they create deep trenches with steep walls in the DM distribution $\sigma(r)$. Because of the singularity of the function $f(r,D)$, represented in Figure 6, the acceleration $a(D)$ would then display two sharp peaks of opposite sign at the inner and outer walls of such a trench. In the middle, the dominant attraction towards the closest inner and outer DM rings would compensate one another. This was said to yield "gravitational nulls". It is thus possible to justify that $a(D)\approx 0$ for these planets, but why should there exist a gap in the distribution of rotating DM rings, precisely where the planets are?

We propose a mechanism that explains this apparently exotic assumption. Any planet attracts DM particles in its immediate vicinity. It creates thus a local void, but DM particles tend to be distributed as evenly as possible, because of mutual scattering. However, they have also to remain on the same orbit to insure dynamic equilibrium. They will thus move towards the planet along their normal orbit. Eventually, there remains a gap in the radial distribution of the DM mass-density distribution $\sigma(r)$ exactly on the orbit of the planet and close to it. Is this true? A visual proof is provided by a picture of Saturn's small moon Daphnis [47] . It is orbiting inside a circular gap of the visible rings. This confirms our proposition that asteroids and objects of the Kuiper belt are embedded in DM rings, rotating around the Sun. Saturn's famous rings confirm also that small objects remained distributed in rotating DM rings.

We can now use Figure 8 to estimate the surface mass-density $\sigma(r)$ of DM in the ecliptic plane. Since we get about the same value for $a(D)$ as the observed one when $\sigma(r)\approx 3$ for $r\approx 20\text{ AU}$ and G = 1, $\sigma\approx(3/G)10^{-10}$

m/s^2. Since $G = 6.8 \times 10^{-11}$ m^3/s^{2}kg, this yields $\sigma \approx 5$ kg/m^2, which is not negligible. It would thus be useful to launch another spacecraft to find out if the onset and other features of Figure 7 are reproduced or not. This could also help to solve other basic problems. The Solar system presents indeed several gravitational anomalies [48] [49] [50] . It is known, for instance, that the distance between the Earth and the Sun is increasing at about 15 m/century [51] . This secular increase of the astronomical unit could be due to a small negative value of the acceleration $a(D)$ when D ≈ 1. This appears in Figure 8, but would mean that the Earth is not sufficiently protected, although it creates also a circular gap in the DM density $\sigma(r)$. The global effect of all greater DM rings is somewhat stronger than that of smaller ones.

4.5. The Flyby Anomaly

If the Earth did really sweep up all DM that was close to its orbit, its center would be surrounded by rotating DM rings. They are invisible and we ignore their actual distribution. Since the Earth captured DM particles that were moving with respect to the center of the Earth and are not subjected to friction, they constitute a rotating DM disk. It is situated in the equatorial plane, since it is gravitationally coupled to the spheroidal shape of the Earth. DM particles will then be in dynamical equilibrium when they are orbiting at the adequate speed with respect to the center of the Earth. Is there any evidence of such a DM disk?

It does exist, but was too unexpected to be considered at least as a possible cause of a small, but detectable anomaly. According to Newtonian gravity, a spacecraft that is approaching a planet from far away, it will turn around it on a hyperbolic path in the reference frame where the planet is at rest. The speed of the spacecraft has the same value V when it enters and leaves the sphere of gravitational influence of the planet. This symmetry is broken in the heliocentric frame, since the planet is orbiting around the Sun. The spacecraft can then acquire a higher velocity. This “gravity assist maneuver” is very useful and the result is exactly predictable, but during the Earth flyby of the Galileo spacecraft in December 1990, there appeared an unexpected velocity increase ΔV. Other flyby anomalies were also detected. The values of ΔV could be positive or negative. The greatest one (13.46 mm/s) was observed for NEAR (Near Earth Asteroid Mission). Only 6 Earth flyby anomalies were known in 2009, but Anderson et al. [52] recognized a trend:

$\Delta V = KV\Delta_1$ where $\Delta_1 = \cos\delta_1 - \cos\delta_2$ (12)

K is a constant, when ΔV is proportional to V. The angles δ_1 and δ_2 are the declinations of the asymptotes for the arriving and departing spacecraft. They are equivalent to latitudes, but a spacecraft should not be affected by the rotation of the Earth. It follows from (12) that it is irrelevant whether the spacecraft approaches the Earth and leaves it above or below the equatorial plane. However, ΔV is positive and greater when the spacecraft is approaching closer to the equatorial plane than when it is leaving. This suggests that there is something in the equatorial that attracts the spacecraft at grazing incidence. The spheroidal shape of the Earth is not sufficient to explain the anomaly. Could it be due to the equatorial DM disk? Figure 9 and Figure 10 provide the answer in terms of a simple model.

Figure 9 shows the hyperbolic path of a spacecraft in the geocentric system and the angles δ_1 and δ_2. A spacecraft of mass m is then not only subjected to the gravitational force GmM/r^2 exerted by the mass M of the Earth, but also to the gravitational attraction of the ensemble of DM rings. For any given declination δ, it is equivalent to the additional attraction exerted by two identical masses μ. They are situated at a distance s from the center of the Earth along the azimuth of the spacecraft (insert of Figure 9). When the spacecraft is far away, its distance r to the central mass has to be replaced for the masses μ by $d_\pm = r \pm s\cos\delta$. The gravitational force acting on the spacecraft is thus

$$F(r) = Gm\left(\frac{M}{r^2} + \frac{\mu}{d_+^2} + \frac{\mu}{d_-^2}\right) \quad \text{where} \quad \frac{1}{d_\pm^2} = \frac{1}{r^2}\left(1 \pm \varepsilon\cos\delta\right)^{-2}$$

For $\varepsilon = s/r \ll 1$ and $m \ll M$, we get a second order correction:

$$F(r) = \frac{Gm}{r^2}\left(M + 3\mu\varepsilon^2\cos^2\delta\right) \quad \text{where} \quad \delta = \delta_1 \text{ or } \delta_2$$

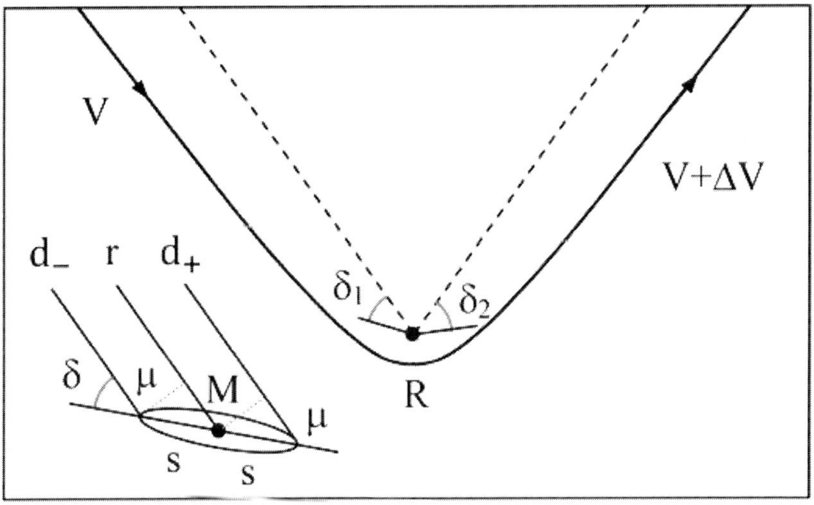

Figure 9. Flyby for a planet with a DM ring (see text).

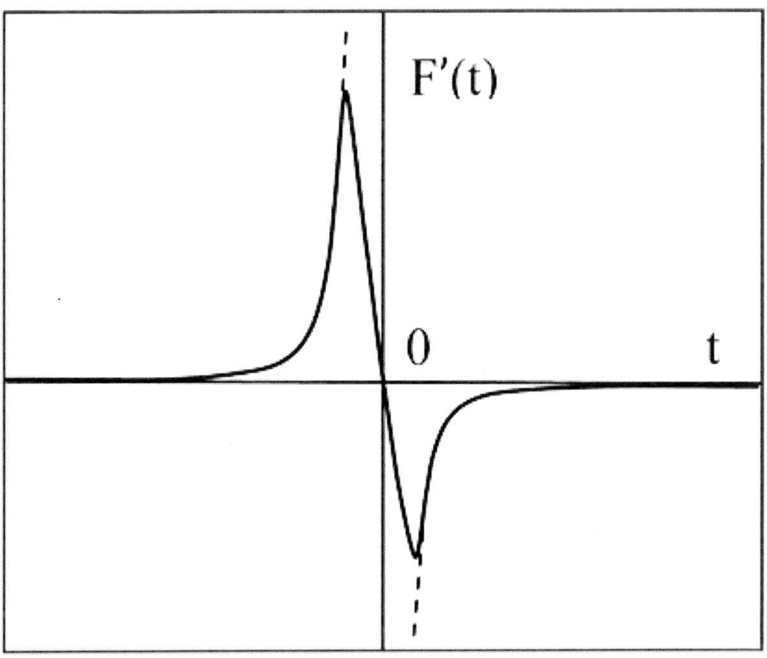

Figure 10. Force causing the flyby anomaly.

When the spacecraft is relatively close to the Earth, it is necessary to consider triangles instead of parallel lines and $d^2 = r^2 + s^2 - 2sr\cos\delta$. However, the spacecraft would be submitted to the same force $F(r)$. It acts like a central force, but the magnitude of the perturbation is

$$F'(t) \approx \frac{Gm}{r^4} 3\mu s^2 \cos^2\delta \quad \text{so that} \quad m\Delta V(t) \approx \int_{-\infty}^{t} F'(t)\mathrm{d}t \qquad (13)$$

The force $F'(t)$ decreases like $1/r^4$ instead of $1/r^2$ and the spacecraft is attracted as long as it is approaching, while it is decelerated when it is leaving. Figure 10 illustrates the variation of $F'(t)$. It is maximal near the point of closest approach, but it changes sign and does not vary in a symmetrical way when δ_1 and δ_2 are different. The global result will be accelerating when $\delta_1 < \delta_2$. The second relation (13) is based on the simplifying assumption that the perturbing central force is acting along the asymptotes of the hyperbolic path of the spacecraft. This is not true near the turning point, but the perturbation will then usually be negligible. The integral can be evaluated, since $r = r(t)$ and $\mathrm{d}t \approx \mathrm{d}r/V$. The modification of the orbital velocity $v(r)$ remains small:

$$\underline{\Delta V} \approx \frac{3G\mu s^2}{V}\int_R^\infty \frac{\mathrm{d}r}{r^4}\Delta_2 = \frac{K'}{VR^3}\Delta_2 \quad \text{where} \quad \Delta_2 = \cos^2\delta_1 - \cos^2\delta_2 \qquad (14)$$

The constant $K' = G\mu s^2$ depends only on intrinsic properties of the DM rings. It is identical for all Earth flybys. The distance R of closest approach is an essential parameter, since there will be no flyby anomalies when $R \gg s$. This did not appear in (12). The dependence on the velocity V is also different. Actually, the flyby anomaly has to increase for a slower spacecraft, since it is exposed during a longer time to the accelerating and decelerating forces. The declinations δ_1 and δ_2 do also have different effects, but Anderson and Nieto [52] deduced already from a small set of measurements that the velocity increase depends on $\cos\delta_1$ and $\cos\delta_2$. Jouannic et al. tried to find "potential causes" of the flyby anomaly by means of a statistical analysis [53] . Their empirical law was identical to (12) when

K is inversely proportional to the mass m of the spacecraft and to the height h of the perigee above the surface of the Earth.

Table 1 is based on data extracted from NASA's Horizon's web interface [54] . Only the values ΔV_{obs} for the reported flyby anomaly were taken from the initial report [52] and for Cassini, from the later compilation [53] . It should be noted that correlations do not yield causal relations, but the height h of the perigee above the surface of the Earth was an essential factor. It accounted at least approximately for the fact that no anomaly was observed for recent Earth flybys, as indicated in the column ΔV_{obs} of Table 1. We calculated the values of ΔV and $\underline{\Delta V}$ by means of (12) and (14), where the constants K and K' were chosen to get agreement with ΔV_{obs} for NEAR. It was not possible to measure the expected flyby anomaly for Juno, since this craft went itself into "save mode".

The proposed theoretical law (14) provides better agreement with observed values than (12), especially when the spacecraft passes far away from the Earth.

The main discrepancy is observed for Rosetta-1. It could result from a more dissymmetric transition from positive to negative values in Figure 10 when V and δ_1 are small. Since $K = G\mu s^2 \approx 3.96\times10^{13}$ km^3 (km/s)^2, it appears that $s^2\mu = 5\times10^{23}\ \text{m}^2\cdot\text{kg}$. We ignore the values of s and μ, but the radius of the Earth is close to 6400 km. For s ≈ 10,000 km, we would get $\mu \approx 5\times10^9$ kg, while the mass of the Earth $M \approx 6\times10^{24}$ kg. Nevertheless, the presence of DM rings rotating around the center of the Earth is detectable.

We add three remarks. 1) The DM disk of the Earth is probably smaller than the orbit of the Moon (r = 385,000 km), but the Moon and satellites that are moving on smaller circular orbits in the equatorial plane of the Earth would create a circular groove in the DM disk. 2) The flyby anomaly and the Pioneer anomaly apply to unprotected spacecraft. 3) Adler tried already to explain the flyby anomaly by considering the possible existence of a DM disk in the equatorial plane of the Earth [55] . However, he assumed that DM would interact with nucleons of the spacecraft, which is negligible [10] . Here, we considered only gravitational interactions.

Table 1. Parameters for Earth flybys and the corresponding values of the observed velocity increase, while ΔV and $\underline{\Delta V}$ are calculated ones.

Spacecraft (and pass)	Flyby (y.m.d)	R (km)	V (km/s)	δ_1 (°dec)	δ_2 (°dec)	ΔV_{obs} (mm/s)	ΔV (mm/s)	$\underline{\Delta V}$ (mm/s)
Galileo-1	90.12.08	7340	8.9	12.8	−34.0	3.92	4.08	2.98
Galileo-2	92.12.08	6682	8.9	34.0	−5.0	−4.6	−4.67	−4.55
NEAR	98.01.23	6910	6.9	21.0	−71.8	13.46	13.46	13.46
Cassini	99.08.18	7574	16.0	12.9	−5.0	−0.5	−1.06	−0.24
Rosetta-1	05.03.04	8338	3.9	2.4	−34.4	1.80	2.13	5.54
Messenger	05.08.02	8714	4.1	−31.4	−31.8	0.02	0.05	0.01
Rosetta-2	07.11.13	11,685	9.5	10.5	18.0	0	0.95	0.16
Epoxi-1	07.12.31	21,943	3.6	4.0	15.7	0	0.05	0.07
Epoxi-2	08.12.29	49,828	3.6	−21.0	59.0	0	0.47	0.05
Rosetta-3	09.11.13	8862	9.4	−18.5	+24.4	0	1.12	0.42
Epoxi-3	09.06.29	1,336,675	1.5	+4.0	−49.0	0	1.61	6×10^{-6}
Epoxi-4	09.12.28	1,325,441	1.1	−40.0	4.0	0	−0.80	-6×10^{-6}
Epoxi-5	10.06.27	36,860	3.5	60.2	−17.0	0	−5.05	−0.15
Juno	13.10.09	6947	10.4	−14.2	39.5	-	6.47	3.90

4.6. Global Positioning Satellites

Ben Harris concluded from a high-precion study of the motions of GPS, Galileo and GLONAS satellites that the mass of the Earth is 0.005% to 0.008% greater than the value established by the International Astronomical Union [56] . It has been stated [57] that he attributed this effect to the presence of a “disk of DM around the equator 191 km thick and 70,000 km across”. The relevant observational data and the calculations were not published, but it is not sufficient to object that “thin disks like that are generally the result of rapid rotation, while gravity tends to make things spherical” [58] . DM rings are possible.

Moreover, GPS satellites are orbiting around the Earth on circular orbits of radius r and specific inclinations (like 55°) with respect to the equatorial plane of the Earth. The declination δ will thus vary and the velocity $v = v(r, \delta)$. According to (13), the additional central force would vary like $\cos^2 \delta$. Since the average value is positive, it accounts for the apparent increase of the mass M of the Earth. If these satellites were passing through the equatorial DM disk, they would even be accelerated just before entering it and decelerated just after leaving it. The velocities $v = v(r, \delta)$ could thus reveal special effects when $\delta = 0$. When no singulariy is observed near the equatorial plane, when r = 26,600 km for instance, this would mean that the DM disk is smaller. A detailed study of the orbital motions of GPS satellites should thus be encouraged.

5. FIRST DIRECT DETECTION OF DM PARTICLES

5.1. The Annual Modulation Signature

In the context of present day uncertainties about the nature and properties of DM particles, any direct detection of DM particles would be very important. Nevertheless, we have to proceed by trial and error. In general, it has been assumed that DM particles can interact with nucleons, but we have shown [10] that this is very rare. Even the most sophisticated system that was based on this assumption did not succeed [59] . Since it was not obvious what method should be used, it was suggested already in 1986 to verify if "detected" signals are genuine or not. They should display an annual modulation [60] . Indeed, the Sun is moving at about 230 km/s with respect to the galactic DM atmosphere, while the Earth is orbiting around the Sun at about 30 km/s. The flux of DM particles should thus vary every year in a specific way. This would yield at least a "significant enhancement of signal-to-background ratio".

Fortunately, the orbit of the Earth around the Sun is not perpendicular to the "wind" of DM particles. It is inclined by about 62° with respect to the galactic disk. Moreover, the flux should be maximal on about June 2nd and minimal close to December 2nd. The detection rate has even to vary according to a perfect cosine function, whatever the nature of DM particles may be. It is required, however, that DM particles can be detected by the chosen method and that other possible causes of annual modulations have been excluded. In terms of these criteria, the DAMA (Dark Matter) experiment, installed in the Gran Sasso Tunnel near Rome, did succeed.

This laboratory is situated at about 1400 m under rocks and NaI(Tl) scintillators were chosen to detect galactic DM particles. First results were published in 2003 and provided already evidence of 7 annual cycles [61] . Using 9 very pure and well-protected crystals, the photomultipliers did yield a modulated counting rate. The ensemble of nearly 100 kg NaI crystals was then replaced by about 250 kg. Every one of the 25 scintillators in a protected 5×5 arrangement was coupled to a pair of photomultipliers. They had to respond in coincidence, to insure detection of DM particles by means of the flash of light that they produce when they are passing through a particular crystal. The initial results were confirmed. Annual modulation for 14 cycles led to a 9.3σ confidence level [62] .

The team leader Rita Bernaby and her collaborators insisted in various publications on the fact that detection occurred in a narrow energy range (2 - 6 keV). At slightly higher energies (6 to 14 keV), the annual modulation did not appear anymore. The energy calibration was regularly controlled by means of x-rays. No systematic process or other side effect could explain the ensemble of observed facts. The signature of a true detection of galactic DM particles is "stringent and unambiguous", but why was this method successful?

5.2. The DM-Electron Interaction

The DAMA team had chosen NaI(Tl) scintillators. Although the nature of DM particles was unknown, it had thus been assumed that a DM particle that passes through a particular crystal might excite there sufficient electrons to get a detectable light flash. Bernaby and collaborators analyzed this process in 2007. They assumed that electrons might be liberated from atoms, as for ionization processes [63] . It seemed important that electrons are not at rest in their initial state, while DM particles were considered as having some great mass (~GeV/c^2). The liberated electrons would then have very low energies, with a threshold effect. Similar and more sophisticated investigations followed in 2014 and afterwards [64] . We want to show here that DM-electron interactions are possible because of STQ (Figure 11) and that detection by NaI scintillators involves not only elastic collisions, but also some solid state physics (Figure 12).

Figure 11 indicates that a nark n_e, which could belong to any DM particle, can for instance be in the $\bar{R}G$ state. Its (u_x, u_y, u_z) quantum numbers are then $(1,-1,0)$, while $u_{ct}=0$ [10] . This nark can change its color by creating a gluon, which is immediately annihilated, by recreating a nark in the initial color state and the antinark of the first emitted one. This yields an N_2 boson, which is the mediator for interactions between DM particles, but also between DM particles and electrons. Indeed, an electon can annihilate the emitted nark to create a boson, which is the supersymmetric partner of an excited electron. Its charge Q = −1 and its mass could be relatively small, like the mass of an electron. This boson merges with the other member of the N_2 boson, to recreate an electron. STQ accounts thus for elastic scattering of an electron by DM particles.

The upper left part of Figure 12 represents a DM particle and its initial momentum vector, as well as the initial momentum of one electon. We assume here identical orientations, but it will turn out that this is irrelevant.

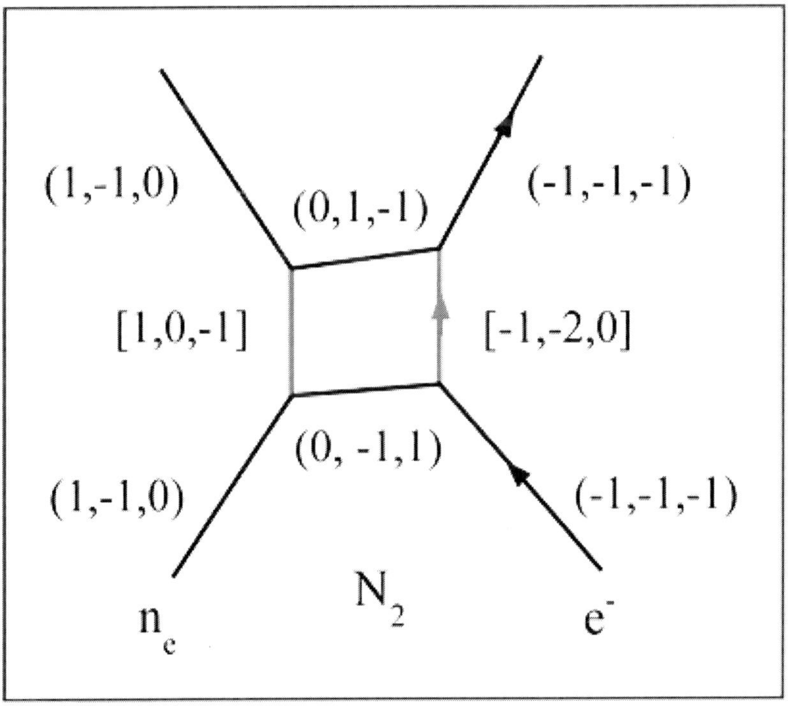

Figure 11. DM-electron interaction is mediated by an N_2 boson.

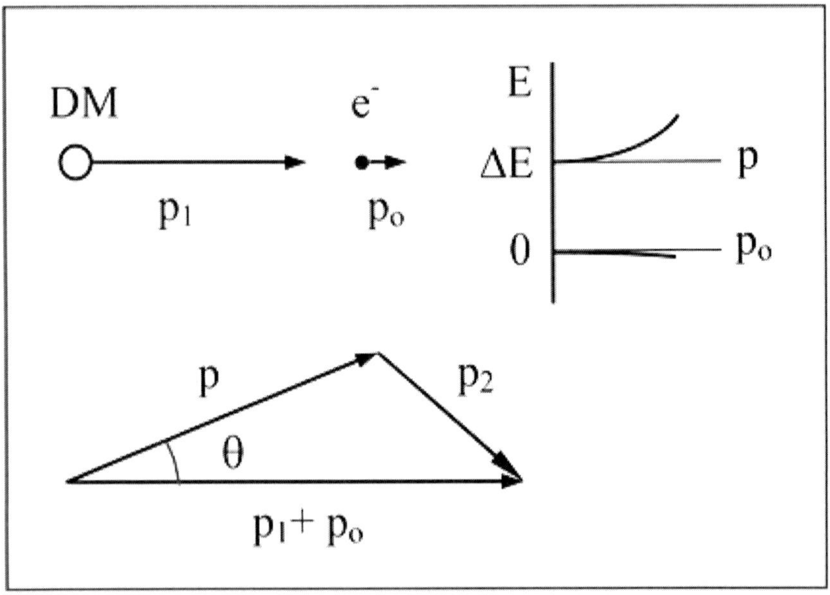

Figure 12. Momenta and possible energies, before and after an elastic DM-electron collision.

The mass M ot the galactic DM particle is unknown, but it is moving at least at 200 km/s with respect to the detector. Its initial momentum p_1 should thus be very great with respect to the momentum p_o of the electron in its initial state. This electron will be scattered by some angle θ and aquire a momentum of magnitude p. The DM particle is also scattered. The lower part of Figure 12 represents the vectorial momentum conservation law. The momentum of the DM particle changed from p_1 to p_2, where

$$p_2^2 = p^2 + (p_1 + p_o)^2 - 2p(p_1 + p_o)\cos\theta \qquad (15)$$

NaI crystals are insulators. They are excellent detectors for gamma rays, which create or liberate rather energetic electrons. For detection, it is sufficient that they excite electrons from the valence band to the conduction band of the NaI crystals. They will then move as free particles and be rapidly trapped in excited states of dispersed Tl atoms. These centers of luminescence allow for electronic transitions that yield detectable photons. The excitation of electrons and light emission occurs very rapidly and yields discrete scintillations, although many excited electrons were involved. The pulse height is even proportional to the energy of the incident particle.

DM particles can only be detected by this method when they transfer sufficient energy to electrons at the upper boundary of the valence band (E = 0) to reach the conduction band. Their energy is then $E = \Delta E + p^2/2m$, as indicated in the upper right part of Figure 12. The minimal excitation energy has thus to exceed the band gap ΔE. Setting $\Delta E = P^2/2m$ and $\mu = m/M$, energy conservation is expressed by

$$P^2 + p^2 = \mu\left(p_1^2 - p_2^2\right) \tag{16}$$

Since $m \square 1$ and $p_1 \square p_o$, it follows from (15) and (16) that

$$p^2 - 2\mu p p_1 \cos\theta + P^2 = 0.$$

Solving for p before considering equally probable angles θ, we get

$$p^2 = \frac{\mu^2}{2} p_1^2 - P^2 \quad \text{or} \quad E_e = \frac{m}{2M} E_1 - \Delta E$$

The kinetic energy of the excited electron $E_e = p^2/2m$ results only from the transfer of a small part of the kinetic energy $E_1 = p_1^2/2M$ of the DM particle. It can excite many electrons, when the band gap ΔE is not too great. This can produce an intense scintillation, comparable to that of x-rays of relatively low energy (2 - 6 keV). For NaI crystals, ΔE = 5.8 eV, while ΔE = 6.3 eV for CsI crystals [65] . The "Korea Invisible Mass Search" (KIMS) tries to detect DM particles by means of CsI(Tl) scintillators. They are less hydroscopic, but seem to be less effective. The flash of light could be weaker for the same type of DM particles or only some of them might be detected. The "Coherent Germanium Neutrino Technology" (CoGeNT) in Minnesota was intended to detect DM particles by means of nuclear recoil, but a modest annular modulation was reported [66] . It can be attributed to DM-electron interactions.

5.3. Detection of Galactic DM Signals

We mentioned and discussed the detection 130 GeV gamma-rays by means of NASA's Fermi telescope and proposed a mechanism for the production of energetic photons according to STQ [10] . However, there is also an x-ray signal at about 3.5 keV. It was detected by means of the XMM-Newton x-ray telescope of the European Space Agency and by NASA's Chandra x-ray observatory [67] . It is a very weak signal, but it appeared in stacked spectra of 73 galaxy clusters with various redshifts. Dark matter could thus be considered as the possible cause. The XMM-Newton telescope detected these x-rays for the Andromeda galaxy and the Perseus cluster, with a 4.3σ confidence level [68] . The 3.5 keV x-ray line was also detected for the center of the Milky Way [69] , but not for the Virgo cluster. The signal was not "clear" enough.

Moreover, x-rays can also result from electronic transitions when highly ionized atoms capture electrons from hydrogen atoms. However, the strongest objection resulted from the measurement of the distribution of x-ray sources at 3.5 keV in the Perseus cluster and near the center of the Milky Way. These sources did not display the radial symmetry, expected for galactic DM density profiles. There were radial and azimuthal irregularities [70] . The spatial distribution of x-rays sources at neighboring energies (between 2 and 5 keV) displayed also irregularities. Plasma emissions, due to charge transfer for highly ionized atoms, are thus very probable for these sources. Moreover, it has now been experimentally proven that that S^{15+} and S^{16+} ions capture electrons and emit x-rays near 3.4 eV [71] . The proposition that the 3.5 keV signal results from DM is questionable.

Nevertheless, skepticism should be omnidirectional. We mention 4 reasons to be prudent. 1) The detected irregular distribution of x-ray sources could result from electronic transitions, while DM signals at 3.5 eV were too weak to establish a map. 2) Since they were absent for the Virgo cluster and are always weak, they could be produced only by a particular type of DM particles. 3) The mass-density distribution $\rho(r)$ is spherically symmetric for DM atmospheres, but it does not distinguish different types of DM particles from one another. 4) We know that they are possible [10] and galactic DM atmospheres could result from gathering clouds of DM particles of different composition. Although mutual scattering of DM particles tends to produce a homogeneous mixture, this takes time and the relevant space is immense.

Could the 3.5 keV x-ray signal be produced by at least one particular type of DM particles? Figure 13 shows a process that could be considered for DM particles of type N_1. They were produced during the Big Bang and should still be present in our universe [10] [11] . They correspond to single narks in the (0,0,0) color-neutral state. Two N_1 particles could thus meet one another and be transformed into a color-neutral gluon. According to the conservation law for u-quantum numbers, this would allow for $[0,0,0] \rightarrow (1,1,1)+(-1,-1,-1)$. The resulting electron-positron pair can yield two photons. We ignore if this process made it possible to detect 3.5 keV photons, but if this were true N_1 DM particles would have this energy. Incidentally, we found the same magnitude for the average mass of galactic DM particles (Section 2.2). It was even suggested by the DAMA team (Section 5.1). Since fusion of DM particles reduces their energy [11] , the possible detection of a galactic DM signal at 3.5 keV merits further studies.

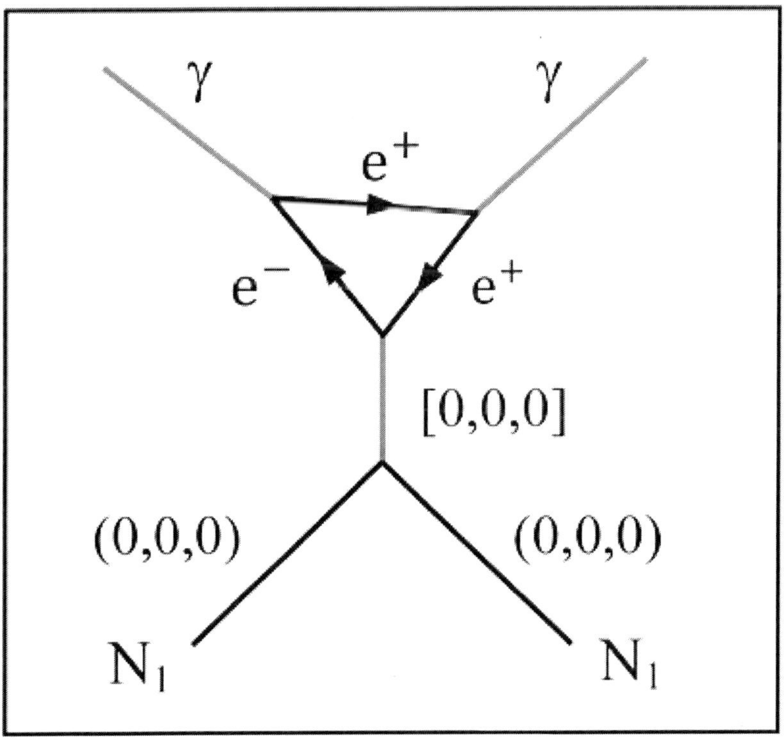

Figure 13. STQ could account for transformations of N_1 dark matter particles into a pair of photons.

6. CONCLUSIONS

The existence of a space-time continuum seemed to be obvious, but this conviction was based on a questionable assumption [10] . We constructed thus a theory of space-time quantization (STQ), where the value a of the ultimate limit for the smallest measurable length was treated as a yet unknown parameter. It appeared that there are no logical inconsistencies when $a \neq 0$ and that this is sufficient to account for the Standard Model of elementary particle physics. The agreement with numerous and varied observations allows us to conclude that STQ is real. Moreover, the Standard Model is generalized and provides then insight into the nature and properties of DM particles. This can also be tested by applying these concepts to cosmology [11] and astrophysical observations.

We did this here in various ways. STQ predicts that DM particles interact by exchanging N_2 bosons and that elastic scattering is predominant for the cosmic DM gas. The resulting pressure is proportional to the density of DM particles and increases thus when the DM gas is compressed by gravitational forces. This leads to the concept of DM atmospheres in hydrostatic equilibrium. It accounts very well for observed properties of DM halos for spiral galaxies. It appeared also that the mass of the central black hole is irrelevant. DM atmospheres could even contain only a small amount of baryonic matter, as has been observed.

The same concepts were applied to the DM atmosphere of the Sun and to explain the stability of the Oort cloud. DM atmospheres are also very important to prevent gravitational collapse of globular star clusters and cosmic filaments, where stars and galaxies remain separated from one another by being embedded in DM atmospheres. Moreover, there are rotating DM rings in the ecliptic plane of the Solar system. They explain the Pioneer anomaly. We proposed a mechanism that explains how planets sweep up all DM that is orbiting at the same distance from the Sun. This does usually protect them from gravitational effects of all DM rings in the ecliptic plane. It explains also why planets have their own DM disk. It is rotating in their equatorial plane and accounts for the flyby anomaly. Properties of this DM disk can also be studied by means of GPS satellites.

We provided already some information concerning the possible production and detection of DM particles [10] . We showed here that STQ allows also for DM-electron interactions. This explains why galactic DM could already be detected by the DAMA collaboration. We advanced also some reasons to reject not yet the possibility that the weak 3.5 keV x-ray signal might be produced by light DM particles. The global conclusion is that STQ and the resulting dark matter theory withstood the test of a detailed and varied confrontation with astrophysical observations. Reality is more astonishing than we thought.

CITE THIS PAPER

Meessen, A. (2017) Astrophysics and Dark Matter Theory. Journal of Modern Physics, 8, 268-298.

REFERENCES

1. Zwicky, F. (1933) Helvetica Physica Acta, 6, 110-127. http://adsabs.harvard.edu/abs/1933AcHPh...6..110Z
2. Rubin, V and Ford, W.K. (1970) The Astrophysical Journal, 159, 379-403. http://adsabs.harvard.edu/full/1970ApJ...159..379R
3. Rubin, V., et al. (1976) The Astronomical Journal, 81, 719-737. http://adsabs.harvard.edu/full/1976AJ.....81..719R
4. Rubin, V., et al. (1980) The Astrophysical Journal, 238, 471-487. http://adsabs.harvard.edu/full/1980ApJ...238..471R7
5. Bosma, A. (1981) The Astronomical Journal, 86, 1791-1824. http://adsabs.harvard.edu/full/1981AJ.....86.1791B
6. Bosma, A. (1981) The Astronomical Journal, 86, 1825 1846. http://adsabs.harvard.edu/full/1981AJ.....86.1825B
7. Brownstein, J.R. and Moffat, J.W. (2005) The Astrophysical Journal, (2006) 636, 721-741.
8. McGaugh, S.S., et al. (2016) Physical Review Letters, 117, 201101, 1-5.
9. Science News (2016) In Rotating Galaxies, Distribution of Normal Matter Precisely Determines Gravitational Acceleration. https://www.sciencedaily.com /releases/2016/09/160921085052.htm
10. Meessen, A. (2017) Journal of Modern Physics, 8, 35-56.
11. Meessen, A. (2017) Journal of Modern Physics, 8, 251-267.
12. Navarro, J.F. Frenk, C.S. and White, S.D. (1996) The Astrophysical Journal, 462, 563-575. http://arxiv.org/abs/astro-ph/9508025
13. Burkert, A. and Silk, J. (1999) On the Structure and Nature of DM Halos. https://cds.cern.ch/record/384623/files/9904159.pdf
14. Einasto, J. (1965) Kinematics and Dynamics of Stellar Systems, Trudy Inst. Astrofiz. Alma-Ata 5, 87.
15. Merritt, D., et al. (2006) The Astronomical Journal, 132, 2685-2700.
16. Di Cintio, A., et al. (2010) Monthly Notices of the Royal Astronomical Society, 000, 1-11, (2013).
17. Spergel, D.N. and Steinhardt, P.J. (2000) Observational Evidence for Self-Interact-ing Cold Dark Matter.
18. Kruezi, L. (2015) The Case for Complex Dark Matter. Interview of James Bullock, Quanta Magazine. https://www.quantamagazine.org/20150820-the-case-for-complex -dark-matter/
19. Erben, T., et al. (2000) A&A, 355, 23.
20. Capelato, H.V., et al. (2008) A&A, 492, 345.
21. van Dokkum, P., et al. (2016) The Astrophysical Journal Letters, 828, 1.
22. Feltzman, R. (2016) A New Class of Galaxy Has Been Discovered, One Made Almost Entirely of Dark Matter. The Washington Post, 25 August 2016.

23. Tully, R.B. and Fischer, J.R. (1977) Astronomy & Astrophysics, 54, 661-673. http://www.wikiwand.com/sv/Tully-Fisher-relationen
24. Read, J.I. (2014) The Local DM Density. JPhysG-100038.R1.
25. Pitjec, N.P. and Pitjeva, E.V. (2013) Astron. Zhurnal or Astronomy Letters, 39, 141-149.
26. Jeans, J.H. (1913) MNRAS, 74, 109-112. http://adsabs.harvard.edu /full/1913MNRAS..74..109J
27. Taylor, M.A., et al. (2015) The Astrophysical Journal, 805, 1-20.
28. Caimmi, R. (2009) Serbian Astronomical Journal, 179, 31-47. http://arxiv. org/pdf/0909.3056.pdf
29. Cappellari, M., et al. (2015) The Astrophysical Journal Letters, 804, L21.
30. Harvey, D., et al. (2015) Science, 347, 1462-1465.
31. National Radio Astronomy Observatory (2014) Birth of Planets Revealed in Astonishing Detail in ALMA's Best Image. https://public.nrao.edu/news /pressreleases/planet-formation-alma
32. Bovaird, T. and Lineweaver, C.H. (2013) MNRS, 435, 1126-1138.
33. Nieto, M.M. (2005) Physical Review D, 72, Article ID: 083004.
34. Anderson, J.D., et al. (1998) Physical Review Letters, 81, 2858-2861.
35. Turyshev, S.G. and Toth, V.T. (2010) Living Reviews in Relativity, 13, 1-173.
36. Katz, J.I. (1998) Physical Review Letters, 83, 1892.
37. Anderson, J.D., et al. (1999) Physical Review Letters, 83, 1893.
38. Bertolami, O., et al. (2008) Space Science Reviews, 151, 75-91 (2010).
39. Francisco, F., et al. (2012) Physics Letters B, 711, 337-346.
40. Rievers, B. (2012) High Precision Modeling of Thermal Perturbations with Applications to Pioneer 10 and Rosetta, Dissert. 1-169. http://elib.suub. uni-bremen.de/edocs/00102500-1.pdf
41. Turyshev, S.G., et al. (2012) Physical Review Letters, 108, Article ID: 241101.
42. Modenini, D. and Tortora, P. (2014) Physical Review D, 90, Article ID: 022004.
43. Turyshev, S.G., et al. (2006) International Journal of Modern Physics D, 15, 56.
44. Moore, G.S.M. and Moore, R.E.M. (2013) Astrophysics and Space Science, 347, 235-247.
45. Iorio, L. and Gioudice, G. (2006) New Astronomy, 11, 600-607.
46. Tangen, K. (2007) Physical Review D, 76, Article ID: 042005.
47. Wilkins, A. (2010) NASA. http://io9.gizmodo.com/5584149/see-the-amazing-moon-inside-saturns-rings
48. Anderson J.M. and Nieto, M.M. (2009) Proceedings of the IAU Symposium, 261.
49. Lämmerzahl, C., Preus, O. and Dittus, H. (2007) Astrophysics and Space Science Library, 349, 75-101. http://www.uv.es/jfgf/Pioneer/0604052.pdf
50. Iorio, L. (2014) International Journal of Modern Physics D, 24, 1530015, 1-35.
51. Krasinsky, G.A. and Brumberg, V.A. (2004) Celestial Mechanics and Dynamical Astronomy, 90, 267-288. http://www.aldebaran.cz/bulletin/2009_24/GAKVAB.pdf
52. Anderson, J.D., et al. (2008) PRL, 100, Article ID: 091102, 1-4. http://virgo. lal.in2p3.fr/NPAC/relativite_fichiers/anderson_2.pdf
53. Jouannic, B., et al. (2015) The Flyby Anomaly: An Investigation into Potential Causes. Proc. 25th Internat.Symp. Space Flight Dynamics, 1-17. http://issfd.org/ 2015/files/downloads/papers/153_Jouannic.pdf
54. NASA, Solar System Dynamics, Horizons Webb-Interface. http://ssd. jpl.nasa.gov/horizons.cgi
55. Adler, S. (2009) Is There Dark Matter in Orbit around the Earth? Institute Adv. Studies. https://www.ias.edu/ideas/2009/adler-dark-matter
56. Harris, R. (2013) Searching for Dark Matter using Navigation Satellites, American Geophysical Union, Fall Meeting. http://adsabs.harvard.edu/ abs/2013AGUFM.G21A0752H
57. www.newscientist.com/article/mg22129503-100-gps-satellites-suggest-earth-is-heavy-with-dark-matter/

58. M. Francis. http://galileospendulum.org/2014/01/02/no-dark-matter-is-not-messing-up-gps-measurements/
59. LUX (Large Underground Xenon) Collaboration (2016) Results from a Search for Dark Matter in the Complete LUX Exposure.
60. Drukier, A., Freese, K. and Spergel, N. (1986) Physical Review D, 33, 3495-3508. http://kicp.uchicago.edu/~collar/drukier.pdf

61. Barnabei, R., et al. (2003) La RivistadelNuovoCimento, 26, 1-73.
62. Barnabei, R., et al. (2014) Advances in High Energy Physics, 2014, Article ID: 605659, 1-10. https://www.hindawi.com/journals/ahep/2014/605659/
63. Barnabei, R., et al. (2008) Physical Review D, 77, Article ID: 023506, 1-15.
64. Roberts, B.M., et al. (2016) P Physical Review D, 93, Article ID: 115037, 1-23.
65. Alkali Halides. http://www.ucl.ac.uk/~ucapahh/research/crystal/homepage.htm
66. Aalseth, C.E., et al. (2014) Search for An Annual Modulation in Three Years of CoGeNT Dark Matter Detector Data.
67. Bulbul, E., et al. (2014) The Astrophysical Journal, 789, 23 p.
68. Boyarsky, A., et al. (2014) Physical Review Letters, 113, Article ID: 251301.
69. Boyarsky, A., et al. (2015) Physical Review Letters, 115, Article ID: 161301.
70. Carlson, E., Jeltema, T. and Profumo, S. (2015) Journal of Cosmology and Astroparticle Physics, 2015, 1-27.
71. Shah, C., et al. (2016) Laboratory Measurements Compellingly Support Charge-Exchange Mechanism for the "Dark Matter" ~3.5 keV X-Ray Line.

CHAPTER 11

Order Statistics of High-Intensity Speckles in Stimulated Brillouin Scattering and Plasma-Induced Laser Beam Smoothing

S Hüller[1,3], A Porzio[1,2] and J Robiche[1]

[1] *Centre de Physique Theorique, CNRS, Ecole Polytechnique, ´ F-91128 Palaiseau, France*

[2] *LAGA, Institut Galilee, Universit ´ e Paris 13, CNRS, F-93430 Villetaneuse, ´ France*

[3] *Centre de Physique Theorique, CNRS, Ecole Polytechnique, ´ F-91128 Palaiseau, France*

ABSTRACT

We have studied plasma-induced smoothing due to stimulated Brillouin scattering (SBS) under the aspect of the extremal statistics of smoothed laser beams. As pointed out in the work by Rose and DuBois (1994 *Phys. Rev. Lett.* **72** 2883), scattered light can be subject to uncontrolled (or even 'explosive') behaviour, associated with a critical gain value for SBS. In this work we show how this critical behaviour can be predicted on the basis of the order statistics of laser speckle fields, and we analyse the transition to uncontrolled behaviour of the laser beam due to the dominance of high intensity speckles.

1. INTRODUCTION

The spreading of optically smoothed laser beams propagating through a plasma has been the subject of theoretical [1–5] and experimental studies [6–8] in recent years. The common interest of these studies was to understand the potential additional smoothing capacities of the plasma that can eventually improve the laser imprint on targets situated behind the plasma layer. That the interaction of the laser beam with the plasma can lead to strong self-focusing (SF) and 'dancing filaments' was demonstrated by simulations in [1]. In this case the strong angular spreading of the beam comes together with loss in temporal coherence due to the non-stationary behaviour of the transmitted light. At lower beam intensities, such that laser speckles at average intensity are not subject to SF, stimulated Brillouin forward scattering (FSBS) was understood as the principal mechanism leading to an additional loss in spatial and temporal coherence in the transmitted light. This loss of coherence in the plasma eventually has useful smoothing capacities. In experiments, mostly using gas jets, with both weakly and strongly ionized plasmas, the smoothing could be identified to be active [7]. In a further step towards higher electron densities, plasmas generated from low-density foam targets were used and the reduction of the laser imprint contrast could be clearly demonstrated [6]. The process leading to additional smoothing is most commonly known in the literature as 'plasma-induced smoothing'.

A critical aspect of this smoothing technique is the potential transition towards uncontrolled behaviour in the scattered light with strong spreading. Uncontrolled characteristics of the beam are, of course, not desirable for the purpose of plasma-induced smoothing. In earlier work by Rose and DuBois [9], the possible explosive behaviour of scattered light from stimulated scattering instabilities has been pointed out and has been associated with a critical gain value of the instability. For the case of stimulated Raman scattering, Rose and Mounaix [10] have recently shown the evidence for the explosive behaviour in spite of diffraction. Mounaix and Divol [11] have shown that the cumulative response of stimulated Brillouin scattering (SBS), based on the independence of speckles (the 'independent hot spot model'), can spread strongly around a critical gain value.

In this paper we show the physical relevance of the high-intensity speckle order statistics for SBS and for the control of the plasma-induced smoothing of optically smoothed laser beams incident on the plasma layer. Since the onset of the latter mechanism is closely related to the amplification of FSBS, the effect of extreme speckles can lead to the poorly controlled spreading of the light beams exiting the plasma layer.

The paper is organized as follows: in section 2 we recall first the model equations for SBS following the model by Elisseev *et al* [12] to determine the solution for the scattered light field response from a single speckle as a function of the standard expression for the gain coefficient [13]. These results will be used to derive the total light field intensity in section 3. For a first layer of speckles (in the propagation direction) of a speckle pattern generated from a random phase plate (RPP), one can make use of a perturbative approach for the scattered light field. In section 5, the modification of the total light intensity due to the scattered light is then evaluated by computing the sum over speckles, using order statistics of speckle fields for both forward (FSBS) and backward SBS (BSBS). From the cumulative response of the scattered light from all speckles, we compute the speckle intensity range corresponding to the speckles which contribute dominantly to this response. Through evaluation of order statistics and by means of the numerical simulations in section 6, we demonstrate that the cumulative response of the scattered light critically depends on the average gain coefficient, associated with the spatial amplification in a single speckle, $\langle G \rangle$. In both sections 5 and 6, a modified speckle distribution for the total field is determined that shows departure from an exponential-type decrease as a function of the speckle intensity around the expectation value of the highest speckle intensity.

2. MODEL EQUATIONS

The system of coupled differential equations for SBS, in the approximation of slowly varying envelopes, can be written as

$$\left(\partial_t + \nu_{\text{iaw}} + \frac{c_s^2}{\omega_{\text{iaw}}}(\vec{k}_0 - \vec{k}_1)\cdot\nabla + \mathrm{i}\frac{c_s^2}{2\,\omega_{\text{iaw}}}\Delta_\perp\right)\frac{n_1}{n_0} = \mathrm{i}\frac{2k_0^2\sin^2(\theta/2)}{\omega_{\text{iaw}}}\beta_{\text{SBS}}E_0E_s^* \tag{1}$$

for the ion acoustic wave (IAW) density perturbation n_1 (with respect to the average electron density n_0) where ν_{iaw} stands for the IAW damping (Landau damping), c_s for the sound speed, $\omega_{\text{iaw}} \equiv 2k_0c_s\sin\theta/2$ for the IAW frequency, β_{SBS} for the coupling coefficient (defined below), $\vec{k}_{0,1}$ for the wave vectors of the incident (0) and scattered (1) electromagnetic wave components, respectively, and with the angle θ between $\vec{k}_0$ and $\vec{k}_1$ and

$$\left(\partial_t + \vec{c}_{g,0}\cdot\nabla + \delta\epsilon_0 + \mathrm{i}\frac{c^2}{2\omega_0}\Delta_\perp\right)E_0 = -\mathrm{i}\frac{\omega_0}{2}\frac{n_0}{n_c}(n_1E_s)/n_0, \tag{2}$$

$$\left(\partial_t + \vec{c}_{g,1}\cdot\nabla + \delta\epsilon_1 + \mathrm{i}\frac{c^2}{2\omega_1}\Delta_\perp\right)E_s = -\mathrm{i}\frac{\omega_1}{2}\frac{n_0}{n_c}\left(n_1^* E_0\right)/n_0, \tag{3}$$

for the electromagnetic incident (pump) and scattered wave component with the (complex-valued) amplitudes E_0 and E_s at frequencies ω_0 and ω_1, respectively, written in paraxial approximation. Here $\vec{c}_{g,0}$ and $\vec{c}_{g,1}$ are the group velocities and $\delta\epsilon_{0,1}$ stands for the difference of the complex dielectric constant $\delta\epsilon_{0,1} = (c^2k^2_0/2\omega_{0,1})[1 - n_e(z)/[n_c(1 + \mathrm{i}\nu_{coll}/\omega_{0,1})] - k^2/k^2_0]$ with respect to the dielectric constant used in the dispersion relation where $k = k_0(1 - n_{ref}/n_c)^{1/2}$ is the wave number for a reference density n_{ref} (here $=n_0$) and ν_{coll} the frequency for collisional absorption. The fields are normalized to a reference value $\hat{E}_0$ which is the peak field value associated with the incoming laser flux. The terms $(\vec{c}_g\cdot\nabla)$ can also be expressed as a function of the angle between $\vec{k}_0$ and $\vec{k}_1$ such that $(\vec{c}_g\cdot\nabla) \simeq c_g\cos\theta\,\partial_z$ holds in the 'paraxial' limit.

Note that in equation (1) the dominating term of $(\vec{k}_0 - \vec{k}_1)\cdot\nabla$ for small angles is given by $\equiv 2k_0\sin(\theta/2)\nabla_\perp$. We define the dimensionless coupling constant at the peak value of the real field, related by $\hat{E} = \hat{E}_0/2$ to the peak field value of the pump (laser) field to which the quantities E_0 and E_s are normalized to, as

$$\beta_{SBS} = \frac{e^2\langle\hat{E}^2\rangle}{m_e m_i c_s^2\omega_0^2} = \frac{1}{4}\frac{\epsilon_0 c\hat{E}_0^2}{n_c c k_B T}$$

with m_e (m_i) denoting the electron (ion) mass, T the effective ($\simeq$ electron) temperature, k_B the Boltzmann constant and ϵ_0 the elementary dielectric constant.

The laser flux ('intensity') is given by

$$I_L = \frac{c\epsilon_0\bar{E}_0^2}{2} = 2c\epsilon\hat{E}^2 = 2\beta_{SBS}n_c c k T$$

which yields, in practical units, the relation

$$I_L = 9.6\times10^{15}\,(\mathrm{W\,cm^{-2}})\beta_{SBS}\frac{n_c}{1.1\times10^{21}\mathrm{cm^{-3}}}T_{keV} \quad \text{or} \quad \beta_{SBS} \simeq 0.01 I_{14}\lambda_\mu^2/T_{keV},$$

where T_{keV} stands for ZT_{eff}/A in units of keV, $\lambda\mu$ for the laser wavelength ($\lambda = 2\pi/k_0$) in microns and I_{14} for the laser intensity in 10^{14} W cm^{-2}.

3. SCATTERING AMPLITUDE

In [12] by Elisseev *et al* it has been shown that the scattered light field along an integration path can be expressed in the form

$$E_s(\theta, \nu t) = E_{seed}\left(1 + \int_0^{\nu t} d\tau \sqrt{\frac{G_l}{\tau}}\mathcal{I}_1\left(2\sqrt{\tau G_l}\right)e^{-\tau}\right), \tag{4}$$

which involves the Bessel function $\mathcal{I}_1$ and the quantity Gl representing the spatial SBS gain coefficient resulting from an integration path trespassing the laser pump wave profile with an angle θ with respect to the propagation direction, see appendix A,

$$G_l(\theta) \equiv \frac{\xi_l(\theta)}{\nu} = \int_{l'=-l}^{l} \frac{\gamma_0^2(l', \theta)}{c_g \nu} dl'. \tag{5}$$

The standard growth rate, γ_0, depends on the angle between the path and the principal propagation axis $\| z$. We recall Elisseev's approach in appendix A and develop there the time-dependent solution in more detail. The time-dependent solution for E_s has the simple and well-known asymptotic solution

$$E_s(\theta, \tau \to \infty) = E_{seed}\exp(G_l), \tag{6}$$

in which E_{seed} stands for the noise level of the scattered light; it establishes for strongly damped IAWs a time interval $\tau > 2Gl$, see appendix A.

For an inhomogeneous pump wave profile, like in laser speckles, the angular dependence, θ, of the integration path has to be taken into account. It is then evident that the expression (5) for a sufficiently long integration path with respect to the speckle correlation length, is nothing else but the SBS gain over a single speckle, which can be written as

$$G_l(\theta) \equiv G_{l_R}\mathcal{F}(\theta, 0)\sin(\theta/2) \tag{7}$$

with $\mathcal{F}$ being a form factor that depends on the shape of the speckle, see appendix B. For the correlation length, one can assume $l_R \sin\theta/2$, with l_R being the Rayleigh length. The peak value of the speckle gain Gl_R is then defined as

$$G_{l_R} \equiv \frac{\hat{\gamma}_0^2}{c_g \nu_{iaw}} 2l_R \sin(\theta/2)|_{\theta=\pi} = \frac{1}{4}\frac{\omega_1}{c_g}\frac{\omega_{iaw}}{\nu_{iaw}}\frac{n_0}{n_c}\beta_{SBS} \tag{8}$$

with $\hat{\gamma}_\theta^2 = \max\{\gamma_0^2(l,\theta) \text{ in } l = [-l_R, l_R]\}$. Here enters the (normalized) peak laser intensity, $I_{sp} \propto \beta_{SBS}$, of the speckle and in which $\omega_{iaw}(\theta) \equiv 2k_0c_s \sin\theta/2$ is the IAW frequency.

The damping for IAWs may similarly depend on the angle, so that the angular dependence of $(\omega/\nu)_{iaw}$ is typically very weak. The latter is the case when linear Landau damping is the dominant damping mechanism, because both the linear IAW Landau damping coefficient as well as the IAW frequency itself are proportional to the IAW wave vector. For viscous damping, usually proportional to the square of the wave number, the angular dependence of $(\omega/\nu)_{iaw}$ has to be considered.

The value of the gain Gl hence depends on the length, $l_{eff}(\theta)$ of the efficient amplification along the path. We will use the expressions for Gl for BSBS and FSBS for further evaluation in the following sections.

3.1. Backward stimulated Brillouin scattering

Expression (8) with $2l_R \sim 2k_0l^2_\perp$ yields the SBS gain for backscattering in a speckle, which is

$$G_{BSBS} \equiv G_{l_R} = \left(\frac{\omega}{\nu}\right)_{iaw} \frac{n_0}{n_c} \frac{\omega_1}{c_{g,1}\lambda_0} \frac{I_{sp}\pi l_\perp^2}{n_c ckT} = \frac{\pi}{2}\left(\frac{\omega}{\nu}\right)_{iaw} \frac{n_0}{n_c} \beta_{SBS} \frac{l_R}{\lambda} \tag{9}$$

for homogeneous plasmas in the strong damping limit. It is well known that the amplification of SBS in an inhomogeneous plasma can be expressed via Rosenbluth's gain coefficient, replacing l_R by the effective interaction length $\sim \nu_{iaw}Lv/c_s$ in a plasma with flow gradient length Lv, see e.g. [14]. The latter is of interest when the contribution of speckles from subsequent plasma layers can be considered to be independent.

3.2. Near forward stimulated Brillouin scattering

For FSBS, Gl, being by the factor $l_{eff}/l_R < 1$ smaller than the value for backscattering, can be computed along the integration path for a given angle θ and the impact parameter $\mathbf{x}_i$, see appendix B. There it is shown that the scattered field over one speckle should take into account all paths with a shift with respect to the central path. In appendix B we compute the angle-dependent form factor that comes into play when integrating over all paths. Following the analysis in appendix B, reporting equation (B.8), the relevant expression for the SBS gain along the central path (setting the impact factor to $\mathbf{x}_i = 0$) can be written as

$$G_{\rm sp} \equiv G_{l_{\rm R}} \mathcal{F}(\theta, {\rm x}_{\rm t} = 0)\ \sin(\theta/2) \simeq G_{l_{\rm R}} \frac{l_\perp}{l_{\rm R}} \frac{\sqrt{\pi}/\cos(\theta/2)}{1+\sqrt{1+(4/\pi)(l_\perp/l_{\rm R})^2 \cot^2\theta}}, \tag{10}$$

with $G_{\rm sp} < G_{l_{\rm R}}(l_\perp/l_{\rm R})(\sqrt{\pi}/2\cos(\theta/2))$ for $\theta < \pi/2$.

It is important to mention that expression (9) is by a numerical factor $f_{\rm num}$ of the order of unity and by the factor relating IAW frequency and damping, $(\omega/\nu)_{\rm iaw}$, equivalent to the ratio $P_{\rm sp}/P_{\rm c}$ between the laser power in a speckle, $P_{\rm sp} = \pi I_{\rm sp} l^2_\perp$ and the SF critical power [15–17],

$$G_{\rm FSBS} \equiv G_{\rm sp} = f_{\rm num} \frac{l_\perp(\theta)}{l_{\rm R}} \left(\frac{\omega}{\nu}\right)_{\rm iaw} \frac{P_{\rm sp}}{P_{\rm c}} \tag{11}$$

with $P_{\rm c} = 2N_{\rm c}(c\lambda_0/\omega_0)(n_{\rm c}/n_0)n_{\rm c}ckT(1 - n_0/2n_{\rm c})$, in which $N_{\rm c}$ assumes the value of $N_{\rm c} = 1.90$ for the example of a Gaussian beam (with a corresponding definition of $l_\perp$, see [18]). This is an important issue for discussion on the dominating process, namely between FSBS and SF, see section 6.

The analysis in appendix B also shows that we can, with good approximation, make use of equation (6) to express the SBS amplification in a speckle by using $G_{\rm sp}$ from equation (10) as the gain value.

4. Total field: pump and scattered fields over numerous speckles

For the perturbative approach that allows separation of the pump and scattered light field contributions to the total field in the plasma, we establish in this section a hierarchy of the different terms contributing to the total light field intensity. In the total field, the complex field components of the pump field E_0 and the scattered light field $E_{\rm s}$ are superposed. The sum over speckles (index j) then reads

$$E_{\rm tot} = \sum_j E_{0j}\, e^{i\Phi_j} + \sum_j E_{sj}\, e^{i\Psi_j}. \tag{12}$$

This superposition in a nonlinear medium, such as the plasma, is valid as long as either the spectral components (in the frequency and/or wave vector) are well separated or the amplitude of the scattered field can be considered small with respect to the pump field.

The square field $|E_{\rm tot}|^2$ has then components that can be ordered in powers of the smallness parameter $\varepsilon \propto |E_{\rm s}/E_0|$ associated with a perturbative approach,

$$|E_{\mathrm{tot}}|^2 = \mathcal{O}(\varepsilon^0) + \mathcal{O}(\varepsilon) + \mathcal{O}(\varepsilon^2) + \ldots, \tag{13}$$

$$\text{with} \quad \mathcal{O}(\varepsilon^0) = \sum_j |E_{0j}|^2 + \sum_j \sum_{k \neq j} E_{0j} E_{0k}^* \, \mathrm{e}^{\mathrm{i}(\Phi_j - \Phi_k)} + \mathrm{c.c.},$$

$$\mathcal{O}(\varepsilon) = \sum_j \sum_k E_{0j} E_{sk}^* \, \mathrm{e}^{-\mathrm{i}(\Phi_j - \Psi_k)} + \mathrm{c.c.},$$

$$\mathcal{O}(\varepsilon^2) = \sum_j |E_{sj}|^2 + \sum_j \sum_{k \neq j} E_{sj} E_{sk}^* \, \mathrm{e}^{\mathrm{i}(\Phi_j - \Psi_k)} + \mathrm{c.c.}$$

For field components due to stimulated scattering, the phase Ψj can be decomposed into time and angle dependent part, being $2k_0c_st \sin\theta/2$ for SBS (where the angle θ is in the interval $[-\pi/2, \pi/2]$) and in a random phase, so that we replace in the following Ψj by $\Psi j + 2k_0c_st \sin\theta/2$.

The first term from equation (13) that contains modifications due to scattering is then

$$\sum_j \sum_k E_{0j} E_{sk}^* \, \mathrm{e}^{\mathrm{i}(\Phi_j - \Psi_k)} + \mathrm{c.c.} = \sum_j E_{0j} E_{sj}^* \, \mathrm{e}^{\mathrm{i}(\Phi_j - \Psi_j)}$$
$$+ \sum \sum_{j \neq k} E_{0j} E_{sk}^* \, \mathrm{e}^{\mathrm{i}(\Phi_j - \Psi_k)} \, \mathrm{e}^{-\mathrm{i}2k_0c_st \sin\theta/2} + \mathrm{c.c.}, \tag{14}$$

wherein Φj and Ψj are uniform random variables in the interval $(-\pi, \pi]$.

Note, however, that for stimulated scattering the phases of the scattered light Ψj have to be considered enslaved to the pump field. Consequently, the phase Ψj of the scattered field excited in speckle j is no longer independent of the phase Φj of the pump field in the same speckle.

The second order terms $\mathcal{O}(\varepsilon^2)$ read analogously

$$\sum_j \sum_k E_{sk} E_{sj}^* \, \mathrm{e}^{\mathrm{i}(\Psi_j - \Psi_k)} + \mathrm{c.c.} = \sum_j |E_{sj}|^2 + \sum \sum_{j \neq k} E_{sj} E_{sk}^* \, \mathrm{e}^{-\mathrm{i}4k_0c_st \sin\theta/2} \, \mathrm{e}^{\mathrm{i}(\Psi_j - \Psi_k)}. \tag{15}$$

Each of the terms can be ordered in powers of the average field intensity of the incident field, $\langle I \rangle \propto \langle |\hat{E}_0|^2 \rangle^{1/2}$ and the relative (normalized) intensity Mj of a speckle, $|E_0j(M)| \propto M^{1/2}j \ \langle I \rangle$. By using speckle statistics, the sums can be replaced by integrals over the speckle distribution

$$\mathrm{d}F(M) = f_{\mathrm{sp}}(M)\mathrm{d}M, \tag{16}$$

for which we specify f_{sp}, the probability density function (pdf), below. This yields for the terms containing E_sj average contributions

$$\left\langle \sum_j E_{0j} E_{sj}^* \right\rangle = \sum_j \int \mathrm{d}F(M) \, |E_0(M)| \, |E_s(\theta, t, M)| \, \mathrm{e}^{\mathrm{i}(\Phi_j - \Psi_j)} \, \mathrm{e}^{-\mathrm{i}2k_0c_st \sin(\theta/2)}$$
$$= \langle I \rangle \mathrm{e}^{-\mathrm{i}2k_0c_st \sin(\theta/2)} \sum_j \langle \mathrm{e}^{\mathrm{i}(\Phi_j - \Psi_j)} \rangle \int \mathrm{d}F(M) M^{1/2} |E_s(\theta, t, M)|, \tag{17}$$

$$\left\langle \sum_j \sum_{k \neq i} E_{0j} E^*_{sk} \right\rangle$$
$$= \sum_j \sum_{k \neq i} \int \mathrm{d}F(M) \int \mathrm{d}F(M') |E_{0j}|\, |E_s(\theta, t, M')| \langle \mathrm{e}^{\mathrm{i}(\Phi_j - \Psi_k)} \rangle\, \mathrm{e}^{-\mathrm{i}2k_0 c_s t \sin(\theta/2)}$$
$$= \langle I \rangle\, \mathrm{e}^{-\mathrm{i}2k_0 c_s t \sin(\theta/2)} \sum_j \sum_{j \neq k} \langle \mathrm{e}^{\mathrm{i}(\Phi_j - \Psi_k)} \rangle \int \mathrm{d}F(M) M^{1/2} \int \mathrm{d}F(M') |E_s(\theta, t, M')|. \tag{18}$$

$$\left\langle \sum_j |E_{sj}|^2 \right\rangle = \int \mathrm{d}F(M) || E_s(\theta, t, M)|^2 \tag{19}$$

and

$$\left\langle \sum_j \sum_{k \neq i} E_{sj} E^*_{sk} \right\rangle = \sum_j \sum_{j \neq k} \int \mathrm{d}F(M) |E_s(\theta, t, M)| \int \mathrm{d}F(M') |E_s(\theta, t, M')| \langle \mathrm{e}^{\mathrm{i}(\Phi_j - \Psi_k)} \rangle. \tag{20}$$

We refer to [18] for the calculation of the distribution of the maxima *Mj*, $f_{\mathrm{sp}}(M) = Z^{-1} \exp\{(1/2)\mathrm{e}^{M - \log n_{\mathrm{sp}}} - M\}$ with $Z = \int_1^{-\log n_{\mathrm{sp}} + \gamma_E} \exp((1/2)\mathrm{e}^{M - \log n_{\mathrm{sp}}} - M)\mathrm{d}M$. The value of Z can be approximated by the integral $\int_{e/2n_{\mathrm{sp}}}^{\gamma/2} \exp(u) u^{-2}\, \mathrm{d}u \simeq 0.736$ n_{sp} with n_{sp} number of speckles (in finite volume) and the Euler Gamma constant $\gamma_F \simeq 0{,}5772$.

Note that we can assume that the term in equation (17), $\sum_{j=1}^{N} \langle \mathrm{e}^{\mathrm{i}(\Phi_j - \Psi_j)} \rangle / N$, is of the order of unity, as the phases Φj and Ψj are relative to the same speckle. The sum $\sum_j \sum_{k \neq i} \langle \mathrm{e}^{\mathrm{i}(\Phi_j - \Psi_k)} \rangle$ in equation (20), in which the phases Φj and Ψk refer to different $j \neq k$, is a finite quantity that we need not compute here as it appears in negligible orders.

Indeed, all terms with $j \neq k$ in the preceding expressions and in equation (13) correspond to products between the scattered field originating from speckle *k* and the pump field from speckle $j \neq k$, for which the correlation is in general weak and assuming a correlation length close to the typical speckle size—an assumption used for independent speckles. The sums over a single index should therefore, in general, be dominant. The terms of equations (17) and (19) can be evaluated in the following via the integral over the speckle distribution functions and powers of the speckle intensity *M*.

4.1. Backward stimulated Brillouin scattering and the role of phase conjugation

For BSBS, in contrast to FSBS discussed below, the components E_0 and E_s are spectrally well-separated in wave vector space and the terms $\sim E_0 j E_s k$ (for any k,j) are rapidly oscillating functions in z with $|2k_0 \sin\theta/2| \simeq 2k_0$ (corresponding to the IAW response). Due to this spectral separation, the backscattered light response is represented by $\propto \sum_{j,k} E_{sj} E^*_{sk}$, dominated by $\sum_j E_{sj} E^*_{sj}$; hence we can use equation (19) to evaluate the response. The requirement for the separation between E_0 and E_s to be valid is thus essentially determined by flux conservation, $|c_{g,0}E_0E^*_0| - |c_{g,1}E_sE^*_s| = \text{const}$, the consequence of which is pump depletion in the case of strong backscattering.

In our evaluation of the contributions from scattering in speckles, equations (16)–(20), we have not accounted for phase conjugation (PC) [19], which is a phenomenon that has particular relevance in the case of backscattering. To account for it, one has to be aware that the scattered light may contain two components, $E_s = \sigma E^*_0 + E_{s,n}$ (with σ real and positive), one of them being proportional to the conjugated pump field E^*_0, while the other component, $E_{s,n}$, is the 'noise' excluding the PC component. The value of σ indicates then the magnitude of the portion of the scattered light seed that is due to PC.

Using this 'orthogonal' decomposition, one can compute the contributions from speckles out of the 'noise' source seed term with the help of expression (19). For PC it is the spatial average of E_0E_s over the dimensions transverse to the propagation direction $\overline{E_0 E_s} = \sigma \overline{|E_0|^2} \propto \sigma \langle I \rangle$ that leads to a spatial amplification, with $\langle I \rangle$ being the average beam intensity. The ensemble average over speckle realizations, as in (19), is therefore unnecessary (for an $\langle I \rangle$ value independent of the RPP realization). The phase conjugated scattered light signal will be amplified at average intensity along z. The total amplification is then either limited by the length of the interaction region or by the finite spatial extent of the resonant coupling (for inhomogeneous plasmas). One can associate a gain value corresponding to PC, $G_{pc}(z) = 2z\hat{\gamma}_0^2/(c_g \nu_{iaw})$ wherein $\hat{\gamma}_0^2$ is evaluated at the average intensity of the beam, so that we can express it in units of the average speckle gain $G_{pc}(z) = 2G_{sp}|_{M=1}\, z/l_R = 2\langle G \rangle_{BSBS}\, z/l_R$. Analogous to the asymptotic solution for scattered light in equation (6), the amplification due to PC hence results in

$$\overline{|E^2_{s,pc}(L)|^2} = \overline{|E^2_0|^2}\sigma^2 e^{G_{pc}} = \sigma^2 \langle I \rangle e^{2(z/l_R)\langle G \rangle_{BSBS}}. \quad (21)$$

4.2. Forward stimulated Brillouin scattering

In the case of FSBS the pump and the scattered light fields are spectrally not strongly separated because the wave number and frequency shifts of the scattered light ($\sim 2k_0 \sin\theta/2$) are small. In a perturbative approach, as used here for a layer limited to about one speckle length, the dominating modification to the pump field can be evaluated from equation (17), while terms from equation (19) should remain small. In section 5 we will hence determine the modification of the total field of speckles due to FSBS by evaluating equation (17).

5. ORDERING CONTRIBUTIONS TO THE TOTAL FIELD FROM SPECKLES

In section 3 we have derived the response of the scattered light from a single speckle, which can be written in the asymptotic limit as equation (6), with the gain coefficient for convective amplification from equation (10). The result for the scattered light collecting the contributions from numerous, independent, single speckles brings about the problem of contributions from extreme speckles to the total field. For this reason we will express, from now onwards, the single speckle gain G_{sp} as a multiple M of the gain of an average speckle

$$G_{sp} \equiv M\langle G\rangle,$$

where $\langle G\rangle$ refers to the gain in a speckle with the average beam intensity $\langle I\rangle$.

In [18] order statistics of speckle fields have been investigated, which will be used in the following.

5.1. Backward stimulated Brillouin scattering from speckles

For BSBS, apart from contributions due to PC, the response function from a single speckle, indexed k, that contributes to the total field via expression (19) can be written as

$$R_{BSBS}(M_k) = \min(1, E_{seed}^2\, e^{2\langle G\rangle_{BSBS} M_k}) = E_{seed}^2 \exp\{2\langle G\rangle_{BSBS} \min(M_k, M_{BSBS}^*)\}, \qquad (22)$$

where Mk is the speckle peak intensity normalized to the average pump field intensity, $\langle G\rangle_{BSBS}$ is the gain of an 'average intensity' speckle and $M^*_{\text{BSBS}} = \langle G\rangle^{-1}_{\text{BSBS}} \ln(E^{-1}_{\text{seed}})$ denotes the 'cutoff' of exponential growth in M_k values after which the response saturates. The latter definition is an auxiliary way to account for pump depletion, which works sufficiently well, but can of course be described in an improved manner, usually leading to an expression which is no longer explicit for $R(M_k)$ in BSBS [20].

In what follows we continue the evaluation of the cumulative response from speckles for the case of FSBS. The derivation for the case of BSBS, not explicitly shown here, is by analogy straightforward and will be discussed again in section 5.4.

5.2. Forward stimulated Brillouin scattering from speckles

For FSBS the response function from a single speckle, indexed k, that contributes to the total field via expression (17) reads similarly to BSBS:

$$R_{\text{FSBS}}(M_k) = \sqrt{M_k}\,\min(1, E_{\text{seed}}\, e^{\langle G\rangle M_k}) = M_k \bar{E}_{\text{seed}} \exp\{\langle G\rangle \min(M_k, M^*)\}, \tag{23}$$

where M_k is the speckle peak intensity normalized to the average pump field intensity and $\langle G\rangle$ is the FSBS gain of an 'average intensity' speckle. The speckle gain defined in equation (11) is hence replaced by $M_k\langle G\rangle$ from now on. Here we choose the seed amplitude as proportional to the speckle field amplitude, namely $E_{\text{seed}} = \bar{E}_{\text{seed}}\sqrt{M_k}$ (and not to the average field amplitude) and allow for pump depletion in order to limit the exponential amplification of the scattered light. Accordingly, the (normalized) speckle intensity M^* defined via $E_{\text{seed}} \exp\{\langle G\rangle M^*\} = \bar{E}_{\text{seed}} M^{*1/2} \exp\{\langle G\rangle M^*\} \equiv 1$ corresponds to the intensity where the scattered light starts to be saturated by pump depletion, hence

$$M^* = \langle G\rangle^{-1} \ln(E^{-1}_{\text{seed}}) = \langle G\rangle^{-1}\left(\ln(\bar{E}^{-1}_{\text{seed}}) - \tfrac{1}{2}\ln M^*\right), \tag{24}$$

the latter being an implicit expression (that can be solved iteratively, $M^* < \langle G\rangle^{-1}[\ln(\bar{E}^{-1}) + \frac{1}{2}\ln(n_{\text{sp}} + \gamma_{\text{E}})]$, with $\gamma_{\text{E}} = 0.5772$ standing for the Euler gamma constant).

The total (or cumulative) response from all n_{sp} speckles can be written as

$$\mathcal{R}_{\text{tot}} = \sum_{k=1}^{n_{\text{sp}}} R(M_k) \equiv \sum_{k=1}^{n_{\text{sp}}} \tilde{R}(M_k)\mathcal{P}^{(G)}(M_k) \tag{25}$$

with $\tilde{R}(M_k) = Z\bar{E}_{\text{seed}} M_k$ and $Z = \sum_k \exp\{\langle G\rangle \min(M_k, M^*)\}$ so that

$$\mathcal{P}^{(G)}(M_k) = Z^{-1} e^{(G)\min(M_k, M^*)}, \tag{26}$$

wherein the values of *Mk* correspond to the values of speckle maxima taken from a speckle field realization.

To the cumulative field response one can associate a typical speckle intensity

$$\langle M \rangle \equiv \frac{\mathcal{R}_{\text{tot}}}{\bar{E}_{\text{seed}} Z}, \tag{27}$$

that indicates the intensity of the speckle population which contributes most to the scattered light field $\mathcal{R}_{\text{tot}}$ as discussed later.

By introducing a speckle order M_1,...*Mk* following the hierarchy $M_1 > M_2 > \ldots > Mk$, strong fluctuation in $\mathcal{R}_{\text{tot}}$ can be expected when the value of $\langle M \rangle \sim M_1 \sim \ln n_{\text{sp}}$.

To evaluate properly the cumulative response from the total field on the basis of statistics, one can make use of order statistics, which allows to take into account the intervals in *M* over which the successive speckle maxima can occur. By defining as *fk*(*M*) the probability density of the *k*th order speckle, the cumulative response can be, in contrast to equation (25), defined as

$$\bar{\mathcal{R}}_{\text{tot}} = \sum_{k=1}^{n_{\text{sp}}} \bar{E}_{\text{seed}} \int_0^\infty M f_k(M) e^{(G)\min(M, M^*)} \, dM, \tag{28}$$

wherein the probability density follows the expression [18] obtained from the marginal probability distribution

$$f_k(M) = \frac{n_{\text{sp}}^k}{\Gamma(k)} e^{-n_{\text{sp}} \exp(-M) - kM} \tag{29}$$

which peaks close to $M \sim \ln(n_{\text{sp}})/k$. It follows from order statistics that the values *Mk* of the most intense maxima cover a large interval in *M*, while smaller, and in our notation higher-indexed maxima, can be well described by a shrinking interval in *M* around their expectation value $\bar{M}_k$ from equation (29).

A technical remark concerning the evaluation of equation (28) using (29) and (31): the interval in *M*on which the function *fk*(*M*) assumes non-negligible values becomes narrower with increasing orders of *k*. For this reason it is not necessary to evaluate the integrals in equation (28) for high orders of *k*, i.e. lower-intensity speckles. This applies for $\langle G \rangle < [k \ln(n_{\text{sp}}/k)/2]^{1/2}$. Practically, for $k > k_>$, the integral in equation (28) can be simply replaced by evaluating the response function for a single speckle at the expectation value $\bar{M}_k \equiv \int M f_k(M)\,\mathrm{d}M$ of the kth order speckle from $\bar{M}_k^{(G)} \equiv \int M f_k^{(G)}(M)\mathrm{d}M$,

$$\tilde{\mathcal{R}}_{\mathrm{tot}} = \sum_{k}^{k_{>}-1} \int_0^{\infty} M f_k(M) \mathrm{e}^{\langle G \rangle \min(M, M_k^{*})} \mathrm{d}M + \sum_{k=k_{>}}^{n_{\mathrm{sp}}} \tilde{R}(\bar{M}_k) \mathcal{P}^{\langle G \rangle}(\bar{M}_k). \tag{30}$$

5.3. Modified probability density of speckles

To find equivalence between equations (28) and (25) we can introduce a new probability density function $f^{\langle G \rangle}_k(M)$ in equation (28) such that

$$f_k^{\langle G \rangle}(M) = (1/N_k^{\langle G \rangle})\, f_k(M)\, \mathrm{e}^{\langle G \rangle \min(M, M^{*})} \tag{31}$$

with the normalization coefficient $N_k^{\langle G \rangle} \equiv \int_0^{\infty} f_k(M) \exp\{\langle G \rangle \min(M, M^{*})\}\mathrm{d}M$. This results in an alternative definition (to expression (27)) of the average speckle intensity,

$$\langle \tilde{M} \rangle = \tilde{\mathcal{R}}_{\mathrm{tot}} / E_{\mathrm{seed}} \tilde{Z}, \tag{32}$$

with $\tilde{Z} \equiv \sum_k N_k^{\langle G \rangle}$. It is important to note that $f^{\langle G \rangle}_k$ contains the factor $\exp\{(\langle G \rangle - k) M\}$ which changes from an exponential decrease to growth when the average gain $\langle G \rangle$ exceeds k, the speckle order index. For this reason, one can observe a critical change in the features of the scattered light in the vicinity of $\langle G \rangle = 1$, coming primarily from the most intense speckle M_1, as illustrated by figures 1–7.

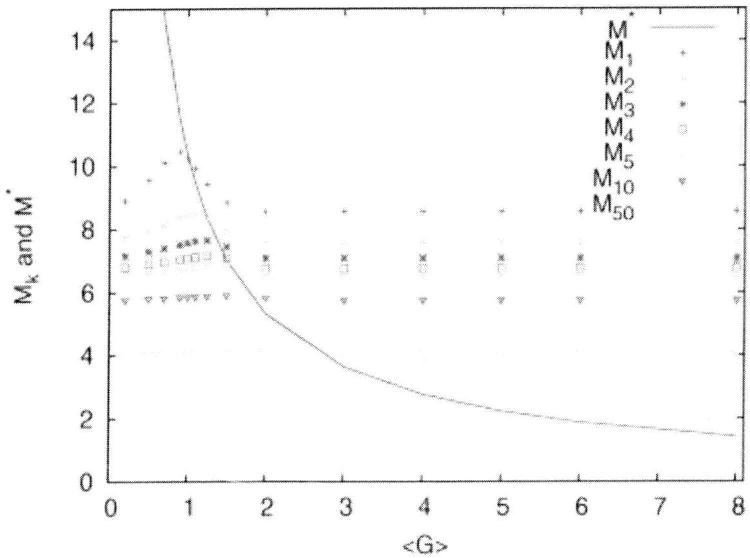

Figure 1. Speckle expectation values computed from equation (35) for speckle orders 1,..., 5, 10 and 50, as well as the value of M^* for $n_{sp} = 3000$ and $\bar{E}_{seed} = 10^{-5}$. Values of $M[G]k < M^*$ are affected by pump depletion.

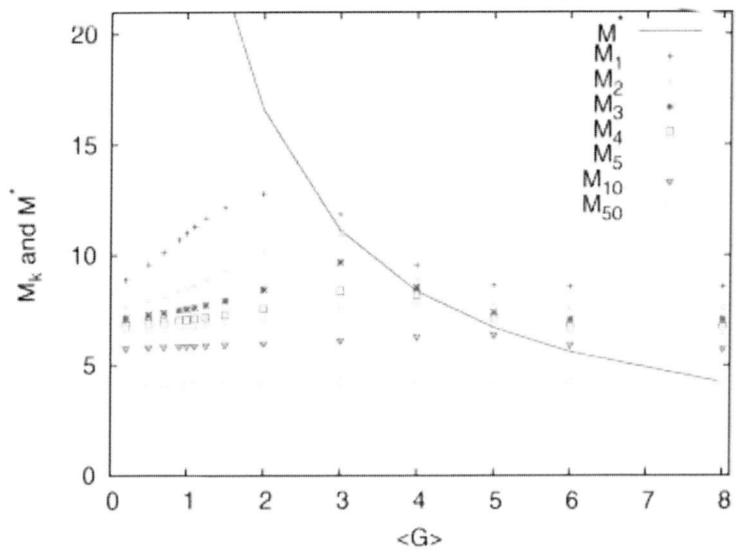

Figure 2. Speckle expectation values computed from equation (35) for speckle orders 1,...,5, 10 and 50, as well as the value of M^* for $n_{sp} = 3000$ and $\bar{E}_{seed} = 10^{-15}$. Values of $M[G]k < M^*$ are affected by pump depletion.

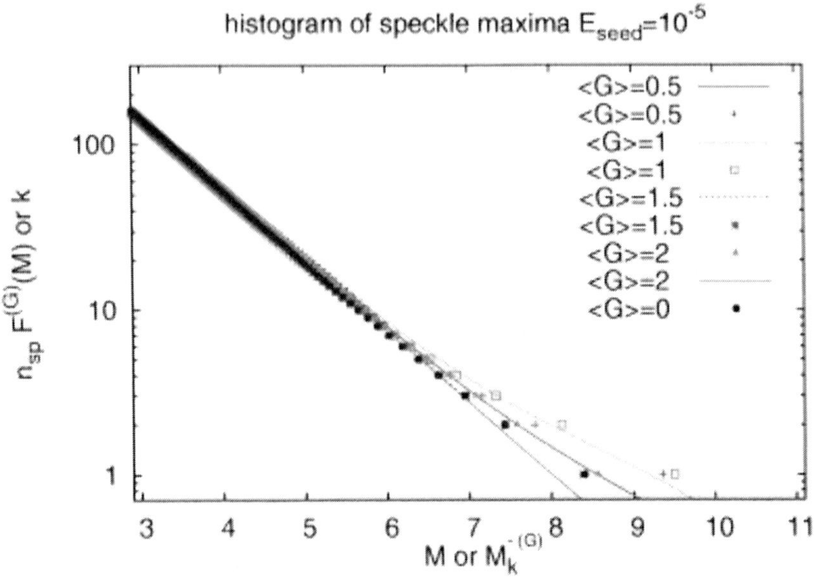

Figure 3. Histogram of the speckle distribution to linear-log scale. Symbols show the speckle number k (ordered from the most intense to the least intense) as a function of its expectation value $\bar{M}_k^{(G)}$ computed from equation (35) for different values of the average gain $\langle G\rangle = 0.5$ (red), 1 (pink), 1.5 (blue), 2 (grey), 0 (black); lines show the speckle distribution $F^{(G)}$ as a function of the normalized speckle intensity M computed from equation (33), in the same colour as for the symbols. Parameters: $n_{sp} = 3000$ and $\bar{E}_{seed} = 10^{-5}$.

Corresponding to $f^{(G)}_k(M)$ we can define a complementary probability distribution function

$$F^{(G)}(M) \equiv \sum_k^{n_{sp}} \int_M^{\infty} f_k^{(G)}(M')\,dM' \tag{33}$$

which is the equivalent to $\int_M^{\infty} f_{sp}(M)$ in equation (16) for $\langle G\rangle = 0$.

To compare $\mathcal{R}_{tot}$ with $\bar{\mathcal{R}}_{tot}$, the values of Mk have to be replaced by their expectation values, either by (i) from equation (29) (see [18])

$$\bar{M}_k = \int_0^{\infty} M f_k\,dM = \bar{M}_{k-1} - \frac{1}{k} \simeq \ln n_{sp} - \ln(k-1) - \frac{1}{2(k-1)} \text{ for } k > 1 \tag{34}$$

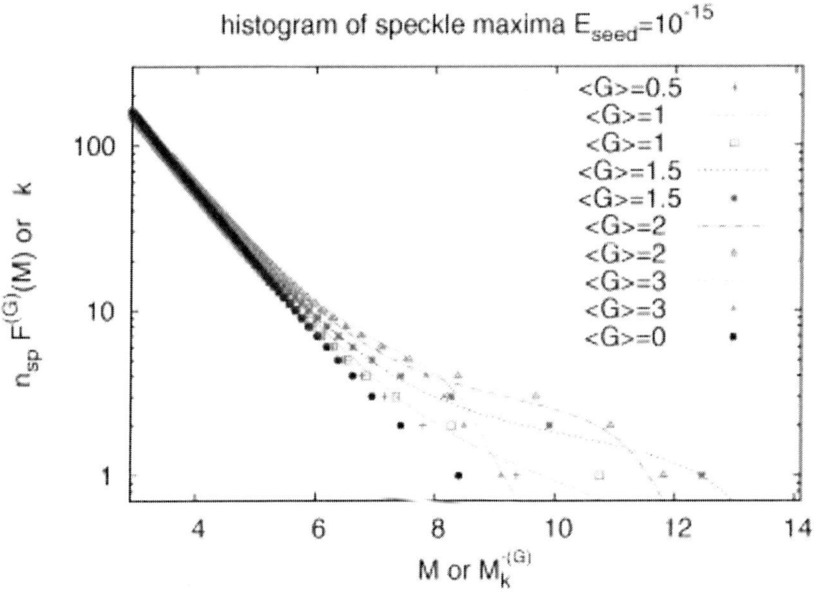

Figure 4. Histogram of the speckle distribution as in figure 3 computed from equations (35) and (33) to linear-logarithmic scale. Symbol index for different values of the average gain $\langle G\rangle$ = 0.5 (red cross), 1 (pink), 1.5 (blue), 2 (red triangle), 3 (grey), 0 (black). Parameters: n_{sp} = 3000 and $\bar{E}_{\mathrm{seed}} = 10^{-15}$.

with $\bar{M}_1 = \ln n_{\mathrm{sp}} + \gamma_{\mathrm{E}}$ or by (ii) from equation (31)

$$\bar{M}_k^{(G)} = \int_0^\infty M f_k^{(G)}(M)\,\mathrm{d}M = \frac{\bar{E}_{\mathrm{seed}} \int_0^{M_k^*} M\ f_k(M)\mathrm{e}^{\langle G\rangle M}\mathrm{d}M + \int_{M_k^*}^\infty M f_k(M)\mathrm{d}M}{\bar{E}_{\mathrm{seed}} \int_0^{M_k^*} f_k(M)\mathrm{e}^{\langle G\rangle M}\mathrm{d}M + \int_{M_k^*}^\infty f_k\mathrm{d}M}. \tag{35}$$

Note that $\bar{M}_k = \bar{M}_k^{(G)}|_{\langle G\rangle=0}$.

The expression shows that due to the modification of the field from scattering, expectation values of the intense speckle maxima will shift $\bar{M}_k$ to higher values $\bar{M}_k^{[G]} > M_k^{[G]}|_{\langle G\rangle=0}$ for gain values around and beyond $\langle G\rangle$ ~ 1. We illustrate this for two cases: one considering n_{sp} = 3000 speckles in the volume and with a seed level for the scattered field of $\bar{E}_{\mathrm{seed}} = 10^{-5}$, and an extreme case (with the same n_{sp}) with a very low

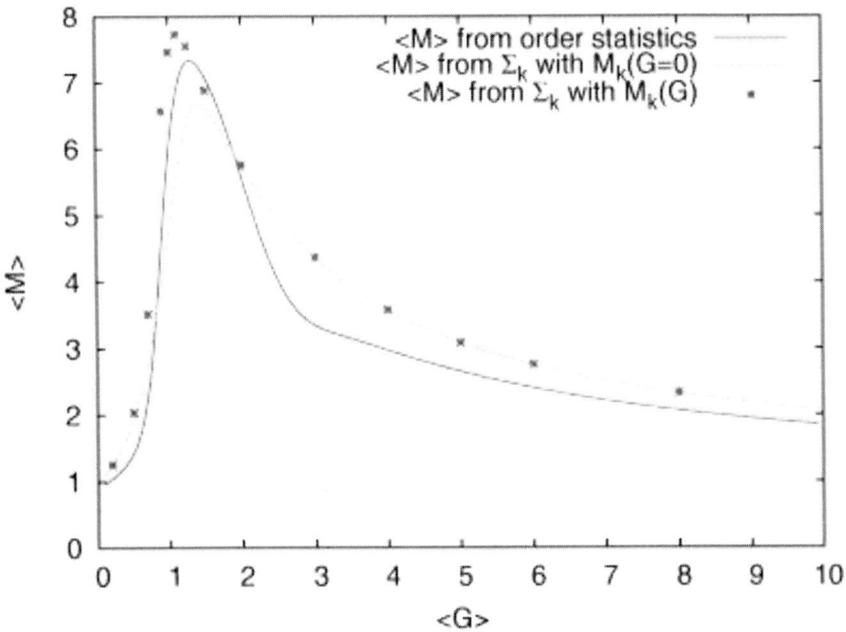

Figure 5. Average speckle intensity $\langle M \rangle$ over $n_{sp} = 3000$ speckles as a function of the gain value $\langle G \rangle$ computed from equation (27) for $\bar{E}_{seed} = 10^{-5}$ using the values of $M(G)k$ of equation (35) (star symbols) and of Mk of equation (34) in equation (25).

seed value, $\bar{E}_{seed} = 10^{-15}$, so that pump depletion of individual speckles practically does not arise. Speckle expectation values computed from equation (35) for speckle orders 1...5, 10 and 50 are shown in figures 1 and 2 together with the value of M^* indicating the onset of pump depletion once $M[G]k > M^*$. For $\bar{E}_{seed} = 10^{-5}$ pump depletion rapidly limits the influence of the most intense speckles for increasing values of $\langle G \rangle > 1$. For the extreme case with a very low seed level, many intense speckles remain unaffected by pump depletion up to average gain values $\langle G \rangle \sim 5$, so that the modification of the values $M[G]k$ as a function of $\langle G \rangle$, with respect to $\langle G \rangle = 0$, covers a wider range than for lower seed levels.

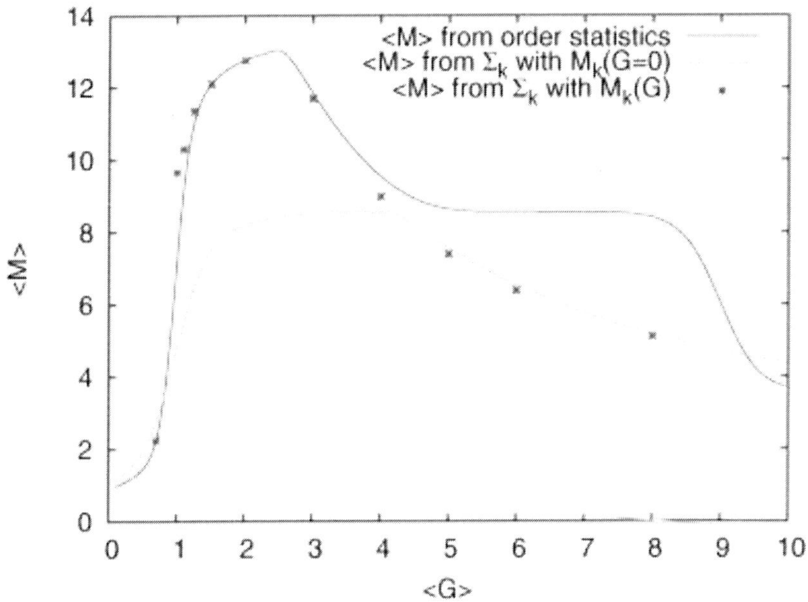

Figure 6. Average speckle intensity $\langle M \rangle$ over $n_{sp} = 3000$ speckles as a function of the gain value $\langle G \rangle$ computed from equation (27) for $\bar{E}_{seed} = 10^{-15}$ using the values of $M^{(G)}k$ of equation (35) (star symbols) and of Mk of equation (34) in equation (25).

The expressions for the speckle distribution equation (33) and for the expectation values for the kth order speckle allow to construct histograms as a function of the speckle peak intensity, as shown in figure 3 for $n_{sp} = 3000$ speckles, $E_{seed} = 10^{-5}$ and (4) for the case with an extremely low seed value $E_{seed} = 10^{-15}$. The histograms from the expectation values are constructed by plotting the speckle order count (from the most to the least intense speckle) as a function of its expectation value $M^{[G]}k$. The histogram for no amplification, $\langle G \rangle = 0$, follows an exponential with a slightly slower decrease for the most intense speckles, as derived in [18]. For the case of $E_{seed} = 10^{-5}$, a strong modification of the histogram in the tail for the most intense speckles is limited to the gain interval $0.5 < \langle G \rangle < 1.5$. In a limited interval of M values, around $M \sim M_1 \sim \ln n_{sp}$, a departure from Gaussian statistics can be observed. The values computed via the expectation values and via equation (33) show good agreement and reproduce the same tendency. The extreme case with a very low seed value

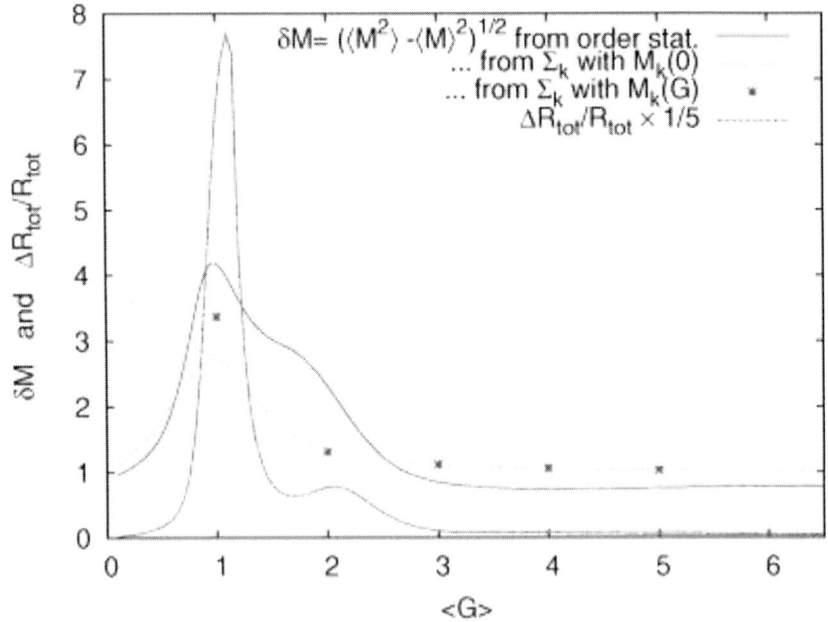

Figure 7. Standard deviation of the average speckle intensity $\delta M = [\langle M^2\rangle - \langle M\rangle^2]^{1/2}$ and $\Delta\mathcal{R}_{\rm tot}/\mathcal{R}_{\rm tot}$ (divided by the factor 5) as a function of the gain value $\langle G\rangle$ for an ensemble of $n_{\rm sp} = 3000$ speckles and $\bar{E}_{\rm seed} = 10^{-5}$ computed from equation (27) using the values of $M^{(G)}_k$ of equation (35) (star symbols) and of M_k of equation (34) in equation (25).

and much less pump depletion in intense speckles, figure 4, shows much stronger departure from $\langle G\rangle = 0$ and for all shown values $\langle G\rangle > 0.5$. The decrease in the tail of the distribution is considerably slower than an exponential one (close to a power-law up to a cutoff value) depending on $\langle G\rangle$.

The quantity $\langle M\rangle$ deduced from equation (27) or (32) of the transmitted light field is a good measure to understand which speckle population the scattered light field originates from. This is illustrated in figures 5 and 6. Three different curves for the value of $\langle M\rangle$, with similar behaviour, are shown. These follow three different methods to evaluate $\langle M\rangle$. In the first case (i) $\langle M\rangle$ results from equation (32) by evaluating equation (28), where the sum over the integrals of all speckle orders is computed; in the second and third cases, (ii) and (iii) respectively, one simply computes the sum according to expression (25) by evaluating $R(M_k)$ at discrete values, namely in case (ii) $M_k = \bar{M}_k(G=0)$ from equation (34) and in case (iii) $M_k = \bar{M}_k(G)$ from equation (35). For all three cases, a rapid transition from $\langle M \rangle \simeq 1$ towards values close to the expectation value of the most intense speckle $\langle M \rangle \sim \bar{M}_1$ can be observed close to $\langle G \rangle \simeq 1$. The values evaluated following method (i) are the most precise, while methods (ii) and (iii) are easier to compute.

The steepness in the transition around $\langle G \rangle = 1$ is more pronounced when accounting properly for order statistics. This quite steep transition, with $d\langle M \rangle / d\langle G \rangle$ being positive and maximum around $\langle G \rangle = 1$, is a clear signature of critical behaviour as also discussed in [9, 11] and more recently shown in [10] for the cumulative backscatter reflectivity of stimulated Raman backscattering as a function of gain. The value of $\langle M \rangle$ shown as a function of the gain value $\langle G \rangle$ indicates which speckle population of the speckle field contributes most to the cumulative response of the scattered field $\mathcal{R}_{tot}$. For the case with an extremely small seed, the average value $\langle M \rangle$ remains considerably greater than 1 up to $\langle G \rangle \sim 5$, while for the case with $\bar{E}_{seed} = 10^{-5}$ due to pump depletion $\langle M \rangle$ converges towards 1 with $\langle G \rangle$.

The fact that for $\langle G \rangle \simeq 1$ the value of $\langle M \rangle$ is close to $\bar{M}_1(\langle G \rangle \simeq 1)$, has also the consequence that the spread of the response of the scattered field from all speckles depends also critically on the variance of M_1 itself. The standard deviation $\delta M = [\langle M^2 \rangle - \langle M \rangle^2]^{1/2}$ of the (normalized) speckle intensity M as a function of $\langle G \rangle$ is shown in figure 7. It is evident that for the regime when the relatively low abundant intense speckles contribute most, i.e. around $\langle G \rangle \sim 1$, the cumulative response varies strongly for different realizations which is translated in an important spread δMaround the average value $\langle M \rangle$. This spread δM can be of the order of $\langle M \rangle /2$. The expectation value of cumulative scattering response $\mathcal{R}_{tot}$ has hence a very low reliability in this regime. As a function of δM one can estimate the standard deviation in $\mathcal{R}_{tot}$, namely $\Delta\mathcal{R}_{tot}/\mathcal{R}_{tot} \simeq |\exp\{\langle G \rangle \min(\langle M \rangle + \delta M, M^*)\} - \exp\{\langle G \rangle \min(\langle M \rangle - \delta M, M^*)\}|/2$ yielding $\sim \exp\{\langle G \rangle \delta M\} - 1$ for $\langle M \rangle < M^*$. For the numbers from the examples given, yielding $\Delta\mathcal{R}_{tot}/\mathcal{R}_{tot} \sim \exp(4)$ for $\langle G \rangle = 1$ and $\delta M \sim 4$, it follows that $\mathcal{R}_{tot}$ may vary strongly from one to another speckle pattern realization.

5.4. Consequences of backward stimulated Brillouin scattering and the presence of phase conjugation

We recall here that the derivation to compute the expectation values $\langle M \rangle$ and the standard deviation δM, here done for the example of FSBS, can analogously be applied to BSBS. Mounaix and Divol [11] have shown in their work, based on simulations using numerous speckle field realizations accompanied by a concise analytical model, that the standard deviation to the average backscatter reflectivity from an ensemble of laser speckles strongly increases for $\langle G \rangle > 0.5$ and peaks at $\langle G \rangle \sim 1.5$ with $\Delta \mathcal{R}_{tot}/\mathcal{R}_{tot} \sim 30$. Our results support these findings based on numerical simulations, as illustrated in figure 7.

For BSBS, the role of PC has to be discussed with respect to the cumulative response of backscattered light from speckles. Rose and Mounaix [10] have investigated this problem for the case of stimulated Raman scattering. As a first measure one has to compare the gain values in a single speckle, $\langle G \rangle M$, to the equivalent gain for PC, $2 \langle G \rangle z/l_R$, involved in equations (22) and (21), respectively. PC could then dominate the backscatter response if its amplification along z is not limited (by a box length or a characteristic length) so that $2z/l_R > \alpha_{pc} + \langle M \rangle$, where $\langle M \rangle$, as defined above, stands for the typical intensity of speckle dominating the backscatter response and where $\alpha_{pc} = \langle G \rangle^{-1} \ln(E_{seed}/\sigma)$ compares the levels between speckle and PC seed. As pointed out in [21], supposing $E_{seed} = \sigma$, PC will be dominant for small speckle gains $\langle G \rangle$ because $\langle M \rangle \sim 1$, as evident from figures 5 and 6. In homogeneous plasmas (media) where hypothetically z/l_R can attain large values, pump depletion due to PC could occur, $2z/l_R > \ln(1/\sigma)$ which would of course question the contribution of the speckle ensemble in the pump-depleted zone.

6. NUMERICAL SIMULATIONS

We have carried out numerical simulations in two spatial dimensions (2D) with the paraxial laser–plasma coupling code HARMONY2D [15, 22]. While the role of intense speckles for gain values around unity for the case of BSBS has clearly been demonstrated via numerical simulations in [11], we concentrate here on the case of FSBS. To simulate this situation, we use a single equation for the electromagnetic light field propagating in the forward direction. Hence, for the simulations we do not separate the light field in a perturbative manner as in the preceding analysis. Equation (2) is hence solved for $E = E_0 + E_s$. The response of the IAWs, equivalent to equation (1), is computed via the fluid response of the plasma under the ponderomotive drive proportional to ∇EE^* which contains the term the $\propto E_0 E^*_s$ driving the IAW resonant with the FSBS process, as well as the term $\nabla_\perp E_0 E^*_0$ responsible for SF and filamentation. The equations in the fluid code account for a wave number-dependent damping so that, as in our model, both the IAW frequency as well as the IAW damping follow the same wave number dependence. In these simulations, no BSBS occurs because the short-wavelength IAWs are not seeded. Both FSBS and SF are in direct competition because of the slower time scale and the longer IAW wavelength with respect to BSBS. This competition is the subject of substantial, and in our opinion not yet conclusive, discussion in the literature [1, 3, 4].

The onset of SF in 2D simulations has been determined in [3, 15], leading to the criterion for speckle power, P_{sp} to SF critical power, P_c

$$p_{2D} \equiv \left(P_{sp}/P_c\right)_{2D} = l^2_{\perp,2D} k^2_{opt} \simeq (\pi^2/2) f^2_\# (n_0/n_c) \beta_{SBS}, \tag{36}$$

where k_{opt} stands for the transversal wave number with optimum growth rate for filamentation and where $l_{\perp,2D} \simeq 0.5 f_\# \lambda$ is the typical radius of a (Gaussian-type) RPP speckle (here in 2D), see [17] and the appendix in [18].

The speckle gain for FSBS, as discussed in the preceding sections, can be estimated using equation (11), $G_{sp,2D} \simeq (0.2/f_\#) G l_R \simeq 0.25\pi(\omega/\nu)_{iaw}(n_0/n_c)\beta_{SBS} f_\#$, where we have assumed $l_{R,2D} \simeq 2.5 f^2_\# \lambda$ [18]. For an average speckle gain of $G_{sp,2D} - \langle G \rangle \equiv 1$, we have $\beta_{SBS} = 0.01 I_{14} \lambda^2 \mu / T_{keV} \simeq 1.27(\nu/\omega)_{iaw}(n_c/n_0) f^{-1}_\#$ and for $p_{2D} \equiv 1$, $\beta_{SBS} \simeq 0.2(n_c/n_0) f^{-2}_\#$.

For $f_\# = 4$, $(\nu/\omega)_{iaw} = 0.03$ and $\beta_{SBS} \simeq 0.025$, both p_{2D} and $G_{sp,2D}$ are close to 1. In this parameter regime the onset of SF should appear together with effects from FSBS due to intense speckles. For higher damping $(\nu/\omega)_{iaw} > 0.07$, SF for an average speckle and thus for the entire beam, should be expected to be the dominant process.

We have chosen to run simulations for which p_{2D} stays below the SF threshold $p_{2D} = 0.5$. Indeed for the regime $P_{sp}/P_c < 1$ it was suggested to introduce a new criterion for the onset of beam spreading effects [4], depending on the IAW damping. To control the average speckle gain for FSBS and to discriminate the behaviour from SF ($\langle G \rangle = G_{sp,2D}$ around 1, namely, $G_{sp,2D} = 0.5$, 1 and 1.5), we varied the IAW damping $(\nu/\omega)_{iaw} = 0.03$, 0.015 and 0.01, respectively. While the onset of SF should not directly depend on IAW damping4, the change of the FSBS gain value governs the behaviour of the plasma response. No additional noise for FSBS was imposed to seed the process. For this reason, both SF and FSBS grow from the ponderomotively-induced density perturbations and the IAW evolving out of these perturbations.

We have chosen a simulation box 100 laser wavelengths (λ) long, corresponding to 2.5 $l_{R,2D}$ for the optical beam aperture $f_{\#} = 4$, and 8200 wavelengths wide. To smooth the incident laser light corresponding via RPPs, the method proposed in [17] was used. For this configuration the simulation volume should contain about 2200–2400 speckles in width, which corresponds to $\log n_{sp} \sim 7.8$.

Figure 8 shows a snapshot of the light field patterns in a part of the interaction volume, for the mentioned cases (a) $\langle G \rangle = 0.5$, (b) =1 and (c) =1.5. The light intensity is displayed in units of the mean intensity. The most significant changes arise clearly from intense speckles, as illustrated in the supplementary movie 1 (available from stacks.iop.org/NJP/15/025003/mmedia). Both enhancement of the peak intensity value as well as lateral oscillations in the rear part can be observed.

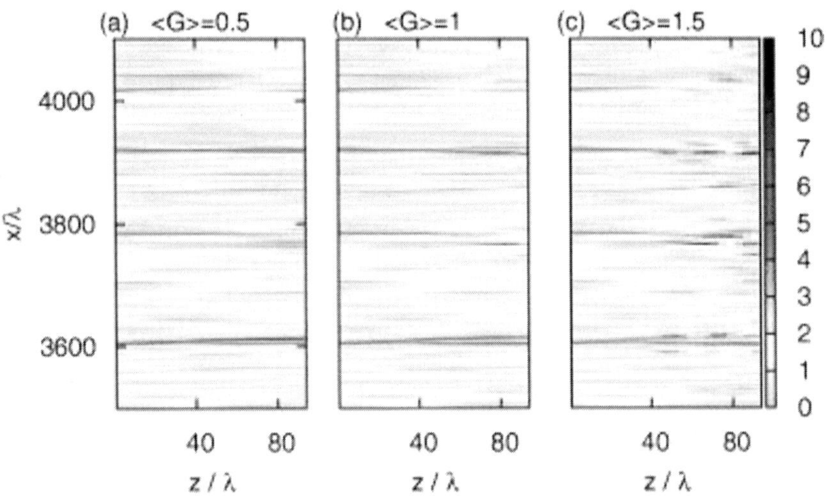

Figure 8. Snapshot of total field intensity, normalized to the average beam intensity (see colour bar), in a part of the interaction volume in $x_\perp$ (out of 0–8200λ), for three cases of IAW damping, $(\nu/\omega)_{iaw} = 0.03$, 0.015 and 0.01, corresponding to (a) $\langle G \rangle = 0.5$, (b) 1 and (c) 1.5, respectively.

From the resulting total field, we have computed a speckle distribution function by determining the local field intensity maxima and integrating around their neighbourhood ($l_{R,2D}/2$ in z), while assuming that the speckle size has not changed substantially (which, of course, can no longer be assumed in the layers subsequent to this first layer of speckles in z). Note that by this method, integrating along the speckle axis, simple speckle intensity enhancement due to SF would not immediately lead to a change of the measured speckle intensity because SF, while enhancing the peak value, would shorten the effective Rayleigh length.

We show the histograms corresponding to this distribution function in figure 9, as a function of the speckle intensity M with respect to the average speckle intensity, and the corresponding pdf in figure 10. The results, here for a single speckle field realization, show the same tendency as the histogram shown for the case of low seed noise in our model based on order statistics, figure 4:

- The value of the most intense speckles observed increases up to $M \sim 12$. In the simulation results, of course, this value depends not only on the gain value, as in the model, but also on the particular realization. An error bar for the value $\langle G \rangle = 1$ indicates the range derived from four realizations. As the standard deviation δM in figure 7 is largest around $\langle G \rangle \sim 1$, the spread of values in the distribution tail can be important.

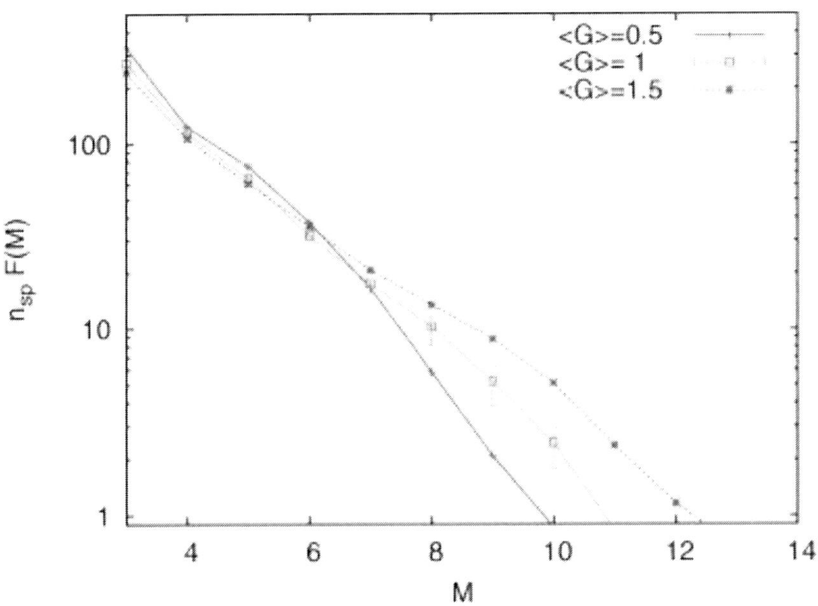

Figure 9. Histogram of the speckle distribution computed from numerical simulations with the code HARMONY2D, in linear–logarithmic scale. Symbol index for different values of the average gain: $\langle G \rangle = 0.5$ (red), 1 (pink), 1.5 (blue). Parameters: $n_{sp} \sim 2300$.

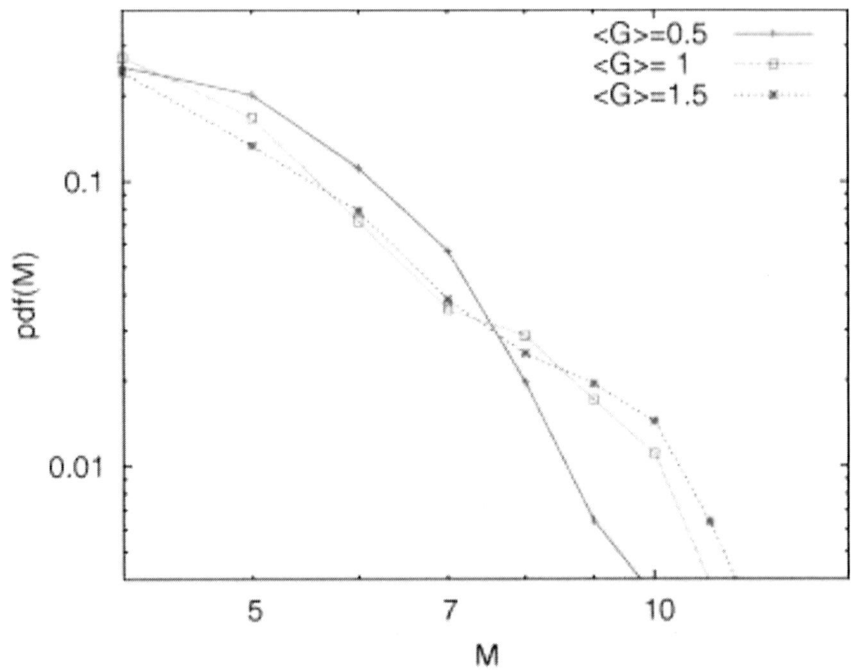

Figure 10. Probability density function of the speckle distribution computed from numerical simulations with the code HARMONY2D, in log–log scale. Symbol index for different values of the average gain: $\langle G \rangle = 0.5$ (red), 1 (pink), 1.5 (blue). Parameters: $n_{sp} \sim 2300$. In the interval $M = 5–10$, the cases for $\langle G \rangle > 1$ show a power law decrease instead of an exponential dependence as still visible for $\langle G \rangle = 0.5$.

- The decrease with the speckle intensity slows down in the vicinity of the expectation value of the most intense speckle associated with the (initial) pump field, $\sim \ln n_{sp} \simeq 8$. The probability density shown in figure 10 in the double-logarithmic scale shows for $\langle G \rangle = 1$ and 1.5 a decrease that is, around $M \sim 8$, closer to a power law dependence (roughly $f(G)_{sp} \sim M^{-3.5\pm0.2}$ in $5 < M < 10$) than an exponential law, indicating a departure from Gaussian statistics in this domain of the distribution function.

7. DISCUSSION AND CONCLUSIONS

Our results show that the magnitude of (forward) scattered light through a plasma layer can exhibit critical behaviour when the gain coefficient $\langle G\rangle$ of FSBS, associated with average laser speckles, increases in the interval from $\langle G\rangle - \epsilon$ to $\langle G\rangle + \epsilon$. The light field can show explosive behaviour because of the dominance of high-intensity speckles. Explosive behaviour has been suggested in earlier work [9], and can be explained by the fact that an exponential increment for the amplification of stimulated scattering, $\sim \exp\{\langle G\rangle M\}$ (with $M \equiv I_{sp}/\langle I\rangle$), compensates the—usually exponential—decrease, $\sim \exp\{-M\}$ in the speckle distribution function [17] occurring for $\langle G\rangle \geq 1$. In more recent work [4] it had been shown that the observed strong spreading of transmitted light behind a plasma layer was related to a criterion different from the onset for SF [16], i.e. P_{sp}/P_c with the ratio between laser power in an average speckle, P_{sp}, and the SF critical power P_c. The modified criterion relates the ratio P_{sp}/P_c to the ratio between the IAW damping and its frequency, $(\nu/\omega)_{iaw}$. The resulting criterion $(P_{sp}/P_c)/(\nu/\omega)_{iaw} > O(1)$ is closely related to our criterion for the SBS gain $\langle G\rangle$.

Our results show that when $\langle G\rangle$ increases to $\langle G\rangle > 1$, the dominance of the high-intensity speckles is crucial: in the mentioned regime, they strongly contribute to the cumulative response in the scattered and in the transmitted light image, although their abundance is small with respect to the total number of speckles in the pattern. This has the consequence that changes in speckle patterns can yield a strongly diverging response of the scattered light. This means that the resulting light field, due to the contributions from scattering, is poorly controlled and not reliable, for example for the purposes of 'plasma-induced smoothing'.

With respect to previous work [9, 11] using speckle distribution functions to compute the cumulative scattering response, we here make use of the order statistics [18] of intense speckles, which allows to take into account the whole intensity range and the order in which each speckle is likely to contribute. By using this method, the impact of the different populations of speckles on the critical behaviour in the cumulative scattering response becomes evident. A good measure for sensitivity of the scattering response on high-intensity speckles is the expectation value of the intensity of speckles that contribute most to the response.

As elaborated in the previous sections, it becomes evident that the closer the expectation value $\langle M\rangle$ approaches the intensity value of the most intense speckle, $\sim \ln n_{sp}$, the less controllable and predictable the light field behind the plasma may be. The standard deviation δM shown in figure 7 is a measure of the interval, around $\langle M \rangle$, over which the intensity of the most contributing speckle population can vary from one realization of a speckle pattern to another. We find that for the regime when the most intense speckles contribute most, i.e. around $\langle G \rangle \sim 1$, the spread in intensity, δM, around $\langle M \rangle$ can be of the order of $\langle M \rangle /2$ itself. Consequently, the expectation value of the cumulative scattering response $\mathcal{R}_{tot}$, as shown also by means of numerical simulations in [11], is therefore of low significance in this regime and has low reliability. The uncertainty in the cumulative response which can be estimated by $\Delta\mathcal{R}_{tot}/\mathcal{R}_{tot} \sim \exp(\langle G \rangle \delta M) - 1$ may assume large values (~30–60 in the examples given).

While a plasma layer can have an extension of many speckles along the laser propagation axis, in the frame of this work, our analysis is restricted to a first layer of the order of one speckle length. We include the influence of pump depletion in our analysis in order to see the saturation of amplification-scattered light in speckles. The information computed by superposition of the incoming laser light with the scattered light from the first layer is already quite significant and important for further scattering in successive layers. It characterizes the light intensity pattern behind this zone. In particular, as shown in section 5, the statistics of the light field exiting the first layer can show a departure from an exponential decrease in speckle intensity, and slower than exponential in the tail of the speckle distribution.

For the purpose of 'plasma-induced smoothing', uncontrolled behaviour that originates already from the first layer of speckles has to be avoided. It is therefore preferential to operate in a regime with plasma parameters such that $\langle G \rangle$ assumes values well beyond $\langle G \rangle \sim 1$ so that the highest intensity speckles do not dominate the scattered light response.

ACKNOWLEDGMENT

This work has been supported by the French 'Agence Nationale de Recherche', in the frame project 'ILPHYGERIE' no. ANR-12-BS04-0006-02.

Appendix A.: Transient solution to the scattered light field

In [12] by Elisseev *et al* the temporal dependence of the scattered light wave amplitude E_s for the case of SBS has been derived by combining the equation of the scattered light field and a heavily damped IAW. In the regime of FSBS, one can neglect the temporal derivative for the scattered light wave in order to simplify the description, and replace the paraxial propagation operator $\vec{c}_{g,1} \cdot \nabla - \mathrm{i}\frac{c^2}{2\omega_1}\Delta_\perp$ by a derivative along the path with $(c^2k_1/\omega_1)\partial l$, where the path-length l diverges with the angle θ from the principal propagation axis $\|z$,

$$c_{g,1}\partial_l E_s = -\mathrm{i}\frac{\omega_1}{2}\frac{n_0}{n_c}\frac{n_1^*}{n_0}E_0, \tag{A.1}$$

furthermore, the assumption of a sufficiently damped IAW allows equation (1) to be reduced to a first-order equation in time, reading

$$(\partial_t + \nu_{\mathrm{iaw}})\frac{n_1^*}{n_0} = -\mathrm{i}c_s k_0 \beta_{\mathrm{SBS}} \sin(\theta/2) E_0^* E_s, \tag{A.2}$$

which eventually yields for $a = (n_1^*/n_0)\exp(-\nu_{\mathrm{iaw}} t)$ the partial differential equation [12]

$$\frac{\partial^2 a}{\partial_t \partial_l} - \frac{\gamma_0^2(l,\theta,\mathbf{x}_i)}{c_g}a = 0, \tag{A.3}$$

and with $\gamma_0(l,\theta,\mathbf{x}_i)$ denoting the local SBS growth rate at the path-length l, in which the additional impact parameter $\mathbf{x}_i$ indicates the shift in the integration path with respect to the path going exactly through the speckle peak. With an intermediate step using the method of Riemann functions, the scattered wave amplitude can be expressed as

$$E_s(\theta,\mathbf{x}_i,\nu t) = E_{\mathrm{seed}}\left(1 + \int_0^{\nu t} \mathrm{d}\tau \sqrt{\frac{\xi_l}{\nu\tau}}\mathcal{I}_1\left(2\sqrt{\tau\xi_l/\nu}\right)\mathrm{e}^{-\tau}\right) \tag{A.4}$$

which involves the Bessel function $\mathcal{I}_1$ and where E_{seed} stands for the noise level of the scattered light. The quantity $\xi l/\nu$ is the spatial amplification coefficient computed over the integration path length $l$and we replace it, see equation (5), in the following by $G_{l,i} \equiv \xi_l/\nu_{\mathrm{iaw}} \int_{l'=-l}^{l} (\gamma_0^2(l',\theta)/c_g\nu)\,\mathrm{d}l'$; we keep here the additional index 'i' to account for the fact that the gain depends on the choice of the integration path with respect to the impact parameter x_i. By partial integration one obtains

$$E_{s,i}(\theta,\mathbf{x}_i,\tau) = E_{\mathrm{seed}}\left(1 + \left[\mathrm{e}^{-\tau}\mathcal{I}_0(2\sqrt{G_{l,i}\tau})\right]_{\tau=0}^{\tau=\nu_{\mathrm{iaw}}t} - \int_0^{\tau=\nu_{\mathrm{iaw}}t}\mathrm{d}\tau'\mathrm{e}^{-\tau'}\mathcal{I}_0(2\sqrt{G_{l,i}\tau'})\right). \tag{A.5}$$

The integral part of equation (A.5) dominates for $\tau = \nu_{\mathrm{iaw}}t > 1$, while the term $\mathrm{e}^{-\tau}\mathcal{I}_0(2\sqrt{\tau G_{l,i}})$ dominates for the early time behaviour, $\tau = \nu_{\mathrm{iaw}}t < 1$.

The asymptotic value of the integral, equation (A.4), for $\nu_{\mathrm{iaw}}t \rightarrow \infty$, results in the simple expression for the exponential amplification of the seeding scattered field amplitude, E_{seed},

$$E_{s,i}(\theta, \mathrm{x}_i, \tau \rightarrow \infty) = E_{\mathrm{seed}} \exp\left(G_{l,i}\right). \tag{A.6}$$

A good approximation for all times (within the model of equations (A.1)–(A.2) not covering the very initial growth, and the strong coupling limit) can be derived applying the saddle point method to the integral part of equation (A.5) by assuming that $\mathcal{I}_0(2\sqrt{G_{l,i}\tau})$ can be replaced by its asymptotic expression for $2\sqrt{G_{l,i}\tau} \gg 1$. This yields, see figure A.1,

$$E_{s,i}(\theta, x_i, \tau) = E_{\mathrm{seed}} \left[e^{-\tau}\mathcal{I}_0(2\sqrt{G_{l,i}\tau}) + \frac{1-e^{-\tau+2\sqrt{G_{l,i}\tau}}}{2\sqrt{\pi G_{l,i}}} \right.$$
$$\left. + \exp(G_{l,i}) \frac{\mathrm{erf}(\sqrt{G_{l,i}}) - \mathrm{erf}(\sqrt{G_{l,i}} - \sqrt{\tau})}{2} \right], \tag{A.7}$$

where the last term asymptotically dominates once $\tau > 2Gl{,}i$ because $(1/2)|\mathrm{erf}(\sqrt{G_{l,i}}) - \mathrm{erf}(\sqrt{G_{l,i}} - \sqrt{\tau})|$ in equation (A.7) transits from =0 for $t = 0$ to =2 for $\tau \gg Gl{,}i$.

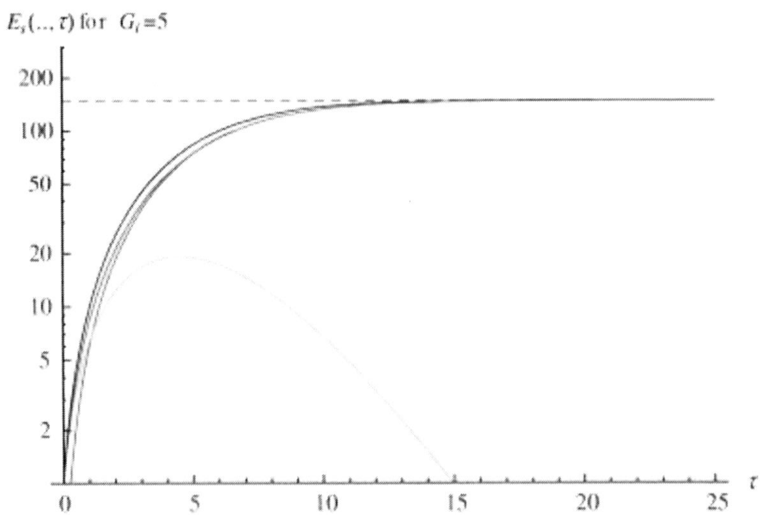

Figure A.1. Solution of the transient solution of the scattered light field amplitude E_S as a function of the normalized time $\tau = \nu_{\mathrm{IAW}}t$ for the example of a speckle with the gain $Gl{,}i = 5$ from equation (A.5) (black line), from equation (A.7) with all terms (red line), from equation (A.7) with the blue line for the limit $\tau > \sqrt{G_i}$ and the green line for the limit $\tau < 1$. The dashed line indicates the asymptotic limit equation (A.6).

We have restricted our analysis in section 5 to the asymptotic regime in time, for $\nu_{iaw}t \gg 1$ of convective amplification of scattered light, characterized by an exponential increment with the speckle gain Gi coefficient. The transient regime, $\tau < 2(Gi\tau)^{1/2}$, implies a transition function, see equation (A.7), depending in fact on the speckle intensity $M \langle I \rangle$, via $Gi = \langle G \rangle M$. The onset of the asymptotic regime is therefore retarded for speckles with increasing peak intensity, yielding the criterion $\tau > 4M \langle G \rangle$, thus $\tau > 4M$ for the most relevant regime discussed here, $\langle G \rangle \sim 1$.

Appendix B.: Speckle shape and form factor

In equation (5) the SBS is expressed for a path through a speckle. For the scattered field from a speckle, all paths with impact parameter x_i contribute, while the contribution from the central one ($x_i = 0$) is likely to be dominant. For this reason, it is useful to express the SBS gain as a product of the gain value Gl depending essentially on the speckle peak intensity, and of a form factor $\mathcal{F}(\theta, x_i)$, $G_l(\theta, x_i) \equiv G_{l_K}\mathcal{F}(\theta, x_i)\ \sin(\theta/2)$,

where $\mathcal{F}$ depends on the shape of the speckle. For a Gaussian-type speckle it reads

$$\mathcal{F}(\theta, x_i) = \int_{l'=-l/l_R}^{l/l_R} \frac{\exp\{-(l_R/l_\perp)(l' \sin\theta - x_i)^2\}}{1 + l'^2 \cos^2\theta} \mathrm{d}l', \tag{B.1}$$

with $l_\perp$ denoting the speckle (correlation) radius in the direction perpendicular to the propagation axis. The integration over l in equation (B.1) converges rapidly when the limits chosen around the speckle focus ($l = 0$) are chosen sufficiently large, $l > 2l_R$. To simplify the analysis, we can extend the limits to infinity so that the form factor can be expressed as a function of the angle θ, and of x_i,

$$\mathcal{F}(\theta, x_i) = \frac{\pi}{2\cos\theta} \left\{ e^{(l_R/l_\perp)^2 z^2} \mathrm{erfc}(\sqrt{(l_R/l_\perp)^2 z^2}) + e^{(l_R/l_\perp)^2 \bar{z}^2} \mathrm{erfc}(\sqrt{(l_R/l_\perp)^2 \bar{z}^2}) \right\} \tag{B.2}$$

with $z = ix_i + \tan\theta$, $\bar{z} = -ix_i + \tan\theta$ and erfc() [= 1 − erf()] standing for the complementary error function. This expression was obtained using the Plancherel–Parseval theorem.

In order to evaluate the asymptotic expression for the scattered field, equation (A.6), it is useful to simplify the expression by developing around $x_i = 0$, which allows eventually a rapidly converging integration in x_i around the speckle peak. The series expansion of equation (B.2) in x_i only yields even orders,

$$\mathcal{F}(\theta, x_i) = \mathcal{F}(\theta, 0)\left(1 - x_i^2/W^2(\theta) + \mathcal{O}(x_i^4)\right) \tag{B.3}$$

with

$$\bar{\mathcal{F}} \equiv \mathcal{F}(\theta, 0) = \frac{\pi}{\cos\theta} e^{(l_R/l_\perp)^2 \tan^2\theta} \mathrm{erfc}(|(l_R/l_\perp)\tan\theta|)$$

and

$$W(\theta) = \frac{l_\perp}{l_R}\tan^{-1}\theta\left[1 - \frac{2}{\sqrt{\pi}}\frac{e^{-(l_R/l_\perp)^2\tan^2\theta}(l_R/l_\perp)\tan\theta}{\mathrm{erfc}(\sqrt{(l_R/l_\perp)^2\tan^2\theta})} + 2\left(\frac{l_R}{l_\perp}\right)^2\tan^2\theta\right]^{-1/2}. \tag{B.4}$$

For $(l_R/l_\perp)^2 \tan^2\theta \gg 1$ this simplifies to $W(\theta) \simeq (\tan\theta l_R/l_\perp)^{-1}$. In a good approximation the function for the form factor is given by
$\mathcal{F}(\theta, 0) \simeq 2\sqrt{\pi}\cos^{-1}\theta/|(l_R/l_\perp)\tan\theta + \sqrt{(l_R/l_\perp)^2\tan^2\theta + 4/\pi}\,|$
behaving as $\mathcal{F}(\theta, 0) \simeq \sqrt{\pi}(l_\perp/l_R)/\sin\theta$ in the limit $(\tan\theta l_R/l_\perp)^2 \gg 1$.

A good approximation that allows further simplification of the evaluation is

$$\mathcal{F}(\theta, x_i) = \mathcal{F}(\theta, 0)e^{-x_i^2/W^2(\theta)}. \tag{B.5}$$

An integration over all values of the impact parameter x_i hence can be performed in this approximation. Expression (A.6) can hence be integrated over x_i in the limit to a neighbouring speckle $x_i \simeq -2\rho_c \dots +2\rho_c$ where $\rho_c \sim l_R W(\theta)$ is the typical correlation width in x_i. For rapidly converging correlation, the integration should not depend considerably on the integration limit beyond $\pm 2\rho_c$ so that it is convenient to choose $\pm\infty$ as the limit for the evaluation,

$$E_{seed} + E_{seed}\int_{x_i=-2\rho_c}^{+2\rho_c}\left(\exp\{G_{l_R}\mathcal{F}(\theta, x_i)\sin(\theta/2)\} - 1\right)dx_i \tag{B.6}$$

$$\simeq E_{\text{seed}} + E_{\text{seed}} \int_{x_i=-\infty}^{+\infty} \left(\exp\{G_{l_R} \mathcal{F}(\theta, 0) \sin(\theta/2)\, e^{-x_i^2/W^2(\theta)}\} - 1 \right) dx_i$$
$$= E_{\text{seed}} + E_{\text{seed}} \sum_{n=1}^{\infty} \frac{G_{\text{sp}}^n}{n!} \int_{x_i=-\infty}^{+\infty} e^{-n\, x_i^2/W^2(\theta)}\, dx_i$$
$$= E_{\text{seed}} \left(1 + \sqrt{\pi} W(\theta) \sum_{n=1}^{\infty} \frac{G_{\text{sp}}^n}{n!\sqrt{n}} \right), \tag{B.7}$$

with, see equation (10), the definition for the effective speckle gain

$$G_{\text{sp}} = G_{l_R} \mathcal{F}(\theta, 0) \sin(\theta/2) \simeq G_{l_R} \frac{l_\perp}{l_R} \frac{2\sqrt{\pi} \sin(\theta/2)/\sin\theta}{1 + \sqrt{1 + (4/\pi)(l_\perp/l_R)^2 \cot^2\theta}}$$
$$< G_{l_R} \frac{l_\perp}{l_R} \frac{\sqrt{\pi}/2}{\cos(\theta/2)} \text{ for } \theta < \pi/2. \tag{B.8}$$

This gain value of G_{sp} is now the relevant value to use in equation (5) and it is used in the analysis starting in section 4.

The expression (B.7) can be approximated with a Bessel function $\mathcal{I}_\alpha$ of order α close to 1 (best fit with $\alpha \simeq 1.15$), or almost as well, by an exponential function $\sim \exp\{G_{\text{sp}}\}$ for not too small values of G_{sp},

$$E_{\text{seed}} \left(1 + \sqrt{\pi} W(\theta) \sum_{n=1}^{\infty} \frac{G_{\text{sp}}^n}{n!\sqrt{n}} \right) \simeq E_{\text{seed}} \left[1 + \pi\, W(\theta) \mathcal{I}_\alpha(G_{\text{sp}}) \right] \simeq W(\theta) \frac{\exp\{G_{\text{sp}}\}}{[G_{\text{sp}}]^{1/2}}. \tag{B.9}$$

FOOTNOTES

For low values of IAW damping, wave trains originating from the density evacuation out of intense speckles can eventually influence the density in neighbouring speckles; however, there is low likelihood that such wave trains arrive in other intense speckles.

REFERENCES

1. Schmitt A J and Afeyan B 1998 Phys. Plasmas **5** 503
2. Myatt J, Pesme D, Hüller S, Maximov A, Rozmus W and Capjack C E 2001 Phys. Rev. Lett. **87** 255003
3. Maximov A V, Ourdev I G, Pesme D, Rozmus W and Tikhoncuk V T 2001 Phys. Plasmas **8** 1319
4. Grech M, Riazuelo G, Pesme D, Weber S and Tikhonchuk V T 2009 Phys. Rev. Lett. **102** 155001
5. Lushnikov P M and Vladimirova N 2010 Opt. Lett. **35** 1965
6. Depierreux S et al 2009 Phys. Rev. Lett. **102** 195005

7. Malka V, Faure F, Hüller S, Tikhonchuk V T, Weber S and Amiranoff F 2003 Phys. Rev. Lett. **90**075002
8. Labaune C, Baton S, Jalinaud T, Baldis H A and Pesme D 1992 Phys. Fluids B **4** 2224
9. Rose H A and DuBois D F 1994 Phys. Rev. Lett. **72** 2883
10. Rose H A and Mounaix P 2011 Phys. Plasmas **18** 042109
11. Mounaix P and Divol L 2002 Phys. Rev. Lett. **89** 165005
12. Elisseev V V, Ourdev I, Rozmus W, Tikhonchuk V T, Capjack C E and Young P E 1997 Phys. Plasmas **4**4333
13. Pesme D 1993 La Fusion Thermonucléaire Inertielle par Laser ed R Dautray and J P Watteau (Paris: Eyrolles) pp 480–96
14. Tikhonchuk V T, Mounaix P and Pesme D 1997 Phys. Plasmas **4** 2658
15. Tikhonchuk V T, Hüller S and Mounaix P 1997 Phys. Plasmas **4** 4369
16. Hüller S, Masson-Laborde P E, Pesme D, Casanova M, Detering F and Maximov A 2006 Phys. Plasmas**13** 022703
17. Chiao R Y, Garmire E and Townes C H 1964 Phys. Rev. Lett. **13** 479
18. Fibich G and Gaeta A L 2000 Opt. Lett. **25** 335
19. Silberberg Y 1990 Opt. Lett. **15** 1282
20. Rose H A and DuBois D F 1993 Phys. Fluids B **5** 3337
21. Hüller S and Porzio A 2010 Laser Part. Beams **28** 463
22. Porzio A and Hüller S 2010 J. Stat. Phys. **138** 1010
23. Zel'dovich B, Ya Pilipetskii N F and Shkunov V V 1982 Sov. Phys.—Usp. **25** 713
24. Tang C L 1966 J. Appl. Phys. **37** 2945
25. William E A 2006 Phys. Plasmas **13** 056310
26. Hüller S, Masson-Laborde P E, Pesme D, Labaune C and Bandulet H 2008 J. Phys.: Conf. Ser. **112**022031

CITATIONS

1. *Harmonic effects on ion-bulk waves and simulation of stimulated ion-bulk-wave scattering in CH plasmas Q S Feng et al 2017 Plasma Physics and Controlled Fusion **59** 085007*
2. *Weibull-type speckle distributions as a result of saturation in stimulated scattering processes S. Hüller and A. Porzio 2015 Laser and Particle Beams **33** 667*
3. *Reduction of stimulated Brillouin backscattering with plasma beam smoothing V. Yahia et al 2015 Physics of Plasmas **22** 042707*
4. *Focus on high energy density physics R Paul Drake and Peter Norreys 2014 New Journal of Physics **16** 065007*

Index